Engineering the Risks of Hazardous Wastes

Engineering the Risks of Hazardous Wastes

Daniel A. Vallero
Duke University, North Carolina Central University

With a contribution by
J. Jeffrey Peirce
Duke University

An imprint of Elsevier Science
www.bh.com

Amsterdam Boston London New York Oxford Paris
San Diego San Francisco Singapore Sydney Tokyo

Butterworth–Heinemann is an imprint of Elsevier Science.

Recognizing the importance of preserving what has been written, Elsevier Science prints its books on acid-free paper whenever possible.

Library of Congress Cataloging-in-Publication Data

Vallero, Daniel A.
Engineering the risks of hazardous wastes / Daniel A. Vallero, with a contribution by J. Jeffrey Peirce.
p. cm.
Includes bibliographical references and index.
ISBN 0-7506-7742-2 (alk. paper)
1. Hazardous wastes–Risk assessment. I. Peirce, J. Jeffrey. II. Title.

TD1050.R57 V35 2003
628.4′2–dc21

2002035609

British Library Cataloguing-in-Publication Data
A catalogue record for this book is available from the British Library.

The publisher offers special discounts on bulk orders of this book.
For information, please contact:

Manager of Special Sales
Elsevier Science
200 Wheeler Road
Burlington, MA 01803
Tel: 781-313-4700
Fax: 781-313-4882

For information on all Butterworth–Heinemann publications available, contact our World Wide Web home page at: http://www.bh.com

10 9 8 7 6 5 4 3 2 1

Printed in the United States of America

For Janis, the love of my life.

For my children, Daniel and Amelia, who are constant reminders of why environmental stewardship is so profoundly important.

To my ever-supportive mother, Berniece.

And, to my late father, Jim, my late uncles Louie, Joe, and Johnnie, and their fellow miners at the Lumaghi Coal Mine in Collinsville, Illinois, who lived with—and may have died from—hazards far beyond anything that I have yet to study in the laboratory.

View of a partially plugged slope entrance to the Lumaghi Coal Company Mine Number 4 in Collinsville, Illinois in 1982. The entrance has since been plugged and backfilled, and the mining site has been remediated under the Illinois' Abandoned Mined Lands Reclamation Program. (Photo used with permission, courtesy of Illinois Department of Natural Resources.)

Contents

Foreword

Hazardous wastes are not a new problem; they have been a major environmental problem for centuries. As people learned to process certain natural materials to yield various metals that were useful in many ways, they soon learned that the residues from the production of the different metals were toxic to biologic life. Plants did not grow as well in soil containing the waste residues from metal production. As time passed, plants no longer grew in the areas where the waste residues were deposited. Over time, the toxic waste volumes grew larger. Rain falling on the waste residues dissolved some of the toxic ions left in the residues, creating liquid runoff that followed the natural terrain to the nearest drainage ditch. Eventually, the toxic liquid moved into adjacent creeks and streams. As the concentrations of toxic ions increased, fish began to die. Eventually, all biologic life in the creek near the waste residue piles disappeared.

Authorities accepted damage to the environment as part of the cost to be paid for the advances in technology that the metal production had created. When the damages became too great, the authorities simply closed down the metal production facilities and had them moved to a new site. The environmental pollution problem began all over again on clean soil. The danger to people who worked in the early metal-processing facilities was recognized by both the authorities and society as a whole. Although the processing facilities were managed by high-level personnel, the operations were carried out by the lowest level of society. The metal-processing operators were considered as expendable for the greater good. Like it or not, decisions were being made regarding the risks to people and the impact of hazardous wastes on the immediate environment.

As populations increased, people occupied all the available land areas. There was simply no place to move to that could be used to process natural materials into useful products with the corresponding toxic waste piles. As the wastes continued to accumulate, something had to be done to prevent significant damage to the natural environment and to the people working and living around the manufacturing plants, which were producing hazardous wastes. It is interesting that some industrial plant managers still believe that wastes are a natural part of manufacturing that society must accept. This view is especially true in developing nations, where financial resources are limited.

The development of the chemical industry produced new sets of toxic waste materials. Some of the toxic waste materials were discharged into the air. Other toxic wastes were discharged as liquid wastes into nearby bodies of water or as solid wastes on the land. Rainfall on the solid waste piles produced additional water pollution problems. Solving the problems of an ever-growing modern society created additional forms of toxic wastes that spread over the entire environment. Although modern technology created the toxic waste problems, modern technology also developed the ability to remove most of the toxic materials from plant wastes. Unfortunately, removing the toxic materials from plant wastes did not improve the final products, but it did increase the cost of the final products. Because plant managers wanted to minimize their product costs and maximize their profits, the real problem for plant managers lay in minimizing waste production since few manufacturing plants could eliminate all wastes. The overall objective was to reduce the discharge of toxic wastes to levels below those determined to pose a significant danger to important life forms in the immediate environment. Society has yet to solve this problem.

Environmentalists changed our vocabulary from *toxic wastes* to *hazardous wastes*, in order to increase the public's awareness of the dangers from toxic waste discharges. The words *hazardous waste* carried a greater significance than the words *toxic wastes*, which were readily accepted as part of normal operations. It is interesting how words can change our perception of the world around us. Ideally, society would like industries to manufacture goods without producing any hazardous wastes. Unfortunately, it is not possible for all industries to produce zero hazardous wastes, but it is possible to minimize the production of hazardous wastes. One of the tools to help minimize the production of hazardous wastes is called risk assessment. Risk assessment techniques are useful in helping people understand the impact of different levels of hazardous wastes being discharged into our current environment and the potential damages that can be expected over time. Risk assessment is essential for environmental pollution control specialists to set waste discharge levels for the different hazardous components.

Once the allowable hazardous waste discharge quantities have been determined, environmental engineers examine the various treatment systems to remove the hazardous contaminants from all the different waste streams. Inorganic contaminants are removed by physicochemical methods, whereas organic contaminants are removed by various biologic treatment systems. Even photochemical methods are useful with specific contaminants. Environmental scientists play a major role in evaluating the different treatment methods used to remove the hazardous contaminants. Environmental engineers take the basic concepts developed by environmental scientists and design the treatment systems required to remove the hazardous contaminants. Sometimes environmental engineers construct large-scale pilot plants to demonstrate which treatment concept offers the best potential for success. Pilot plants can help illuminate potential problems that might arise in full-scale treatment systems.

Once the engineers have selected the optimum process, they must design the full-scale treatment system and select all of the mechanical equipment to be used. The engineers supervise construction of the full-size treatment units and placement of all mechanical equipment. Once the construction phase is complete, the environmental engineers start the treatment facilities and demonstrate the effectiveness of the treatment systems to remove the desired amounts of contaminants. The final step for the environmental engineer is training the treatment plant operators to ensure correct operation of the waste treatment facilities. Responsibility for day-to-day operations rests with the treatment plant operators and their immediate supervisor. Removal of the hazardous components from the various waste streams and their proper processing for return to the environment is a major responsibility that requires competent, dedicated individuals who can meet the challenges.

The author has chosen to focus this book on the most critical phase of hazardous waste engineering, the engineering of risk management for various types of hazardous contaminants. Understanding risk management is critical to the control of hazardous waste materials for environmental engineers. As one of Dr. Vallero's professors at the University of Kansas and as a colleague at Duke University, it has been a special pleasure to watch his professional development and growth over the years. All older university professors will recognize the special pride faculty members take in seeing the products of a former student's efforts. Knowledge is built on a solid foundation and constructed from the sum of a lifetime of experience.

If you learn to manage the hazardous waste risks, you will also learn to manage life's risks. Knowledge is not designed to be kept in a single box, but rather is designed to grow and blossom throughout the full expanse of life.

Thank you very much, Dan.

Ross E. McKinney
Adjunct Professor, Duke University
Professor Emeritus, Kansas University

Preface

The control and management of hazardous wastes are truly among the most important challenges of our times. Environmental engineers play crucial roles in reducing the amount of hazardous substances produced, treating hazardous wastes to reduce their toxicity, and applying sound engineering controls to reduce or eliminate exposures to these wastes. The calling of engineers is broad. We design the facilities that generate the chemicals that, under the wrong circumstances, become hazards. Once the wastes are released, we are asked to design and operate the containment and treatment facilities to deal with them. We are the professionals who are most frequently called on to address these wastes once they are released into the environment.

There are seldom, if ever, single solutions to environmental problems, especially hazardous wastes. "Everything matters in environmental engineering."[1] To deal with hazardous waste, the environmental engineer must have a command of the physical sciences. It would be imprudent to respond to the release of a chemical without first ascertaining its physical and chemical properties (e.g., first responders are well aware of the consequences of spraying water onto a strong oxidizer or failing to contain an organic compound that has a very high vapor pressure). The physical characteristics of the environment must also be known. For example, how quickly will a spilled substance traverse the vadose zone? What is the recharge rate of the aquifer? The natural and biologic sciences are also requisites to comprehensive hazardous waste management. How toxic is the substance to humans or sensitive species? Does it accumulate in the food chain?

The environmental engineer must also consider the sometimes less obvious fields of the social sciences and the humanities. Is one solution more cost effective than another? How were these costs determined? How does one (or, more important, *should* one) place a value on a human life, or the lost aesthetics, or the fears of nearby residents? These are not solely theoretical constructs. Environmental engineers are confronted daily with the controversies of real and perceived hazards. The well-trained engineer is prepared academically and professionally to incorporate the many engineering disciplines (and those of the social sciences and humanities) to be truly responsive to the challenges arising in the emotionally charged milieu of hazardous waste engineering. How we as environmental engineers decide what is important and how well we confront the problems of hazardous waste will dictate the public's perception of our success as environmental engineers for decades to come.

What This Book Is About

This book provides approaches for incorporating risk assessment and management into hazardous waste engineering decisions. It is intended to be the primary text for an undergraduate-level course in hazardous waste engineering and management, as well as the primary text for an undergraduate or graduate engineering and science course devoted to environmental risk assessment and management, with a particular emphasis on the risks posed by hazardous wastes. To cover the material in one semester, students should be grounded in basic physics and chemistry and somewhat familiar with fluid mechanics and the basic concepts of environmental engineering. This book can also be a supplemental or complementary text for a graduate-level hazardous waste engineering seminar, where a specific focus on risks is desired (I advocate that *any* hazardous waste engineering course include a risk module).

The book is also a reference for the practicing engineer and environmental scientist with an interest in risk assessment. Whether it is used as a textbook or as a reference, the book is designed to provide risk assessment insights to complement the physical and natural science considerations covered in a hazardous waste handbook.

The book can also be useful to a more general audience. It is a resource for an interested and informed reader, yet the reader does not necessarily have to be a practitioner in the field of hazardous waste site remediation or a risk expert. For example, all industry-related jargon and any terminology not widely applied outside of the environmental engineering profession is defined in context. Callout examples, case study discussions, and definitions appear throughout the text to clarify important engineering and risk concepts. This approach is necessary even within the field because environmental engineering has an eclectic mix of perspectives. Therefore, environmental consultants, public interest groups, and neighborhood groups should find this book useful, understandable, and beneficial as they address or learn more about the various aspects of hazardous waste issues.

What This Book Is Not About

This book is not a hazardous waste handbook *per se*. There are many excellent resources out there and probably more coming.[2] This book is a companion to such handbooks and manuals. I address several hazardous waste issues and, in doing so, I consider hazardous waste technologies and processes. In fact, Professor Peirce's excellent contribution to this book (Chapter 4) provides some real-life engineering approaches for engineers to intervene to address hazardous wastes; however, it should not be inferred that this is an exhaustive treatise on hazardous waste engineering techniques. Similarly, this book is not a physics or chemistry text, although I draw heavily from these fields in Chapter 3 and many of the case studies. I use these materials in a senior environmental chemistry course that I teach at North Carolina Central University. Finally, this book is not a philosophy or ethics text; however, hazardous waste engineering is rich in moral and ethical issues and lessons, so I would be remiss not to point these issues out along the way.

In all matters surrounding hazardous wastes, the specific circumstances must dictate the appropriate engineering approach. One size does not fit all! Therefore, this text does not prescribe specific remedies for any single problem. The problem must be considered in light of the scientific, engineering, societal, and legal aspects of each hazardous waste problem. Thus, the appropriate response will vary in each circumstance, depending on the particulars.

September 11, 2001

Like so many other endeavors of 2001, my research and thought processes related to this book were drastically changed following the attacks of September 11, 2001. The book is very different from what it would have been had the United States not been attacked. First, because I have been personally involved in the environmental monitoring around the World Trade Center (WTC), I have used some of the lessons learned from the environmental emergency response to write Chapter 6. The WTC has provided important lessons to environmental engineering that may be applied to more general hazardous waste projects. Second, I have become more aware of the important new roles for environmental engineers in large-scale emergency response efforts. The book now includes new insights regarding how risk-assessment techniques can assist environmental engineers with their new responsibilities to protect our public health. Finally, I recommend a higher profile for environmental engineers as members of the civil engineering community. For example, much that has been written about the roles of civil engineers in responding to September 11 has been devoted to structural considerations for existing and planned buildings and infrastructures.[3] This concern is certainly paramount, but it is not the only one for civil engineers. All engineers who specialize in environmental concerns are also key players in emergency response. In fact, many of the questions and concerns that have arisen as people begin to return to their homes near Ground Zero are related to human health risks, such as exposure to asbestos, lead, or other hazardous substances.

My major goal in writing this book is to help engineers do a complete job of addressing hazardous waste issues and problems. I believe that the risk-assessment paradigm provides several lessons for engineers. As Bill Lowrance said more than 25 years ago:

> We must hope that society at large will come to appreciate the capabilities and inherent limitations of science and technology; and we must hope that those in the technical world will come to appreciate the nonrational nature and great subtlety of social decisions. The risks are changing. Menaces are upon us. Time is short. Decisions have to be made.... May discussion of these troublesome issues be temperate, imaginative, and effective.[4]

My hope is that this book will contribute to this discussion and help to prepare current and future environmental engineers for the challenges that await us.

Acknowledgments

I would like to thank Professor Jeff Peirce for his significant contribution to Chapter 4. Jeff and I first conceived this book following our chemical fate and movement lecture series in his Hazardous Waste Engineering course at Duke University. During these lectures, I began to emphasize a risk-based engineering approach to hazardous waste management.

I am also indebted to many engineers and other professionals who over the years have helped me to elucidate many of the science and engineering concepts covered in this book. The insights of several U.S. Environmental Protection Agency engineers and scientists are reflected in these pages, especially those of Gary Foley, Bill Rice, Len Stockburger, Eric Swartz, Jack Suggs, Ralph Langemeier, Alan Vette, and Laura Webb. I was honored to work with these experts and the other members of the World Trade Center air-monitoring team.

I would also like to express my appreciation for the ongoing advice and wisdom shared by my research colleagues. I especially want to note Ross McKinney (biologic systems), formerly at the University of Kansas and recently retired from Duke; Aarne Vesilind (biosolids and engineering ethics), formerly at Duke and now at Bucknell University; Captain Jerry Farnsworth (degradation and transport of pesticides) at West Point Academy; Bob Lewis (organic chemistry) at the EPA's National Exposure Research Laboratory; Yoram Cohen (compartmental modeling) at UCLA; Paul Lioy (human exposure) of the Environmental and Occupation Health Sciences Institute at Rutgers University; Miguel Medina (hydrologic modeling) at Duke; Yolanda Banks Anderson (environmental justice) at NCCU; and Seymour Mauskopf (history of science and engineering) at Duke.

The editors at Butterworth–Heinemann (now Elsevier Science) have been quite helpful. I have been particularly impressed by the keen eyes of Christine Kloiber and Kyle Sarofeen.

Finally, I would like to thank my wife Janis for her patience and wisdom during the writing of this book.

CHAPTER 1

An Engineering Perspective on the Risks of Hazardous Wastes

How Engineers Can Help Reduce the Risks Posed by Hazardous Wastes

Protecting people and the environment from hazardous wastes presents an enormous challenge to environmental engineers. The engineering solutions to hazardous waste problems can be approached in myriad ways, but all of the solutions consist of applications of physics that are common to all engineers. The engineering solutions also include the applications of chemistry, which is familiar to most engineers. Biology is another key part of the hazardous waste engineer's repertoire, especially the applications of microbiologic principles in the treatment of wastes.

A unique aspect of hazardous waste engineering, however, is the importance of the social sciences in addressing problems. These issues include important considerations such as the psychology and economics of risks (e.g., what do people perceive as risks and how does the engineer incorporate these perceptions into proposed remedial actions?). When engineers address hazards, they must remember that the concept of risk is a human phenomenon. One cannot engage in hazardous waste engineering without a firm grasp of the human concept of risk. Therefore, in this book, we approach these wastes by combining the many disciplines into an engineering approach that draws on two perspectives: environmental engineering and risk assessment.

The field of environmental engineering emerged centuries ago, but the descriptive title of environmental engineer came into widespread use only in the last half of the 20th century. In the 1960s, academic institutions began organizing their curricula and research programs under this new moniker, usually as a specialty within civil engineering. In many instances, the field of sanitary engineering was renamed and reconstituted to become environmental engineering. This was more than simple semantics, however, because environmental engineers were increasingly called on to go beyond the design of water supplies, wastewater treatment facilities, and sanitary landfills; they were being asked to play additional, increasingly important roles in protecting public health and ecosystems.

With these new public health and environmental protection charges and mandates came the need to assess human health and environmental risks. This responsibility is not unique to environmental engineering because all engineers are called on to consider risks in their careers. The structural engineer must be aware of and be able to quantify to some degree of satisfaction the risks associated with a structure during its usable life. What is the risk of a building collapsing under various scenarios? The recent events of September 11, 2001, for example, have caused civil engineers to consider risks of collapse that were not previously forecast.[1] The chemical engineer must be cognizant of the risks associated with the synthesis of certain chemicals in reactors and even the use of those chemicals after synthesis. The biomechanical engineer must consider the risk of failure of implanted devices designed to improve the quality of life. Are the devices improving the ability of the user at the expense of some other life activity?

In a sense, the "go or no-go" decision for most engineering designs is based on some sort of risk-reward paradigm, with the need to have costs and risk heavily outweighed by some societal good.[2] Similarly, environmental engineers must consider all possible outcomes, planned or otherwise, of designs. In contrast to most engineers' common concerns about outcome risks, however, the environmental engineer is entirely driven by risks. Whereas other fields of engineering must consider risks as part of their design, the environmental engineer's whole purpose is to address and ameliorate risks.

For the past three or four decades, North Americans have called for continuously decreasing risks in their daily lives. The National Academy of Sciences[3] has attributed this trend at least partly to the economic development in the Western Hemisphere following World War II. This is intellectually akin to Maslow's hierarchy of needs,[4] which states that people worry about higher-level needs only after basic physiologic needs are met. The rapid economic development in the United States and Western Europe allowed for a more thoughtful analysis of possible chronic and long-term environmental consequences. Before, such risks were relegated behind concerns about infectious diseases (e.g., tuberculosis, cholera, influenza, and yellow fever)[5] and insufficient dietary intakes.

History of Hazardous Waste Engineering

Over the past three millennia, humans have generated wastes in exponentially increasing volumes. As societies have attempted to control the environment, greater amounts of wastes have been produced. Engineering has its roots in these attempts at control. The first written documentation of a municipality attempting to control solid wastes was that of Knossos, Crete's burial program for solid wastes produced by the Minoan civilization (a precursor of the modern landfill where waste was buried in layers intermittently covered by soil).[6] In the millennium before Christ, the city-state of Athens required that citizens be responsible for the refuse and garbage they produced, and that they transport the wastes at least 1,500 meters from the city walls for disposal. The ancient Greeks and Romans also addressed the need for potable water supplies.[7] Vitruvius, for example, recognized in the 1st century B.C. that water would become polluted in stationary ponds left to evaporate, a process we now refer to as *eutrophication*. He also noted the generation of "poisonous vapors," which was probably methane generation from the anaerobic, reduced conditions of eutrophic water bodies. Ironically, Vitruvius may well have avoided recommending that the neurotoxic lead be used for water supplies, not because of its toxicity (unknown until the 20th century) but because bronze could better withstand the pressures on the closed pipe systems that were used to move water relatively long distances.[8]

The innovation of the incineration of wastes was led by Europeans, especially Britain and Germany, in the 19th century. The first municipal garbage incineration program was established in Nottingham, England, in 1874, followed in a couple of decades by Britain's first "waste-to-energy" incinerator in the 1890s.[9]

For centuries, humans had been able to move on and leave their wastes behind. Later, wastes were deposited in dumps on land that was sufficiently out of the way and for which there was no perceived value. Municipal refuse was taken to sites in the middle of woodlands. Industrial wastes were disposed of on company property. These wastes, much of which would now be categorized as hazardous, were simply stored above ground—in pits, ponds, and lagoons—or buried under thin layers of soil. In the 1950s and 1960s, initiatives to eliminate open dumps called on engineers to begin designing sanitary landfills. These engineered systems were a response to public health concerns, but possibly more important, to the need to stem the exponential growth of land being dedicated to waste disposal.

Why Engineers Should Care about Hazardous Wastes

We read about environmental and health hazards constantly; we see television reports about concerned neighborhoods or newly discovered industrial diseases; and we hear news reports about chemical spills and releases on the radio. On the Internet, websites are dedicated to a particular class of compounds (e.g., chlorinated organic pesticides or mercury compounds) or to the particular diseases associated with chemical exposures (e.g., endocrine system dysfunction).

People have good reason to be concerned. The public is exposed to measurable concentrations of carcinogens in their drinking water. Effluents from inadequately treated wastewater cause human beings and wildlife to be exposed to substances that behave like hormones or that interfere with the immune and neurologic systems. Urban neighborhoods are dotted with abandoned waste sites and formerly industrialized areas, known as *brownfields*, which have left behind residual contamination. So-called toxic clouds have wafted across oceans and continents. Local officials must decide whether the possible leaching of contaminants from landfills is less of a problem than the potential release of heavy metals, dioxins, furans, and polycyclic aromatic hydrocarbons from improper incineration.

Engineers are called on to protect people and their environment from the potential damages caused by hazardous substances. We are increasingly asked to apply the latest science and technology to prevent and remove the risks concomitant with hazards. This is our public mandate. Philosophers call this our *credat emptor* ("Let the buyer trust"). Unlike the *caveat emptor* ("Let the buyer beware"), the public entrusts its professionals to make wise decisions in their interests and to follow through with design and implementation of solutions to these problems.

This book is about hazardous waste engineering. In particular, it is about the risks imposed by hazardous wastes on individuals and society, and how engineers can confront these risks. The wastes themselves are simply manifestations of economics of society and of lifestyle decisions. Hazardous wastes are merely combinations and mixtures of a small set of elements. The really bad wastes are those that have been arranged, either intentionally or by accident, in a certain way that causes us harm. The carbon, hydrogen, oxygen, and chlorine atoms of the dioxin molecule, for example, could have been sugar and salt (if we added sodium to the mix) under other conditions.

By anthropomorphizing chemicals (e.g., "chemical X is bad, chemical Y is good"), we may lose sight of the fact that the wastes and possible exposures associated with these chemicals actually result from human decisions. Individual human beings and their institutions have been found guilty of both sins of commission and omission (i.e., doing wrong or failing to do what is right). Engineers and the institutions they represent are accountable. In all of their decisions, engineers are accountable to the public, to the companies and agencies that employ them, and to the profession of engineering. A decision to ignore potential problems is indeed a decision. Thus the risks presented by hazardous wastes can include actions knowingly taken by individuals, companies, and governments, as well as any decision that discounts the possibility that hazardous wastes may be generated by the operations. If the past 25 years of legal precedents, the size of fines and penalties, and the enormous cleanup costs accrued is any indication, the public can be no less tolerant of decisions made out of ignorance than they are of those made with intent.

A seminal case study in hazardous wastes was that of Love Canal in upstate New York. The case involved many public and private parties who shared the blame for the contamination of groundwater and exposure of humans to toxic substances. Some, possibly most, of these parties may have been ignorant of the possible chain of events that led to the chemical exposures and health effects in the neighborhoods surrounding the waste site. The decisions by governments, corporations, school boards, and individuals in totality led to a public health disaster. Some of these decisions were outright travesties and breaches of public trust. Others may have been innocently made in ignorance (or even benevolence, such as the attempt to build a school on donated land, which tragically led to the exposure of children to dangerous chemicals). The bottom line is that people were exposed to these substances. Cancer, reproductive toxicity, neurologic disorders, and other health effects resulted from exposures, no matter the intent of the decision makers. Neither the public nor the attorneys and company shareholders accept ignorance on the part of engineers as an excuse for designs and operations that lead to hazardous waste–related exposure and risks.

Case Study: The Case of Love Canal, New York

Addressing the so-called conventional pollutants, such as particle matter, carbon monoxide, and sulfur dioxide in the air, and pathogenic bacteria, biochemical oxygen demand (BOD), and nutrients in water, and moving from open dumps to sanitary landfills occupied much of the attention and concern of environmental engineers up to the 1970s. In the middle part of that decade, however, concerns about toxic and hazardous pollutants began to capture the awareness of both the general public and environmental engineers. No single event epitomized this new concern and crystallized the need for hazardous waste engineering more than the Love Canal controversy.

Love Canal was the key event that led to the passage of national hazardous waste laws, especially the Comprehensive Environmental Response, Compensation and Liability Act (Superfund) and the Resource Conservation and Recovery Acts at the end of the 1970s.

The Love Canal Hazardous Waste Site[10]

A fence now surrounds the infamous 16-acre Love Canal hazardous waste landfill in upstate New York, adjacent to the Niagara River. William T. Love excavated his namesake canal on the site in the 1890s to create a hydroelectric power project, but his dream never materialized. In 1942, Hooker Chemicals and Plastics (now Occidental Chemical Corporation [OCC]) used the landfill to dispose of more than 21,000 tons of chemical wastes that included halogenated organics, pesticides, chlororobenzenes, and dioxins. Disposal stopped in 1952, and the landfill was covered with soil in 1953, the same year that the deed for the site property was ceded to the Niagara Falls Board of Education (NFBE). Subsequently, the area near the covered landfill underwent extensive residential growth. An elementary school and numerous single-family houses were constructed on what was to become the Love Canal hazardous waste site.

Complaints about odors and residues emanating from the abandoned landfill began to be reported in the 1960s, with the frequency and intensity of complaints increasing during the 1970s. The water table rose in the 1970s, and, with the rise, contaminated groundwater was able to migrate near the ground surface and began to leach into basements and structures. Engineering studies found that several toxic chemicals had migrated from the original landfill disposal site to the residential area.

Three miles upstream from the site, contaminated runoff water was also found to be draining into the Niagara River at the intake tunnels for the Niagara Falls water treatment plant. Dioxins and other contaminants had migrated from the landfill into the sewers that drained into feeder creeks. President Jimmy Carter declared an environmental emergency for the Love Canal area in 1978 and again in 1980, which resulted in evacuating nearly 950 families from the 10 square blocks surrounding the landfill. The Federal Emergency Management Agency (FEMA) coordinated the home purchases and residential relocations. In 1980, the neighborhoods on-site became the Emergency Declaration Area (EDA). The EDA covered about 350 acres and was divided into seven separate smaller areas of concern. Approximately 10,000 people are located within one mile of the site, and 70,000 people live within three miles.

The remedial action at the site includes a lining of clay and synthetic materials that covers a total area of 40 acres, along with a barrier drainage system and a leachate collection and treatment system. A consent decree settlement went into effect on December 21, 1995, for the United States to recover costs from OCC and the U.S. Army, which also was found to have contributed to the waste disposal problems. As part of the settlement, OCC and the U.S. Army agreed to pay for the costs resulting from the federal government's response and remediation of the site. This amounted to $129 million for the company. In addition, OCC has also agreed to reimburse certain other costs, including federal oversight costs, and to pay natural resource damages claims. The Army agreed to reimburse $8 million of the response costs. Another $3 million of the settlement funds is dedicated to the Agency for Toxic Substances and Disease Registry to conduct a comprehensive health study from the Love Canal Health Registry. The New York State Department of Health is conducting this study.

Response and Remedial Actions

The site was remediated in several phases, including one initial set of emergency actions and six long-term remedial action phases:

1. Landfill containment with leachate collection, treatment, and disposal.
2. Excavation and interim storage of the sewer and creek sediments.
3. Final treatment and disposal of the sewer and creek sediments and other wastes.
4. Remediation of the 93rd Street School soils.
5. EDA home maintenance and technical assistance by the agency implementing the Love Canal Land Use Master Plan.
6. Buyout of homes and other properties in the EDA.

Three additional smaller remedial actions were taken in 1993: (1) the Frontier Avenue Sewer remediation, (2) the removal of EDA soil, and (3) the repair of a portion of the Love Canal landfill cap.

In 1987, the United States Environmental Protection Agency (U.S. EPA) selected a remedy to destroy and dispose of the dioxin-contaminated sediments from the sewers and creeks, consisting of the following steps:

1. Construction of an on-site facility to dewater and contain the sediments.
2. Construction of a separate facility to treat the dewatered contaminants through high-temperature thermal destruction.
3. Thermal treatment of the residuals stored at the site from the leachate treatment facility and other associated Love Canal waste materials.
4. On-site disposal of any nonhazardous residuals from the thermal treatment or incineration process.
5. Off-site EPA-approved thermal treatment and/or land disposal of the stored Love Canal waste materials.

The sewer and creek sediments and other waste materials were subsequently shipped off-site for final disposal; this remedial action was completed in March 2000.

The 93rd Street School property remediation consisted of excavating 7,500 cubic yards of contaminated soil adjacent to the school and conducting on-site solidification and stabilization. This remedy was reevaluated in 1991, so that the subsequent selected remedy was excavation and off-site disposal of the contaminated soils. This remedial action was completed in 1992. The school building was razed, and the land will be kept vacant.

The amount of material moved and treated during the Love Canal remediation was massive, as the following statistics attest:

- Removed more than 62,000 tons of sewer and creek sediment wastes.
- Collected 76,000 liters of dense nonaqueous phase liquids (DNAPLs).
- Filtered 48,000 liters of DNAPLs.
- Handled 11,000 kg carbon filter wastes.
- Treated about 12 million liters of groundwater per year.

Homeowners have now repopulated the habitable areas of the Love Canal EDA. More than 260 formerly abandoned homes make up a new neighborhood.

The company, OCC, is responsible for continued operation and maintenance of the leachate treatment facility and monitoring. The groundwater on the site is monitored continuously using monitoring wells installed throughout the area. Annual monitoring results have demonstrated that the engineering actions of containment, leachate collection, and treatment are operating as designed.

What Is Our Focus?

In this text, we draw from our experiences in the public, private, and research segments of engineering to give practical means for identifying potential and actual waste problems, preventing future problems, correcting existing problems, and developing comprehensive engineering systems to manage hazardous wastes. This is not a cookbook; rather, it is our attempt to provide practical solutions to the growing problems of hazardous wastes. In particular, we concern ourselves with the risks presented by hazardous wastes and how we as engineers must deal with substances to reduce the risks. Note that we use the term *reduce* rather than *eliminate*. Rarely is it possible to eliminate a very dangerous substance completely from the environment, but it usually is practical to reduce the amount and reduce exposures. In an industrial society, however, the risks of any process from all chemicals cannot be zero. Thus the engineer is increasingly called on to advise decision makers about how to minimize the risks as much as possible.

What Human Values Are Important in Hazardous Waste Decisions?

Engineers are often presented with the challenge of optimizing a set of variables to manage a risk in a manner that provides the greatest number of benefits and reduces the monetary and nonmonetary costs of a project. An engineer can rarely design a system that completely eliminates all risk. In the field of hazardous waste engineering, this is never possible. Under a given set of conditions, any substance is hazardous. The reason for calling in the engineers is to address a problem, whether to ameliorate an existing problem, prevent a future problem, or design systems to render other's wastes "safe." Any option available to the engineer presents risks. Cleaning a site presents short-term risks to the workers and the nearby public during the remediation. Various treatment processes present unique hazards and exposure scenarios that must be evaluated to select the best approach. Intervention is only justifiable if the risks from "no action," the status quo, are greater than the risks associated with any engineering solution.

Our distinguished colleague Henry Petroski has written adeptly on the subject of failure and how engineers must be aware of potential problems, some that have been disastrous, when pushing the envelopes of science and engineering. In his book *To Design Is Human,* Professor Petroski cautions us to be bold, even though we know that there is plenty of self-doubt and always the potential for failure. This approach certainly holds for engineering solutions to hazardous waste problems. The engineer's daunting task is to select from all available solutions the best one, but even assigning what is best depends on human and societal values. The values are a mixture of those held by the profession as expressed by the codes of ethics, building and construction regulations, environmental rules, and other professional mandates. Even the most sound approach (e.g., zero exposure) to a problem is not successful without the concurrence of key decision makers and the incorporation of societal values.

For example, a 50-foot-tall, 10-foot-thick wall surrounding an abandoned waste site may reduce the risk to the public health to nearly zero, but the public would be unlikely to consider this engineering solution acceptable because of its obnoxious aesthetic appeal, its potential effect on property values, and the knowledge that the problem still exists, albeit within the crypt! Ultimately, the values of the individual professionals and the potentially affected public must be integral to the selected action in order to remedy a hazardous waste problem.

Unfortunately, the engineer and the public sometimes work from competing values (e.g., the best scientific approach versus a more socially acceptable approach). This dilemma calls for collaboration and, often, compromise. Even deciding which engineering and scientific approach should be applied to address a waste problem is not always completely clear. In the 1960s and 1970s, much controversy existed over whether environmental engineering—or sanitary engineering as it was called then—should be a set of chemical and physical steps or an emulation of what goes on in natural biologic systems. Ross McKinney, a bioremediation pioneer, was fond of telling his students in the 1980s "to look under their feet" for the solution to environmental problems. He meant that literally, as one could adapt and acclimate soil bacteria like *Pseudomonas* spp. to break down complex organic compounds into simpler, less toxic compounds. The predominant culture in previous decades was one of instantaneous results and abiotic chemistry as the solution to many evils (reminiscent of the advice given to Ben in the movie *The Graduate* to invest in plastics). Professor McKinney, however, was among the first to combine his background in microbiology at MIT with subsequent engineering expertise to help establish biologic treatment methods, which now predominate in the treatment of wastewater. These techniques have also carried over to hazardous waste treatment, where microbial processes are key components of detoxification.

Timeliness and responsiveness are also important human values. The extent and probability of exposure to a hazard may increase with the time elapsed before a remedy is devised. When an abandoned waste site is first discovered, the appropriate, immediate solution may be simply separating people and wildlife from the site using a barrier. Even a simple earth berm may be an overdesign at this point, when a cyclone fence that is sufficiently tall would prevent entry to the site. This is what often occurs at the so-called Superfund sites, where fences and other barriers are erected within a buffer zone around the site, giving engineers and scientists time to begin plans to monitor and to propose remedies. Before the passage of the Comprehensive Environmental Response, Compensation and Liability Act (CERCLA), the Superfund[11] law, preventing exposure to abandoned hazardous wastes was difficult from a legal perspective.

An example of this challenge occurred in 1977 in Kansas City, Kansas, where a trucking company was hired by a local lead smelter to haul away dross from the metal refining process. Unfortunately for the smelter's residential neighbors, the wastes did not find their way to adequate disposal, but were deposited in large piles along the streets. No direct federal, state, or local laws allowed for swift action to prevent exposure to the waste piles, which were found to contain the toxic heavy metals cadmium and nickel, and the metalloid arsenic. The engineers, scientists, and attorneys in the regional office of the U.S. Environmental Protection Agency (U.S. EPA) decided that something had to be done, so they researched the provisions of the 1899 Rivers and Harbors Act and the Federal Water Pollution Control Act Amendments of 1972 to determine whether the piles could be deemed a "spill" into the waters of the United States. In fact, a rivulet near the piles was found to flow to a small stream that entered the Kansas River. The attorneys considered this to be a spill, albeit a slowly moving one, approaching the U.S. waters, so immediate removal of the piles and legal actions under the water rules ensued. The need for such legal gymnastics was obviated by the Superfund law's provisions of timely, emergency responses to reduce the opportunities for exposure to hazardous wastes.

The major lesson learned from these experiences is that the engineer should be bold and creative in searching for and developing solutions to hazardous waste problems, while being sufficiently attentive to the possibility of failure. The corporate client, as represented by the company's field engineers and CEO, and the public, as represented by elected and appointed officials, will demand acceptable risks. But costs, time, and other expressions of the values held by the public will not allow zero-risk solutions. In the words of Professor Petroski, "While it is theoretically possible to make the number representing risk as close to zero as desired, human nature in its collective and individual manifestations seems to work against achieving such a risk-free society."[12]

Hazardous waste management decisions will remain important to the public and will be emblematic of professional engineers' success or failure for decades and centuries. The public considers a wide range of factors beyond the math and science of risk and reward. Any hazardous waste solution must be based on strong engineering principles, but this approach is not sufficient for public acceptance. We will explore ways to incorporate other societal factors into hazardous waste management decisions.

What Is Hazardous Waste, Anyway?

Both words in the term *hazardous wastes* are crucial to engineers. *Hazard* is a component of risk. A hazard is expressed as the potential of unacceptable outcome (see Table 1-1). For chemicals, the most important hazard is the potential for disease or death (measured by epidemiologists as morbidity and mortality, respectively). So, the hazards to human health are referred to collectively in the medical and environmental sciences as toxicity. Toxicology is the study of these health outcomes and their potential causes.

TABLE 1-1
Four Types of Hazards Important to Hazardous Wastes, as Defined by the Resource Conservation and Recovery Act

Hazard Type	*Criteria*	*Physical/Chemical Classes Included*
Corrosivity	A substance with an ability to destroy tissue by chemical reactions.	Acids, bases, and salts of strong acids and strong bases. The waste dissolves metals, other materials, or burns the skin. Examples include rust removers, waste acid, alkaline cleaning fluids, and waste battery fluids. Corrosive wastes have a pH of < 2.0 or > 12.5. The U.S. EPA waste code for corrosive wastes is D002.

(continued)

TABLE 1-1 *(continued)*

Hazard Type	*Criteria*	*Physical/Chemical Classes Included*
Ignitability	A substance that readily oxidizes by burning.	Any substance that spontaneously combusts at 54.3°C in air or at any temperature in water, or any strong oxidizer. Examples are paint and coating wastes, some degreasers, and other solvents. The U.S. EPA waste code for ignitable wastes is D001.
Reactivity	A substance that can react, detonate, or decompose explosively at environmental temperatures and pressures.	A reaction usually requires a strong initiator (e.g., an explosive like TNT, trinitrotoluene), confined heat (e.g., salt peter in gunpowder), or explosive reactions with water (e.g., Na). A reactive waste is unstable and can rapidly or violently react with water or other substances. Examples include wastes from cyanide-based plating operations, bleaches, waste oxidizers, and waste explosives. The U.S. EPA waste code for reactive wastes is D003.
Toxicity	A substance that causes harm to organisms. Acutely toxic substances elicit harm soon after exposure (e.g., highly toxic pesticides causing neurologic damage within hours after exposure). Chronically toxic substances elicit harm after a long period of exposure (e.g., carcinogens, immunosuppressants, endocrine disruptors, and chronic neurotoxins).	Toxic chemicals include pesticides, heavy metals, and mobile or volatile compounds that migrate readily, as determined by the Toxicity Characteristic Leaching Procedure (TCLP), or a TC waste. TC wastes are designated with waste codes D004 through D043. (See Table 1-3.)

Risk is a function of the hazard and exposure. The term *hazard* refers exclusively to the chemical of concern. What are the intrinsic characteristics of the chemical or mixture of chemicals in the waste that can cause harm? The threshold level[13] of chemical is the lowest amount needed to induce harmful effects in an organism. In addition to this inherent toxicity of the compound (e.g., cyanide and dioxin are highly and acutely toxic, whereas the metal iron is usually only acutely toxic in high doses), the hazard is also influenced by factors such as (1) a chemical's mobility (how quickly does it move through the environment or across cellular membranes), (2) its persistence (remaining in the environment for years without being altered, for example, a chlorinated compound is inherently more difficult to break down than its nonhalogenated counterpart), and (3) its likelihood to accumulate in living tissue (e.g., mercury and lead can build up in tissue over years and decades with long-term exposures).

The hazard term can be expressed as a gradient. Dose is the amount (often mass) of a chemical administered to an organism (so-called applied dose), the amount of the chemical that enters the organism (internal dose), the amount of the chemical that is absorbed by an organism over a certain time interval (absorbed dose), or the amount of the chemical or its metabolites that reaches a particular target organ (biologically effective dose), such as the amount of a hepatotoxin (liver-damaging chemical) that reaches the liver. Theoretically, the higher the concentration of a hazardous substance that an organism contacts, the greater the expected adverse outcome. The classic demonstration of this gradient is the so-called dose-response curve (Figure 1-1). If one increases the amount of the substance, a greater incidence of the adverse outcome would be expected.

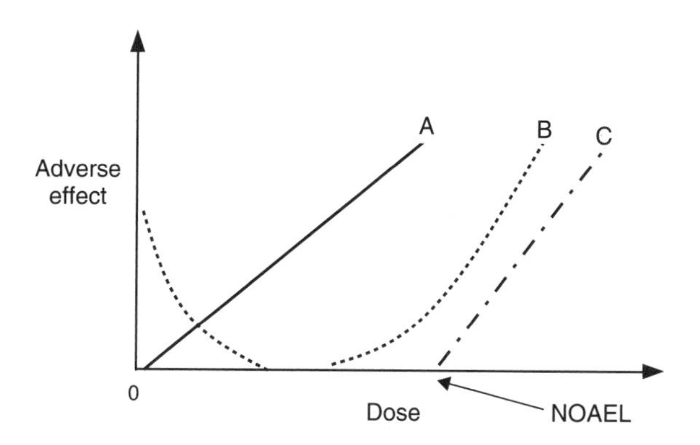

FIGURE 1-1. Three prototypical dose-response curves. Curve A represents the no-threshold curve, which expects a response (e.g., cancer) even if exposed to a single molecule (this is the most conservative curve). Curve B represents the essential nutrient dose-response relationship and includes essential metals, such as trivalent chromium or selenium, where an organism is harmed at the low dose due to a deficiency (left side) and at the high dose due to toxicity (right side). Curve C represents toxicity above a certain threshold (noncancer). This threshold curve expects a dose at the low end where no disease is present. Just below this threshold is the no observed adverse effect level (NOAEL).

The three curves in Figure 1-1 represent those generally found for toxic chemicals.[14] Curve A is the classic cancer dose-response curve. Regulatory agencies generally subscribe to the precautionary principle that any amount of exposure to a cancer-causing agent may result in an expression of cancer at the cellular level. Thus the curve intercepts the x-axis at 0. Metals can be toxic at high levels, but several are essential to the development and metabolism of organisms. Thus Curve B represents an essential chemical (i.e., a nutrient) that will cause dysfunction at low levels (below the minimum intake needed for growth and metabolism) and toxicity at high levels. The segment of Curve B that runs along the x-axis is the optimal range of an essential substance. Curve C is the classic noncancer dose-response curve. The steepness of the three curves represents the potency or severity of the toxicity. For example, Curve C is steeper than Curve A, so the adverse outcome (disease) caused by chemical in Curve C is more potent than that of the chemical in Curve A. This simply means that the response rate is higher; however, if the diseases in question are cancer (Curve A) and a relatively less important disease for Curve C, such as short-lived headaches, then the steepness simply represents a higher incidence of the disease, not greater importance.

The shape and slope of the curve are formed according to available data. Several uncertainties are associated with these data. The dose-response relationship is often based on comparative biology from animal studies. These are usually high-dose, short-duration (at least compared to a human lifetime) studies. From these animal data, models are constructed and applied to estimate the dose-response relationship that may be expected in humans. Thus the curve may be separated into two regions (Figure 1-2). When environmental exposures do not fall within the range of observation, extrapolations must be made to establish a dose relationship. Generally, extrapolations are made from high to low doses, from animal to human responses, and from one route of exposure to another. The first step in establishing a dose-response relationship is to assess the data from empirical observations. To complete the dose-response curve, extrapolations are made either by modeling or by employing a default procedure based on information about the chemical's biochemical characteristics.[15]

Dose-response models may be biologically based, with parameters calculated from curve-fitting of data. If data are sufficient to support a biologically based model specific to a chemical, and significant resources are available, then this is usually the model of choice. Biologically based models require large amounts of data.

Case-specific models employ model parameters and information gathered from studies specific to a particular chemical. Often, however, neither the biologically based nor the case-specific model is selected because the necessary data or the significant costs cannot be justified.

Curve-fitting is another approach used to estimate dose-response relationships for chemicals. Such models are used when response data in the

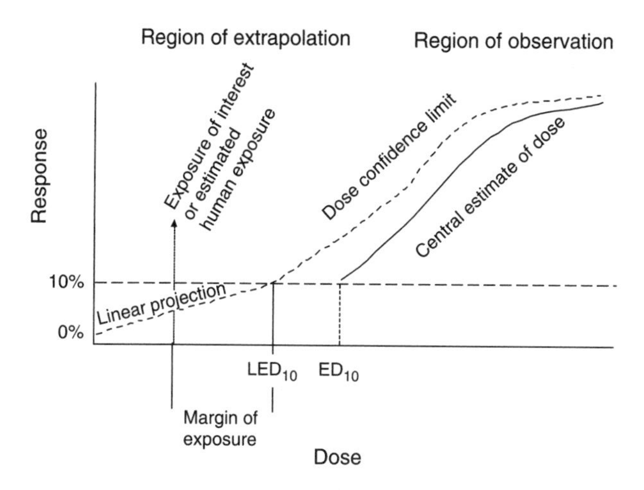

FIGURE 1-2. Dose-response curves showing the two major regions of data availability. (*Source:* Based on discussions with the U.S. Environmental Protection Agency.)

observed range are available. A so-called "point of departure" for extrapolation is estimated from the curve. The point of departure is a point that is either a data point or an estimated point that can be considered to be in the range of observation, without the need for much extrapolation. The LED_{10} in Figure 1-2 is the lower 95% confidence limit on a dose associated with 10% extra risk. This is an example of such a point and, in fact, is often the standard point of departure. The central estimate in Figure 1-2 of the ED_{10} (the estimate of a 10% increased response) also may be used to describe a relative hazard and potency ranking.

Risk is calculated by multiplying the slope of the dose-response curve by the actual contact with the substance (i.e., exposure). If either term is zero, the risk is zero. The risk associated with even the most toxic substance is zero if there is no exposure. If there is an extremely toxic substance on the planet Jupiter, one's risk on Earth is zero. The risk will only increase if the substance finds its way to Earth or if we find our way to Jupiter. Similarly, a nontoxic substance—if there is such a substance—will never elicit a risk because the toxicity is zero; however, the reality of risk is always within these extremes. The engineer is challenged to reduce risks at both ends, by decreasing the toxicity of a substance and by eliminating, or at least limiting, the exposures to the substance.

Hazardous waste is specifically defined by the federal government. Section 1004(5) of the Resource Conservation and Recovery Act (RCRA) defines a hazardous waste to be a solid waste that may "pose a substantial present or potential threat to human health and the environment when improperly treated, stored, transported, or otherwise managed." The RCRA made the U.S. EPA responsible for defining which specific solid wastes would be considered hazardous waste either by identifying the characteristics of a hazardous waste or by listing particular hazardous wastes. Thus, a solid waste is hazardous if:[16]

1. *The waste is officially listed as a hazardous waste on one of the four U.S. EPA groupings* (Note: The engineer should check frequently whether any of the wastes of concern have been listed because the lists are updated periodically by the federal government, as new data and research are published.):

    - *F List*. Chemicals that are generated via nonspecific sources by chemical manufacturing plants to produce a large segment of chemicals. A solvent must comprise at least 10% of the waste before use.
    - *K List*. Wastes from 17 specific industries that use specific chemical processes (e.g., veterinarian drug or wood preservative manufacturing). The processes included on the K List are specifically defined by regulation, so the engineer involved in work related to chemical manufacturing processes is well advised to investigate all past, present, and possible processes to determine whether they fall into this list.
    - *P List*. Acutely hazardous, technical-grade (i.e., approximately 100% composition and sole active ingredient) chemicals discarded by commercial operations.
    - *U List*. Toxic, but not acutely hazardous, technical-grade chemicals discarded by commercial operations, which are also classified as corrosive, ignitable, reactive, or toxic (see Table 1-1).

2. *Based on testing, the waste is found to be corrosive, ignitable, reactive, or toxic* (see Table 1-1).
3. *The generator of the waste reports and declares that the waste is hazardous based on its proprietary information or other knowledge about the waste*. (Note: It is always good ethics and good business practice to exercise full disclosure in matters related to potential hazards, including those for chemicals that are not listed per se by the enforcement agencies. Full disclosure is also sound professional practice because it would be embarrassing and potentially damaging to an engineer's career if information were available to the company documenting a hazard, but this was not disclosed until legal proceedings.)[17]

Mixtures of any listed hazardous waste with other wastes will require that the engineer manage all of the mixture as a listed hazardous waste. Spills of listed waste that impact soils and other unconsolidated material are also regulated as the listed hazardous waste. If a listed hazardous waste is spilled, the engineer must immediately notify the appropriate state agency or the U.S. EPA to determine how best to manage the impacted material that contains the listed waste. The so-called characteristic wastes may not appear on one of the EPA lists, but they are considered hazardous if they exhibit one or more of the characteristics described in Table 1-1. The quantity of the waste also matters. Generators of small quantities of hazardous wastes can be treated less stringently than large-quantity generators (see Appendix 3).

Other classifications have been applied to hazardous wastes. For example, biologically based criteria have been used to characterize the hazard and ability of chemicals to reach and affect organisms (see Table 1-2). This text is concerned primarily with human health hazards and risks; however, the engineer should be aware of the risks to other receptors, especially those associated with ecosystems (see Sidebar Discussion: Ecological Risk Assessment in Chapter 2).

TABLE 1-2
Biologically-Based Classification Criteria for Hazardous Waste

Criterion	*Description*
Bioconcentration	The process by which living organisms concentrate a chemical to levels exceeding the surrounding environmental media (e.g., water, air, soil, or sediment).
Lethal Dose (LD)	A dose of a chemical calculated to expect a certain percentage of a population of an organism (e.g., minnow) exposed through a route other than respiration (dose units are mg [chemical] kg^{-1} body weight). The most common metric from a bioassay is the lethal dose 50 (LD_{50}), wherein 50% of a population exposed to a chemical is killed.
Lethal Concentration (LC)	A calculated concentration of a chemical in the air that, when respired for four hours (i.e., exposure duration = 4 h) by a population of an organism (e.g., rat) will kill a certain percentage of that population. The most common metric from a bioassay is the lethal concentration 50 (LC_{50}), wherein 50% of a population exposed to a chemical is killed. (Air concentration units are mg [chemical] L^{-1} air.)
Phytotoxicity	The chemical's ability to elicit biochemical reactions that harm flora (plant life).

Source: P. Aarne Vesilind, J. Jeffrey Peirce, and Ruth F. Weiner, *Environmental Engineering*, 3rd edition (Boston, MA: Butterworth-Heinemann, 1993).

Toxicity Testing

The U.S. EPA also developed standard approaches and set criteria to determine whether waste exhibited any of the hazardous characteristics. The testing procedures are generally defined and described in the *Test Methods for Evaluating Solid Waste* (SW-846).[18] The extraction procedure (EP) was the original test developed by the EPA to establish whether a waste was hazardous by virtue of its toxicity. Because the RCRA defines a hazardous waste as a waste that presents a threat to human health and the environment when the waste is improperly managed, the government identified the set of assumptions that would allow for the means for a waste to be disposed if the waste is not subject to controls as mandated by Subtitle C of the RCRA. This so-called "mismanagement scenario" was designed to simulate a plausible worst case of mismanagement. Under a worst-case scenario, a potentially hazardous waste is assumed to be disposed along with municipal solid waste in a landfill with actively decomposing substances overlying an aquifer. When the government developed the mismanagement scenario, it recognized that not all wastes would be managed in this manner but that a dependable set of assumptions would be needed to ensure that the hazardous waste definition is implemented. So the U.S. EPA used a conservative approach.

The conservative assumption of mismanagement drove the EP. This led to selecting drinking water that has leached from a landfill as the most likely pathway for human exposure. So, the EP defined the toxicity of a waste by measuring the potential for finding toxic substances in the waste that have leached and migrated to contaminate groundwater and surface water (and ultimately sources of potable water).

The specific EP called for the analysis of a liquid waste or liquid waste extract to see if it contained unacceptably high concentrations of any of 14 toxic constituents identified in the National Interim Primary Drinking Water Standards,[19] because at the time that the EP was being developed these were the only official health-based federal standards available.

Following the worst-case scenario, the solid waste (following particle size reduction, if necessary) was extracted using organic acids (acids likely to be found in a landfill containing decomposing municipal wastes). To simulate the likely dilution and degradation of the toxic constituents as they would migrate from the landfill to a water source, the drinking water standards were multiplied by a dilution and attenuation factor (DAF) equal to 100, which the government considered to represent a substantial hazard.

The Hazardous and Solid Waste Amendments of 1984 (HSWA) redirected the government to broaden the toxicity characteristic (TC) and to reevaluate the EP, especially to see if the EP adequately addressed the mobility of toxic chemicals under highly variable environmental conditions. The U.S. Congress was specifically concerned that the leaching medium being used was not sufficiently aggressive to identify a wide range of hazardous wastes, but mainly focused on metals (particularly in their "elemental" or zero-valence form) and did not give enough attention to wastes that contain hazardous organic compounds. So in 1986, a new procedure was developed.

The Toxicity Characteristic Leaching Procedure (TCLP) was designed to provide replicable results for organic compounds and to yield the same type of results for inorganic substances as those from the original EP test. The government added 25 organic compounds for testing (see Table 1-3). These additions were based on the availability of chronic toxicity reference levels. The U.S. EPA applied a subsurface fate and transport model to confirm whether the DAF of 100 was still adequate. The TCLP begins with the same mismanagement assumptions as those that established the EP. The test procedure is the same as that of the EP, except that the TCLP allows the use of two extraction media. The specific medium used in the test is dictated by the alkalinity of the solid waste. The liquid extracted from the waste is analyzed for the 39 listed toxic constituents in Table 1-3, and the concentration of each contaminant is compared to the TCLP standards specific to each contaminant.

The concept of risk is expressed as the likelihood (statistical probability) that harm will occur when a receptor (e.g., human or a part of an ecosystem) is exposed to that hazard. So, an example of a toxic hazard is a carcinogen (a cancer-causing chemical), and an example of a toxic risk is the likelihood that a certain population will have an incidence of a particular type of cancer after being exposed to that carcinogen (e.g., the population risk that one person out of 1 million will develop lung cancer when exposed to a certain dose of carcinogen X for a certain period (we will consider several examples of these exposure and risk calculations in Chapter 5).

Other hazards besides toxicity are also important to hazardous waste engineering. The outcome may relate to environmental quality, such as an ecosystem stress, loss of important habitats, and decreases in the size of the population of sensitive species. Outcomes related to public and personal safety are also important to engineers. These may include a substance's potential to ignite, its corrosiveness, it flammability, or its explosiveness. Finally, a substance may be a public welfare hazard that damages property values or physical materials, expressed for example as its corrosiveness or acidity. The so-called hazard may be inherent to the substance, but more than likely, the hazard depends on the situation and conditions in which the exposure may occur. The substance is most hazardous when several dangerous conditions exist simultaneously; witness the hazard to firefighters using water to extinguish flames in the presence of barrels of oxidizers.

The word *wastes* means that the substances of concern to hazardous waste engineers have no apparent value; however, there is certainly no unanimity in a substance's value. In fact, the adage "One man's trash is another man's treasure" holds true in hazardous waste management and engineering. The successes of waste exchanges and clearinghouses provide evidence of the situational usefulness and harmfulness of chemical substances.

TABLE 1-3
Toxicity Characteristic Chemical Constituent Regulatory Levels for 39 Hazardous Chemicals.

Contaminant	*Regulatory Level (mg/L)*	*EPA Number*
Arsenic	5.0	D004
Barium	100.0	D005
Cadmium	1.0	D006
Chromium	5.0	D007
Lead	5.0	D008
Mercury	0.2	D009
Selenium	1.0	D010
Silver	5.0	D011
Endrin	0.02	D012
Lindane	0.4	D013
Methoxychlor	10.0	D014
Toxaphene	0.5	D015
2,4-D	10.0	D016
2,4,5 TP (Silvex)	1.0	D017
Benzene	0.5	D018
Carbon tetrachloride	0.5	D019
Chlordane	0.03	D020
Chlorobenzene	100.0	D021
Chloroform	6.0	D022
o-Cresol	200.0	D023
m-Cresol	200.0	D024
p-Cresol	200.0	D025
Cresol	200.0	D026
1,4-Dichlorobenzene	7.5	D027
1,2-Dichloroethane	0.5	D028
1,1-Dichloroethylene	0.7	D029
2,4-Dinitrotoluene	0.13	D030
Heptachlor (and its hydroxide)	0.008	D031
Hexachlorobutadiene	0.5	D033
Hexachloroethane	3.0	D034
Methyl ethyl ketone	200.0	D035
Nitrobenzene	2.0	D036
Pentachlorophenol	100.0	D037
Pyridine	5.0	D038
Tetrachloroethylene	0.7	D039
Trichloroethylene	0.5	D040
2,4,5-Trichlorophenol	400.0	D041
2,4,6-Trichlorophenol	2.0	D042
Vinyl chloride	0.2	D043

The challenge to the engineer is to remove or modify the characteristics of a substance that render it hazardous or to relocate the substance to a situation where it has value. An example of the former would be the dehalogenation of chlorinated benzenes to transform them into compounds that can be used as solvents in manufacturing or laboratories. An example of the latter is the so-called adopt-a-chemical programs in laboratories, where solvents and reagents left over in one laboratory are made available to other laboratories.[20]

Risk assessment sounds like a technical term. It can be, but risk assessment is really something people do constantly. Human beings decide throughout each day whether the risk from particular behaviors is acceptable or whether the potential benefits of a behavior do not sufficiently outweigh the hazards associated with that behavior. Classic examples may include one's decision whether to drink coffee that contains the alkaloid caffeine. The benefits include the morning jump-start, but the potential hazards include induced cardiovascular changes in the short term and possible longer-term hazards from chronic caffeine intake.

Note that the foregoing example includes a no-action alternative, along with several other actions. One may choose not to drink coffee or tea. Other examples may include other actions, with concomitant risk. As mentioned earlier, however, the no-action alternative is not always innocuous. For example, if one knows that exercise is beneficial but does not act on this knowledge, the potential for adverse cardiovascular problems is increased. If one does not ingest an optimal amount of vitamins and minerals, disease resistance may be jeopardized. If one always stays home to avoid the crowds, no social interaction is possible and the psyche suffers. This is a microcosm of the engineer's challenge. The engineer must take an action only if it provides the optimal solution to the hazardous waste problem, while avoiding unwarranted financial costs, without causing unnecessary disruption to normal activities, and in a manner that is socially acceptable to the community.

Thus the engineer is faced with an enormous responsibility to represent the client in a manner that the handling and management of hazardous wastes is conducted with due diligence. This book has been written to be one of the available tools to all engineers, whether hazardous waste is their principal charge or whether there is even a remote possibility that hazardous wastes may be generated as an ancillary outcome of the planning, design, construction, and operation of any project.

CHAPTER 2

Entering the Risk Era

How Engineers Can Manage Hazardous Waste Risks

Several paradigms have been proposed to deal with environmental risk assessment and management. Risk assessment is a process during which information is analyzed to determine the extent that an environmental hazard might cause harm to exposed persons and ecosystems.[1] Risk assessment of an existing problem or of a decision about possible actions requires an understanding of (1) a chemical hazard; (2) the adverse outcomes associated with that hazard; (3) the possible exposures of people, ecosystems, and materials to those chemicals; and (4) a means of compiling this information to determine the overall risks of a chemical in the environment (Figure 2-1). Risk management lays out the approaches needed to address a risk. This includes the scientific and engineering basis that underpins the risk assessment, plus it must add socioeconomic, political, legal, spiritual, ethical, and other human values to frame the risk decision. Finally, the risk must be communicated to the parties who stand to be affected by the decision. The risk paradigm indicated in Figure 2-1 applied to these decisions may be a series of contingent steps, a single set of feedback mechanisms and simultaneous activities, or permutations of both. The important message is that risk assessments must be reliable (based on sound scientific methods and data) and thorough (see Sidebar: Cleaning up a Hazardous Waste Site).

Discussion: Cleaning up a Hazardous Waste Site

The U.S. EPA has established a set of steps to determine the potential for a release of contaminants from a hazardous waste site. These steps are known as the Superfund cleanup process. The first step is a preliminary assessment/site inspection (PA/SI), from which the site is ranked in the agency's Hazard Ranking System (HRS). The HRS is a process that

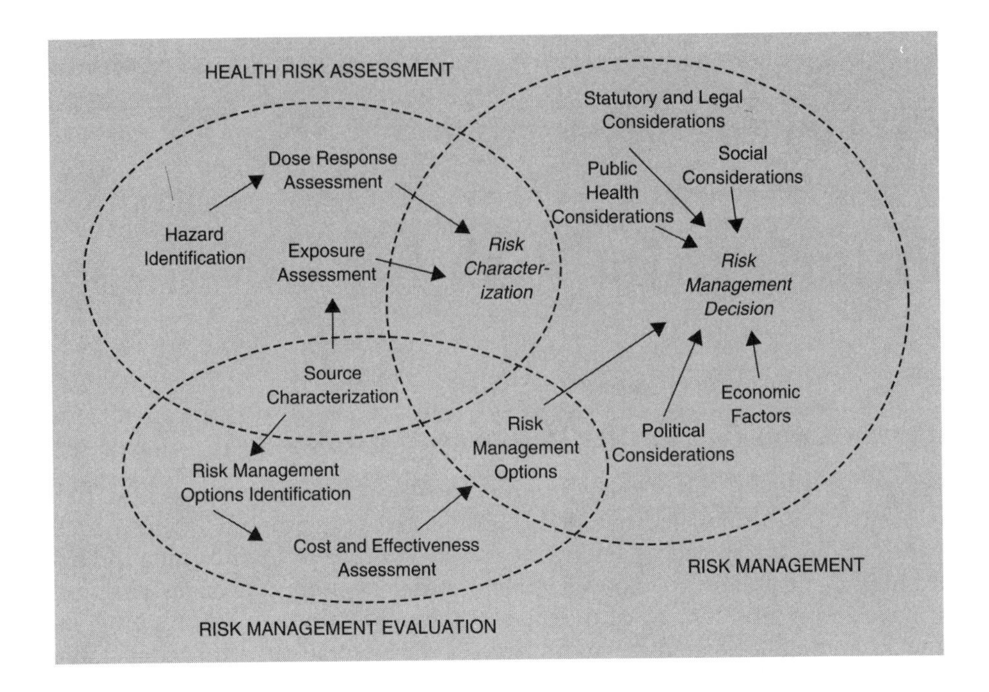

FIGURE 2-1. The risk-assessment paradigm. (*Source:* U.S. Environmental Protection Agency.)

screens the threats of each site to determine if the site should be listed on the National Priority Listing (NPL), which is a list of the most serious sites identified for possible long-term cleanup, and what the rank of a listed site should be. Following the initial investigation, a formal remedial investigation/feasibility study (RI/FS) is conducted to assess the nature and the extent of contamination. The next formal step is the record of decision (ROD), which describes the various possible alternatives for cleanup to be used at an NPL site. Next, a remedial design/remedial action (RD/RA) plan is prepared and implemented. The RD/RA plan specifies which remedies will be undertaken at the site and lays out all plans for meeting cleanup standards for all environmental media. The construction completion step identifies the activities that were completed to achieve cleanup. After completion of all actions identified in the RD/RA plan, a program for operation and maintenance (O&M) is carried out to ensure that all actions are as effective as expected and that the measures are operating properly and according to the plan. Finally, after cleanup and demonstrated success, the site may be deleted from the NPL.

The engineer may be called on as a consultant to a company or to a government agency to lead remediation efforts at a hazardous waste site. Although all sites are unique, several steps must be taken for any hazardous waste facility. First, the location of the site should be clearly specified, including the formal address and geodetic coordinates. The history of the site, including present and all past owners and operators, should be documented. The search for this background information should include both formal (e.g., public records) and informal (e.g., newspapers) documentation.

The main businesses that have operated on the site, as well as any ancillary interests, should be documented and investigated. For example, in the famous Times Beach, Missouri, incident, the operator's main business was an oiling operation to control dust and to pave roads. Unfortunately, the operator also ran an ancillary waste oil hauling and disposal business. The combination of these two businesses (i.e., spraying waste oil that had been contaminated with dioxins) led to the widespread problem and numerous Superfund sites in Missouri, including the relocation of the entire town of Times Beach.

The investigation at this point should include *all* past and present owners and operators. Any decisions regarding *de minimus* interests will be made later (by the government agencies and attorneys). At this point, one should search for every potentially responsible party (PRP). A particularly important part of this review is to document all sales of the property or any parts of the property. Also, all commercial, manufacturing, and transportation concerns should be known because these may indicate the types of wastes that have been generated or handled at the site. Even an interest of short duration can be important, if this interest produced highly persistent and toxic substances that may still be on-site, or that may have migrated off-site. The investigation should also determine whether any attempts were made to dispose of wastes from operations, either on-site or, through manifest reports, whether any wastes were shipped off-site. A detailed account should be given of all waste reporting, including air emission and water discharge permits, voluntary audits that include tests like the toxicity characteristic leaching procedure (TCLP), and compare these results to benchmark levels, especially to determine if any of the concentrations of contaminants exceed the U.S. EPA hazardous waste limit (40 CFR 261). For example, the TLCP limit for lead (Pb) is 5 mg L^{-1}. Any concentration above this federal limit in the soil or sand on the site must be reported.

Initial monitoring and chemical testing should be conducted to target those contaminants that may have resulted from operations over previous years or decades. A more general surveillance is also needed to identify a broader suite of contaminants. This is particularly important in soil and groundwater because their rates of migration (Q) are slow compared to the rates usually found in air and surface water transport. Thus the likelihood of finding remnant compounds is greater in soil and groundwater. Also, in addition to parent chemical compounds, chemical degradation products should also be targeted because decades may have passed since the waste was buried, spilled, or released into the environment.

An important part of the preliminary investigation is identifying possible exposures, both human and environmental. For example, the investigation should document the proximity of the site to schools, parks, water supplies, residential neighborhoods, shopping areas, and businesses.

One means of efficiently implementing a hazardous waste remedial plan is for the present owners (and past owners, for that matter) to work voluntarily with government health and environmental agencies. States often have voluntary action programs that can be an effective means of expediting the process, such as the Delaware Department of Natural Resources and Environmental Control's (DNREC) Voluntary Cleanup Program Agreement (VCP).

Through the VCP Agreement companies can participate in, and even lead, the remedial investigation and feasibility study (RI/FS) consistent with a state-approved work plan (which can be drafted by their consulting engineer). The feasibility study (FS) delineates potential remedial alternatives, comparing the cost effectiveness to assess each alternative approach's ability to mitigate potential risks associated with the contamination. The FS also includes a field study to retrieve and chemically analyze (at a state-approved laboratory) water and soil samples from all environmental media on the site. Soil and vadose zone contamination will likely require that test pits be excavated to determine the type and extent of contamination. Samples from the pit are collected for laboratory analysis to determine general chemical composition (e.g., a so-called total analyte list) and TCLP levels (that indicate leaching, i.e., the rate of movement of the contaminants).

An iterative approach may be appropriate as the data are derived. If the results from the screening (e.g., total analytical tests) and the leaching tests indicate that the site's main problem is with one or just a few contaminants, then a more focused approach to cleanup may be in order. For example, if preliminary investigation indicated that for most of the site's history a metal foundry was in operation, then the first focus should be on metals. If no other contaminants are identified in the subsequent investigation, a remedial action that best contains metals may be in order. If a clay layer is identified at the site from test pit activities and extends laterally beneath the foundry's more porous overburden material, the clay layer should be sampled to see if any screening levels have been exceeded. If groundwater contamination has not been found beneath the metal-laden material, an interim removal action may be appropriate, followed by a metal treatment process for any soil or environmental media laden with metal wastes. For example, metal-laden waste has recently been treated by applying a buffered phosphate and stabilizing chemicals to inhibit Pb leaching and migration.

During and after remediation, water and soil environmental performance standards must be met and confirmed by sampling and analysis. Post-stabilization sampling and TCLP analytical methods should be used to assess contaminant leaching (e.g., to ensure that Pb does not violate the federal standard of 5 mg L^{-1}). Confirmation samples must be analyzed to verify complete removal of contaminated soil and media in the lateral and vertical extent within the site.

The remediation efforts subject to the terms of the VCP should be clearly delineated in the final plan for remedial action, such as the total surface area of the site to be cleaned up and the total volume of waste to be decontaminated. In Delaware, for example, the VCP document is the DNREC's Final Plan of Remedial Action issued under the Delaware Hazardous Substance Cleanup Act (HSCA) and Delaware's regulations governing hazardous waste cleanup.

At a minimum, a remedial action is evaluated on the basis of the current and proposed land use around the site; applicable local, state, and federal laws and regulations; and a risk assessment that specifically addresses the hazards and possible exposures at or near the site. Any proposed plan should summarize the environmental assessment and the potential risks to public health and the environment posed by the site. The plan should clearly delineate all remedial alternatives that have been considered. It should also include data and information on the background and history of the property, the results of the previous investigations, and the objectives of the remedial actions. Because this document is official, the state environmental agency must abide by federal and state requirements for public notice, as well as provide a sufficient public comment period (about 20 days).

The final plan must address all comments. The Final Plan of Remedial Action must clearly designate the selected remedial action, which will include the target cleanup values for the contaminants, as well as all monitoring that will be undertaken during and after the remediation. It must include both quantitative (e.g., to mitigate risks posed by metal-laden material with total Pb > 1,000 mg kg^{-1} and TCLP Pb $\geq$ 5.0 mg L^{-1}) and qualitative (e.g., control measures and management to ensure limited exposures during cleanup) objectives. The plan should also include a discussion on planned and potential uses of the site following remediation (e.g., will it be zoned for industrial use or changed to another land use?). The plan should distinguish between interim and final actions, as well as interim and final cleanup standards. The proposed plan and the final plan then constitute the remedial decision record.

The ultimate goal of the remediation is to ensure that all hazardous material on the site has either been removed or rendered nonhazardous through treatment and stabilization. The nonhazardous, stabilized material can then be properly disposed of (e.g., in a nonhazardous waste landfill).

This risk-assessment/management/communication process may appear to be logical and based on common sense, and it is, but it has only recently been enunciated.[2] In fact, it has only recently become codified. When William D. Ruckelshaus returned to the post of Administrator of the U.S. Environmental Protection Agency (EPA) in 1983, he called for the separation of risk assessment and risk management in EPA decisions.[3] The EPA had been correctly accused of sometimes making decisions in the absence of sound science. It is important to remember that the EPA's reputation at the time of Ruckelshaus's return was at an all-time low. The public was losing confidence in its ability to protect the environment. Hazardous wastes were in the news daily.[4] Commonly used products were found to contain previously unknown contaminants with long names, like polychlorinated biphenyls (PCBs).[5] People were calling for advocacy from governments, businesses, and policy makers to take actions to deal with these wastes in socially just ways.[6]

In 1983, the National Research Council[7] proposed a four-component process that has since been adopted by most public health institutions and agencies in the United States:

1. *Hazard identification.* The process of identifying the chemical substances of concern, and the compilation, review, and evaluation of relevant data concerning the hazardous properties of the substances, especially their toxicity.
2. *Dose-response evaluation.* Assessment of the relationship between dose and response for each chemical of potential concern, often ascertained from animal studies.
3. *Exposure assessment.* Identification of the ways people and organisms are exposed to the chemical substances, including the exposure pathways, the transport, transformation and fate of chemicals in the environment, and the estimation of the magnitude and duration of chemical exposure for the potential exposure pathways.
4. *Risk characterization.* Calculation of numerical estimates of risks for each chemical substance through each route of exposure, employing the dose-response information and the exposure estimate calculations.

Generally, health risk assessments will provide several estimates:

- Individual risks based on a statistical "central tendency" exposure within defined geographic areas, expressed both as averages across the area and at the location of maximum chemical concentrations within each area.
- Risks to potentially highly exposed or susceptible subgroups, including infants and young children, within the general population.
- Risks associated with certain behavior patterns and activities that could lead to elevated exposures, such as subsistence fishing near contaminated surface waters.
- Individual risks based on high-end exposure to particular subgroups of a population that are considered to be highly exposed; this can account for some of the variability in exposure within an exposed group.
- Cumulative risks to the population in the vicinity of a source, such as people living or working in a plume created by an incinerator's stack emissions.

This risk-assessment approach provides an estimation of risk to specific segments of the population, taking into consideration the site-specific activity patterns, certain unique qualities of the individuals in each subgroup, and the actual locations of individuals within these subgroups.

How Toxicity Is Calculated and Applied to Risk

Risk assessments require a means to describe toxicity.[8] Reference dose (RfD), reference concentration (RfC), toxicity slope factors, and unit risk values are used to determine potential toxic effects (RfD and RfC) or the possibility of excess cancers (slope factors and unit risks). These values are used to calculate risks from specific pathways, especially oral, inhalation, and dermal.

To begin to evaluate whether a substance is toxic, scientists rely on data from animals (comparative biology) and human studies. These studies usually fall under the heading of epidemiology, the study of diseases in human population. An important consideration is whether exposure to certain chemicals causes disease. The best that science usually can do in this regard is to provide enough weight-of-evidence to support or reject a suspicion that a substance causes a disease. The medical research and epidemiologic communities use several criteria to determine the strength of an argument for causality, but the first well-articulated criteria were Hill's Causal Criteria[9] (see Table 2-1). Regarding risk assessment, some of Hill's criteria are more important than others.

TABLE 2-1
Hill's Criteria for Causality

Factors to be considered in determining whether exposure to a chemical elicits an effect:	
Criterion	*Description*
1: Strength of Association	For a chemical exposure to cause an effect, the exposure must be associated with that effect. Strong associations provide more certain evidence of causality than weak associations. Common epidemiologic metrics used in association include risk ratio, odds ratio, and standardized mortality ratio.
2: Consistency	If the chemical exposure is consistently associated with an effect under different studies using diverse methods of study of assorted populations under varying circumstances by different investigators, the link to causality is stronger. For example, if the carcinogenic effects of Chemical X are found in mutagenicity studies, mouse and Rhesus monkey experiments, and human epidemiologic studies, there is greater consistency between Chemical X and cancer than if only one of these studies showed the effect.
3: Specificity	The specificity criterion holds that the cause should lead to only one disease and that the disease should result from only this single cause. This criterion appears to be based in the germ theory of microbiology, where a specific strain of bacteria and viruses elicits a specific disease. This is rarely the case in studying most chronic diseases, because a chemical can be associated with cancers in numerous organs, and the same chemical may elicit cancer, hormonal, immunologic, and neural dysfunctions.
4: Temporality	Timing of exposure is critical to causality. This criterion requires that exposure to the chemical must precede the effect. For example, in a retrospective study, the researcher must be certain that the manifestation of a disease was not already present before the exposure to the chemical. If the disease were present prior to the exposure, it may not mean that the chemical in question is not a cause, but it does mean that it is not the sole cause of the disease (see "Specificity").

TABLE 2-1 *(continued)*

5: Biologic Gradient	This is another essential criterion for chemical risks. In fact, this is known as the dose-response step in risk assessment. If the level, intensity, duration, or total level of chemical exposure is increased then a concomitant, progressive increase should occur in the toxic effect.
6: Plausibility	Generally, an association needs to follow a well-defined explanation based on a known biologic system; however, paradigm shifts in the understanding of key scientific concepts do change. A noteworthy example is the change in the latter part of the 20th century of the understanding of how the endocrine, immune, and neural systems function, from the view that these are exclusive systems to today's perspective that in many ways they constitute an integrated chemical and electrical set of signals in an organism.*
7: Coherence	The criterion of coherence suggests that all available evidence concerning the natural history and biology of the disease should stick together (cohere) to form a cohesive whole. By that, the proposed causal relationship should not conflict or contradict information from experimental, laboratory, epidemiologic, theory, or other knowledge sources.
8: Experimentation	Experimental evidence in support of a causal hypothesis may come in the form of community and clinical trials, in vitro laboratory experiments, animal models, and natural experiments.
9: Analogy	The term *analogy* implies a similarity in some respects among things that are otherwise different. It is thus considered one of the weaker forms of evidence.

*For example, Candace Pert, a pioneer in endorphin research, has espoused the concept of mind/body, with all the systems interconnected, rather than separate and independent systems.

Comparison Values

The Agency for Toxic Substances and Disease Registry (ATSDR) is an agency of the U.S. Public Health Service and is required by law to conduct a public health assessment at each of the sites on the EPA's National Priorities List.[10]

These health assessments attempt to see if people are being exposed to hazardous substances. If appropriate, the ATSDR also conducts public health assessments when petitioned by concerned individuals. If approved by the ATSDR, state agencies may also conduct public health assessments.

The ATSDR establishes comparison values for different media (e.g., drinking water) to determine if measured concentrations are safe under default exposure conditions. Comparison values are used as screening values in the preliminary identification of contaminants of concern at a hazardous waste site. A contaminant of concern is a chemical found at the site that the health professionals select to be analyzed for potential human health effects. These contaminants will vary according to the type of industrial or commercial enterprise previously or currently operating at the site. For example, Table 2-2 lists the contaminants of concern at a waste combustion facility. Combustion facilities are associated with two major types of contaminants: (1) those that are part of the feedstock coming into the facility, and (2) those that are actually produced by the combustion process. Pollutants brought to an incinerator may be released to the air via residues or fugitive emissions and to the soil and water from spills and leaking containers or vehicles. Products of incomplete combustion (PICs) are generated when the incinerator is not properly operated. The goal of incineration is to produce carbon dioxide and water from the complex organic compounds (see Chapter 3); however, with improper temperatures, pressures, sorbents, or other key components of the combustion operation, by-products are produced. The list of organic compounds in Table 2-2 includes both types of contaminants of concern. For example, it includes pesticides, such as chlordane, aldrin, and lindane, that are brought to the combustor awaiting incineration, and PICs such as benzo(a)pyrene, dioxins, furans, and hexaclorobenzene, which are produced from other organic compounds. The toxic metals can be concentrated thermodynamically and released in much higher concentrations than they are found in the feedstock; however, metals and toxic inorganic compounds (such as hydrogen cyanide) may also be found in relatively high concentrations before thermal treatment.

A chemical is selected as a contaminant of concern because its maximum concentration in air, water, or soil at the site exceeds one of the ATSDR's comparison values, but they are not toxicity thresholds. The occurrence of any health effects associated with a contaminant of concern depends on the actual on-site, specific conditions and the individual lifestyle and genetic factors that affect the route, magnitude, and duration of actual exposure, but not simply the measured environmental concentrations.

Screening values based on noncancer effects are obtained by dividing the threshold levels (e.g., the no observable adverse effects level [NOAEL] or lowest observable adverse effects level [LOAEL]) determined in animal or, less often, human studies by cumulative safety margins (i.e., safety factors, uncertainty factors, or modifying factors). Cumulative safety factors generally range from 10 to 10^3. Carcinogen screening levels, however, are usually derived by linear extrapolations from data obtained from studies wherein animals are exposed at high doses, because human cancer incidence data for very low levels of exposure do not exist. The resulting screening values (i.e., environmental media evaluation guides and cancer risk evaluation guides) can be used to predict the health risk associated with low-level exposures in humans.

TABLE 2-2
Contaminants of Concern at a Waste Combustion Facility

Products of Incomplete Combustion (PICs) and Residual Organic Compounds		
Acenaphthene	Butylbenzylphthalate	Dichlorobenzidine, 3,3′-
Acenaphthylene	Carbon disulfide	Dichlorobiphenyl
Acetaldehyde	Carbon tetrachloride	Dichlorodifluoromethane
Acetone	Chlordane	Dichloroethane, 1,1-
Acetophenone	Chloro-3-methylphenol, 4-	Dichloroethane, 1,2-
Acrolein	Chloroacetophenone, 2-	Dichloroethene, 1,1-
Acrylonitrile	Chloroaniline, p-	Dichloroethylene, trans-1,2-
Anthracene	Chlorobenzene	Dichlorofluoromethane
Benzaldehyde	Chlorobenzilate	Dichlorophenol, 2,4-
Benzene	Chloroethane	Dichloropropane, 1,2-
Benzoic acid	Chloroform	Dichloropropene, cis-1,3-
Benzotrichloride	Chloromethane	Dichloropropene, trans-1,3-
Benzo(a)anthracene	Chloronaphthalene, beta	Diethylphthalate
Benzo(a)pyrene	Chlorophenol, 2-	Dimethoxybenzidine, 3,3′-
Benzo(b)fluoranthene	Chlorodiphenylether, 4-	Dimethylphenol, 2,4-
Benzo(e)pyrene	Chloropropane, 2-	Dimethylphthalate
Benzo(g,h,i)perylene	Chrysene	Di-n-butylphthalate
Benzo(j)fluoranthene	Cresol, m-Cresol,	Di-n-octyl phthalate
Benzo(k)fluoranthene	o-Cresol,	Dinitritoluene, 2,6-
Benzyl chloride	p-Crotonaldehyde	Dinitro-2-methylphenol, 4,6-
Biphenyl	Cumene	Dinitrobenzene, 1,2-
Bis(2-chloroethoxy) methane	2,4-D	Dinitrobenzene, 1,3-
Bis(2-chloroethyl)ether	4,4′-DDE	Dinitrobenzene, 1,4-
Bis(2-chloroisopropyl)ether	Dibenz(a,h)anthracene	Dinitrophenol, 2,4-
Bis(2-ethylhexyl)phthalate	Dibenz(a,h)fluoranthene	Dinitrotoluene, 2,4-
Bromochloromethane	Dibromo-3-chloropropane,1,2-	Dioxane, 1,4-
Bromodichloromethane	Dibromochloromethane	Ethyl methacrylate
Bromoethene	Dichloro-2-butene, cis-1,4-	Ethylbenzene
Bromoform	Dichloro-2-butene, trans-1,4-	Ethylene dibromide
Bromomethane	Dichlorobenzene, 1,2-	Ethylene oxide
Bromodiphenylether, p-	Dichlorobenzene, 1,3-	Ethylene thiourea
Butadiene, 1,3-	Dichlorobenzene, 1,4-	
Butanone, 2- (Methyl ethyl keytone)		

(continued)

TABLE 2-2 *(continued)*

Products of Incomplete Combustion (PICs) and Residual Organic Compounds		
Fluoranthene	Methyl-2-Pentanone, 4-	Styrene
Fluorene	Monochlorobiphenyl	Tetrachlorobenzene, 1,2,4,5-
Formaldehyde	Naphthalene	Tetrachlorobiphenyl
Furfural	Nitroaniline, 2-	Tetrachloroethane, 1,1,1,2-
Heptachlor	Nitroaniline, 3-	Tetrachloroethane, 1,1,2,2-
Heptachlorobiphenyl	Nitroaniline, 4-	Tetrachloroethene
Hexachlorobenzene	Nitrobenzene	Tetrachlorophenol, 2,3,4,6-
Hexachlorobiphenyl	Nitrophenol, 2-	Toluene
Hexachlorobutadiene	Nitrophenol, 4-	Toluidine, o-Toluidine, p-Trichloro-1,2,2-Trifluoroethane, 1,1,2-
Hexachlorocyclohexane, alpha-	N-Nitroso-di-n-butylamine	Trichlorobenzene, 1,2,4-
Hexachlorocyclohexane, beta-	N-Nitroso-di-n-propylamine	Trichlorobiphenyl
Hexachlorocyclohexane, gamma- ("lindane")	N-Nitrosodiphenyl amine	Trichloroethane, 1,1,1-
Hexachloro cyclopentadiene	Nonachlorobiphenyl	Trichloroethane, 1,1,2-
Hexachloroethane	Octachlorobiphenyl	Trichloroethene
Hexachlorophene	Pentachlorobenzene	Trichlorofluoromethane
Hexane, n-Hexanone, 2-	Pentachlorobiphenyl	Trichlorophenol, 2,4,5-
Hexanone, 3-	Pentachloronitro benzene	Trichlorophenol, 2,4,6-
Indeno(1,2,3-cd)pyrene	Pentachlorophenol	Trichloropropane, 1,2,3-
Isophorone	Phenanthrene	Vinyl acetate
Maleic hydrazide	Phenol	Vinyl chloride
Methoxychlor	Phosgene	Xylene, m-
Methylene bromide	Propionaldehyde	Xylene,o-
Methylene chloride	Pyrene	Xylene,p-
Methylnaphthalene, 2-	Quinoline	
Methyl-tert-butyl ether	Quinone	
	Safrole	

Dioxin Congeners	*Furan Congeners*
2,3,7,8-Tetrachlorodibenzo-para-dioxin	2,3,7,8-Tetrachorodibenzofuran
1,2,3,7,8-Pentachlorodibenzo-para-dioxin	1,2,3,7,8-Pentachorodibenzofuran
1,2,3,4,7,8-Hexachlorodibenzo-para-dioxin	2,3,4,7,8-Pentachorodibenzofuran
1,2,3,6,7,8-Hexachlorodibenzo-para-dioxin	1,2,3,4,7,8-Hexachorodibenzofuran
1,2,3,7,8,9-Hexachlorodibenzo-para-dioxin	1,2,3,6,7,8-Hexachorodibenzofuran
1,2,3,4,6,7,8-Heptachlorodibenzo-para-dioxin	1,2,3,7,8,9-Hexachorodibenzofuran
Octachlorodibenzo-para-dioxin	2,3,4,6,7,8-Hexachorodibenzofuran
	1,2,3,4,6,7,8-Heptachorodibenzofuran
	1,2,3,4,7,8,9-Heptachorodibenzofuran
	Octachorodibenzofuran

TABLE 2-2 *(continued)*

Metals	
Aluminum	Lead
Antimony	Mercury (inorganic and organic)
Arsenic	Nickel
Barium	Selenium
Beryllium	Silver
Cadmium	Thallium
Chromium (hexavalent and trivalent)	Zinc
Copper	

Acid Gases	*Particulate Matter*
Hydrogen chloride	Respirable particles ($\leq 10\ \mu m$ diameter)
Total nitrogen oxides (NO_X)	Fine particles ($\leq 2.5\ \mu m$ diameter)
Total sulfur oxides (SO_X)	

Fugitive Volatile Organic Compounds		
Acetone	Cyclohexane	Hydrazine
Acetonitrile	Cyclohexanone	Indeno(1,2,3-cd)pyrene
Acetophenone	Dibenz(a,h)anthracene	Isobutanol
Acetylaminofluorene, 2-	Dibromoethane, 1,2-	Isopropanol
Acrylonitrile	Dichlorobenzene	Isosafrole
Alcohols	Dichlorodifluoroethane	Maleic anhydride
Aliphatic hydrocarbons	Dichlorodifluoromethane	Methanol
Aniline	Dichloroethane, 1,1-	Methyl methacrylate
Benzene	Dichloroethene	Methylbutadiene, 1-
Benzenedicarboxylic acid, 1,2-	Diethyl stilbestrol	Methylcholanthrene, 3-
Benzidine	Diethylphthalate	Methyl isobutyl ketone
Benzoquinone, p-Benzo(a) pyrene	Dimethyl sulfate	Naphthalene
Butanol	Dimethylamine	Naphthylamine, 1-
Butanone, 2-	Dimethylbenzidine, 3,3′-	Naphthylamine, 2-
Butyl acetate	Dimethylhydrazine	Nitrobenzene
Calcium chromate	Dimethylphenol, 2,6-	Nitrophenol, 4-
Carbon	Dimethylphthalate	Nitropropane, 2-
Carbon disulfide	Dinitrotoluene	N-nitrosodiethanol amine
Carbon tetrachloride	Dioxane, 1,4-	N-nitrosodiethylamine
Chlorobenzene	Epichlorohydrin	N-nitrosodi-n-butyl amine
Chloroform	Ethanol	N-nitrosopyrolidine
Chlorinated paraffin, oil, wax	Ethoxyethanol, 2-	Phenol
Chrysene	Ethyl acrylate	Phthalic anhydride
Creosote (coal tar)	Ethylbenzene	Picoline, 2-
Cresol	Fluoranthene	Pyridine
Crotonaldehyde	Formaldehyde	Resorcinol
Cumene	Formic acid	
	Furfural	
	Heptane	

(continued)

TABLE 2-2 *(continued)*

Fugitive Volatile Organic Compounds		
Tetrachlorobenzene, 1,2,4,5-	Toluene diisocyanate	Trichloroethene
Tetrachloroethane, 1,1,1,2-	Toluenediamine	Trichlorofluoromethane 1,1,2-
Tetrachloroethene	Trichloro-1,2,2,-TFE,	Xylene
Tetrahydofuran	Trichlorobenzene	
Toluene	Trichloroethane, 1,1,1-	
Fugitive Ash Emissions		
Arsenic		Nickel
Barium		Selenium
Cadmium		Silver
Lead		Cyanide

The types of comparison values that the ATSDR uses include:

CREGs = cancer risk evaluation guides
MRLs = minimal risk levels
EMEGs = environmental media evaluation guides
RMEGs = reference dose media evaluations

Cancer risk evaluation guides (CREGs) are estimated contaminant concentrations in air, soil, or water that are expected to cause no greater than one excess cancer in 1 million persons (risk $\leq 10^{-6}$) exposed over a lifetime. CREGs are calculated from the EPA's cancer slope factors (see discussion later in this chapter).

Minimal risk levels (MRLs) are estimates of daily human exposure to chemicals (i.e., doses [mg $kg^{-1}day^{-1}$]) that are not expected to be associated with any appreciable risk of noncancer effects over a specified duration of exposure (see Appendix 2). MRLs are derived for acute (14 or fewer days), intermediate (15 to 364 days), and chronic (more than 365 days) exposures. The ATSDR publishes these MRLs in its toxicologic profiles for specific chemicals. The MRLs are substance-specific estimates that are meant to be used as screening levels for hazardous waste sites; however, the ATSDR does not intend that the MRLs be used to define cleanup or action levels.

The toxicologic profile is a document about a specific substance in which scientists interpret available information on the chemical to specify hazardous exposure levels. The profile also identifies knowledge gaps and uncertainties about the chemical. When a toxicologic profile is developed, MRLs are derived when meaningful data exist to associate a chemical with an effect in a target organ(s) or when scientific information can identify the most sensitive health effect for a specific duration for a given route of exposure to a chemical.

The environmental media evaluation guides (EMEGs) are concentrations of a chemical in the air, soil, or water below which noncancer effects are not expected to be associated with exposures over a specified duration of exposure. The EMEGs are derived from the MRLs by including a default factor for body weights and ingestion rates. Separate EMEGs are given for acute (14 days or fewer), intermediate (15 to 364 days), and chronic (more than 365 days) exposures.

Reference Dose

The RfD[11] is used to determine the systemic effects to human beings exposed to toxic substances, representing the highest allowable daily exposure associated with a noncancerous disease. It is calculated from the threshold value, uncertainty, and modifying factors:

$$\mathrm{RfD} = \frac{(\mathrm{NOAEL})}{(\mathrm{UF}_{1\ldots n}) \times (\mathrm{MF}_{1\ldots n})} \tag{2-1}$$

where,

RfD = Reference dose (mg kg^{-1} d^{-1})
$UF_{1\ldots n}$ = Uncertainty factors related to the exposed population and chemical characteristics (dimensionless, usually factors of 10)
$MF_{1\ldots n}$ = Modifying factors, which reflect the results of qualitative assessments of the studies used to determine the threshold values (dimensionless, usually factors of 10)

The uncertainty factors address the robustness and quality of data used to derive the RfD, especially to be protective of sensitive populations (e.g., children and the elderly). It also addresses extrapolation of animal data from comparative biologic studies to humans, accounting for differences in dose-response curves among different species. An uncertainty factor can also be applied when the studies on which the RfD is based are conducted with various study designs (e.g., if an acute or subchronic exposure is administered to determine the NOAEL, but the RfD is addressing a chronic disease, or if a fundamental study used an LOAEL as the threshold value, requiring that the NOAEL be extrapolated from the LOAEL). The modifying factors address the uncertainties associated with the quality of data used to derive the threshold values, mainly from qualitative, scientific assessments of the data.

For airborne chemicals, a reference concentration (RfC) is used in the same way as the RfD (i.e., the RfC is an estimate of the daily inhalation exposure that is likely to be without appreciable risk of adverse effects during a lifetime). The oral chronic RfD is used with administered oral doses under long-term exposures (i.e., exposure duration longer than 7 years), while the oral subchronic RfD is applied for shorter exposures of 2 weeks to 7 years. The inhalation chronic RfD applies to long-term exposures and is derived from an inhalation chronic RfC. Likewise, the subchronic inhalation RfD is derived from the inhalation subchronic RfC. The dermal chronic RfD and subchronic RfD relate to absorbed doses under chronic exposures and subchronic exposures, respectively.

Minimal Risk Levels

Like the RfDs, the minimal risk levels (MRLs) are based only on noncancer diseases, and the NOAEL/UF approach is used to derive MRLs for hazardous chemicals. An MRL is established below the level expected to induce any adverse health effects in the most sensitive human subpopulations. Presently, a suitable method for deriving dermal MRLs is still not available.

As mentioned, MRLs are intended to serve as a screening tool to help engineers and other public health experts decide where to direct their attention in waste activities. The MRLs can also serve as a way of identifying hazardous waste sites that are not expected to cause serious adverse health effects; however, like other risk metrics, all MRLs include uncertainties in the precision of available toxicologic data for sensitive groups such as children, the elderly, and immunologically compromised persons, as well as the fact that many MRLs are derived from animal data, rather than from human epidemiology. That is why the ATSDR requires that safety factors be built into the MRLs, making them more protective than explicitly called for by the data, sometimes two orders of magnitude below the levels found to be nontoxic to laboratory animals. When adequate information is available, physiologically based pharmacokinetic (PBPK) modeling and benchmark dose (BMD) modeling can complement the NOAEL/UF in deriving the MRLs.

Recently, more than 100 inhalation MRLs, nearly 200 oral MRLs, and 6 dermal MRLs have been set. A listing of the published MRLs by route and duration of exposure is provided in Appendix 2.

Hazard Index

The risk of a population contracting a noncancerous disease is calculated from the maximum dose and a predetermined acceptable daily intake (likely from the same studies used to derive the RfD and RfC). This is the hazard index,[12] which is calculated as:

$$\text{Risk} = \frac{\text{MDD}}{\text{ADI}} \quad (2\text{-}2)$$

where,

Risk = Population risk (dimensionless)
MDD = Maximum dose of a chemical received on any given day during an exposure period (mg kg^{-1})

ADI = Allowable daily intake, the daily dose that is likely not to elicit a noncarcinogenic effect over an extended exposure period (mg kg^{-1})

Cancer Slope Factor

Unlike the reference dose, which provides a "safe" level of exposure, cancer risk assessments generally assume no threshold. Thus, the NOAEL and LOAEL are meaningless for cancer risk. Instead, cancer slope factors are used to calculate the estimated probability of increased cancer incidence over a person's lifetime (the so-called excess lifetime cancer risk, ELCR). Like the reference doses, slope factors follow exposure pathways.

The cancer slope factor is an upper-bound estimate of the likelihood that a carcinogenic response will occur per unit intake of a chemical over a 70-year lifetime exposure. It is a measure of the potency of a chemical carcinogen that is derived by applying a mathematical model to extrapolate from the relatively high doses given to experimental animals (or, less often, from human studies, most often from occupational exposures). The models extrapolate the lower exposure levels expected for human contact in the environment. Numerous models are employed to extrapolate from high-dose to low-dose conditions. The approach used by the U.S. EPA to develop SF values is the linearized multistage model, which is generally believed to be conservative and precautionary, meaning that it is likely to overpredict the true potency of a chemical.

Table 2-3 illustrates the toxicity values[13] for several pesticides that have been applied in the State of New York. Note that only malathion and permethrin are designated as carcinogens and that permethrin is an order of magnitude more carcinogenic than malathion (i.e., the slope factor is 1.84×10^{-2} $mg^{-1}kg\text{-}d^{-1}$ for permethrin versus 1.52×10^{-3} $mg^{-1}kg\text{-}d^{-1}$ for malathion).

Cancer Classifications

Based on the scientific weight-of-evidence available for the hazardous chemical, the U.S. EPA classifies the chemical's cancer-causing potential. Carcinogens fall into the following classifications (in descending order of strength of weight-of-evidence):

- "A" Carcinogen: The chemical is a human carcinogen.
- "B" Carcinogen: The chemical is a probable human carcinogen, with two subclasses:
    - B1: Chemicals that have limited human data from epidemiologic studies supporting their carcinogenicity.

TABLE 2-3
Toxicity Values for Six Pesticides Used in the Northeastern United States

	Non-Cancer Hazards									*Cancer Risk*	
	Acute (short-term exposure duration)			*Subchronic (intermediate exposure duration)*			*Chronic (long-term exposure duration)*			*(the additional probability of contracting cancer over a lifetime)*	
Active Ingredient	*RfD Skin (mg/kg-d)*	*RfD Ingestion (mg/kg-d)*	*RfC Inhalation (mg/L)*	*RfD Skin (mg/kg-d)*	*RfD Ingestion (mg/kg-d)*	*RfC Inhalation (mg/L)*	*RfD Skin (mg/kg-d)*	*RfD Ingestion (mg/kg-d)*	*RfC Inhalation (mg/L)*	*Cancer Slope Factor (CSF) $(mg/kg\text{-}d)^{-1}$*	*Unit Risk Factor (UR), Dust Inhalation $(\mu g/m^3)^{-1}$*
Malathion	0.5	0.60	0.0001	0.5	0.024	0.0001	0.024	0.024	0.0001	0.00152	0.000000434
Naled	0.01	0.01	0.0000023	0.01	0.01	0.0000023	0.002	0.002	0.0000023	NC	NC
Permethrin	1.5	0.26	0.0025	1.5	0.155	0.0025	0.05	0.05	0.00025	0.0184	0.00000626
Resmethnin	10	0.1	0.0001	10	0.1	0.0001	0.03	0.03	0.00001	NC	NC
Sumithnin	10	0.7	0.0029	10	0.7	0.0029	0.071	0.071	0.00029	NE	NE
Propenyl Butoxide	10.0	2.0	0.00074	10	0.0175	0.00074	0.0175	0.0175	0.00007	NE	NE

Source: U.S. Environmental Protection Agency
Notes: RfC = Reference Concentration
RfD = Reference Dose
CSF = Cancer Slope Factor
UR = Unit Risk Factor
NC = No evidence of carcinogenicity
NE = Limited evidence of carcinogenicity; no CFS established
(mg/kg-d) = milligram of elucide active ingredient per kilogram human body weight per day
(mg/L) = milligram of active ingredient per liter of air
$(mg/kg\text{-}day)^{-1}$ = Risk per milligram of active ingredient per kilogram human body weight per day
$(\mu g/m^3)^{-1}$ = Risk per microgram of active ingredient per cubic meter of air; microgram in one millionth of a gram

    - B2: Chemicals for which there is sufficient evidence from animal studies but for which there is inadequate or no evidence from human epidemiologic studies.

- "C" Carcinogen: The chemical is a possible human carcinogen.
- "D" Chemical: The chemical is not classifiable regarding human carcinogenicity.
- "E" Chemical: There is evidence that the chemical does not induce cancer in humans.

If a compound can exist in several forms (see Chapter 3 for the discussion on isomers and congeners), then the toxicity of all these forms may be grouped together using the toxic equivalency method (TEF). In other words, the engineer is more concerned about the contaminant in all of its forms, rather than each species. For example, the chlorinated dioxins have 75 different forms and there are 135 different chlorinated furans, simply by the number and arrangement of chlorine atoms on the molecules. The compounds can be separated into groups that have the same number of chlorine atoms attached to the furan or dioxin ring. Each form varies in its chemical, physical, and toxicologic characteristics. The most toxic form is the 2,3,7,8-tetrachlorodibenzo-*p*-dioxin (TCDD) isomer. Isomers with configurations similar to the 2,3,7,8 arrangement (see Chapter 3) are also considered to have higher toxicity than the dioxins and furans with different chlorine atom arrangements. It is best that the individual toxicity of each chemical in a mixture like hazardous waste is known, but this is usually quite costly and time-consuming or even possible to determine. So, the TEF (Table 2-4) provides an aggregate means of estimating the risks associated with exposure to mixtures of the chlorinated dioxins and furans, as well as other highly toxic groups, such as the PCBs and polycyclic aromatic hydrocarbons (PAHs). These standards can serve as benchmarks for hazardous waste cleanup target levels.

Arguably, dioxins represent the most important group of hazardous chemicals for which TEFs are calculated. The U.S. EPA classifies the toxicity of each individual isomer in the mixture by assigning each form a TEF as it compares to the most toxic form of dioxin. An example of recommended standards for each major pathway or environmental medium is that set by the Province of Ontario, Canada (see Table 2-5). Chlorinated compounds share certain similar chemical structures and biologic characteristics. The 2,3,7,8-tetrachlorodibenzo-*p*-dioxin (TCDD) is the most studied and the most highly toxic of these compounds. Scientists consider that dioxins cause toxic effects in similar ways. Dioxins usually exist as mixtures of congeners in the environment. Using TEFs, the toxicity of a mixture can be expressed in terms of its toxicity equivalents (TEQs), which represents the amount of TCDD that would be required to equal the combined toxic effect of all the dioxin-like compounds found in that mixture. In this approach, the concentration

TABLE 2-4
Toxic Equivalency Factors for the Most Toxic Dioxin and Furan Congeners

	Toxic Equivalency Factor
Dioxin Congeners	
2,3,7,8-Tetrachlorodibenzo-para-dioxin	1.0
1,2,3,7,8-Pentachlorodibenzo-para-dioxin	0.5
1,2,3,4,7,8-Hexachlorodibenzo-para-dioxin	0.1
1,2,3,6,7,8-Hexachlorodibenzo-para-dioxin	0.1
1,2,3,7,8,9-Hexachlorodibenzo-para-dioxin	0.1
1,2,3,4,6,7,8-Heptachlorodibenzo-para-dioxin	0.01
Octachlorodibenzo-para-dioxin	0.001
Furan Congeners	
2,3,7,8-Tetrachorodibenzofuran	0.1
1,2,3,7,8-Pentachorodibenzofuran	0.5
2,3,4,7,8-Pentachorodibenzofuran	0.05
1,2,3,4,7,8-Hexachorodibenzofuran	0.1
1,2,3,6,7,8-Hexachorodibenzofuran	0.1
1,2,3,7,8,9-Hexachorodibenzofuran	0.1
2,3,4,6,7,8-Hexachorodibenzofuran	0.1
1,2,3,4,6,7,8-Heptachorodibenzofuran	0.01
1,2,3,4,7,8,9-Heptachorodibenzofuran	0.01
Octachorodibenzofuran	0.001

TABLE 2-5
Ontario, Canada's Environmental Standards for Dioxins and Furans, Based on Toxic Equivalence (TEQ) to 2,3,7,8-TCDD Congener

Air:	Ambient Air Quality Criterion (24 hour)–5 picograms TEQ / cubic meter.
Drinking Water:	Interim Maximum Allowable Concentration–15 picograms TEQ / liter.
Surface Water:	A Canadian Water Quality Guideline is in preparation.
Surface Soils:	Residential Soil Remediation Criterion–1,000 picograms TEQ / gram.
	Agricultural Soil Remediation Criterion–10 picograms TEQ / gram.

of each dioxin is multiplied by its respective TEF. The measurement and analysis of dioxins is complicated and difficult.

There are several ways of expressing limitations in analysis and detection (see Table 2-7). If a result is reported as nondetected, the U.S. EPA conservatively sets it to one-half of the detection level.[14] The products of the concentrations and their respective TEFs are then summed to arrive at a single TCDD TEQ value for the complex mixtures of dioxins in the sample.

TABLE 2-6
Commonly Used Human Exposure Factors*

Exposure Factor	*Adult Male*	*Adult Female*	*Child*†
Weight (kg)	70	60	15–40
Total fluids ingested ($l\ d^{-1}$)	2	1.4	1.0
Surface area of skin, without clothing (m^2)	1.8	1.6	0.9
Surface area of skin, wearing clothes (m^2)	0.1–0.3	0.1–0.3	0.05–0.15
Respiration/ventilation rate ($l\ min^{-1}$) – Resting	7.5	6.0	5.0
Respiration/ventilation rate ($l\ min^{-1}$) – Light activity	20	19	13
Volume of air breathed ($m^3\ d^{-1}$)	23	21	15

*These factors are updated periodically by the U.S. EPA in the *Exposure Factor Handbook* at www.epa.gov/ncea/exposfac.htm.

†The definition of child is highly contested in risk assessment. The *Exposure Factors Handbook* uses these values for children between the ages of 3 and 12 years.

The TEF method is determined by first identifying all of the isomeric forms in the mixture, multiplying the concentrations of each isomer by its corresponding TEF factor.[15] For dioxins, the products are summed to obtain the total 2,3,7,8-TCDD equivalents in the mixture. From these summed equivalents the human exposures and risks are calculated as total dioxin/furan exposures and total dioxin/furan risks. (The same procedure is used for calculating total PCB or PAH exposures and risks.)

For the past 30 years, several different sets of TEFs have been used to evaluate mixtures of dioxin compounds. No uniform set of TEFs presently exists, but two are currently in use: (1) the International set and (2) the World Health Organization (WHO) set. Both the International approach and the WHO approach include a total of 17 dioxin and furan compounds (see Chapter 3 for descriptions of these compounds). The WHO approach for developing TEF values differs from the International approach for three compounds, two of which would not alone significantly change any TCDD TEQ value. For one compound, however, a pentachlorinated dioxin, the TEF for the WHO 1998 approach is twice as high (1 versus 0.5) as that of the International approach. In other words, in the International approach, the pentachlorinated dioxin is considered to have half the potency of TCDD, but WHO considers the two compounds to have equal potencies (TEF = 1). This is not simply a hypothetical problem because if a waste has high amounts of a pentachlorinated dioxin, it is going to be considered much more hazardous using the WHO method. This issue arose in reporting dioxin results during the World Trade Center environmental response, where it was decided that the International approach would be applied to dioxin findings.

Estimating Exposure to Hazardous Waste[16]

Armed with an understanding of the concept of hazard and of its health metric, toxicity, we are now ready to explore the second part of the risk equation, exposure. Human exposure is defined as the contact of a chemical with a person's surfaces, or exposure boundary. An exposure assessment is the actual calculation or estimation of the amount of the chemical agent that comes in contact with a human population. The assessment characterizes the intensity, frequency, and duration of contact and should measure or model the rates at which the chemical crosses the boundary (chemical intake or uptake rates), the routes by which the chemical enters the person.

An exposure pathway is the physical course that a chemical takes from the source to the organism exposed. Important chemical pathways include air, drinking water, food, or soil. A route of exposure is the manner that a chemical enters an organism after contact (e.g., by dermal, ingestion, or inhalation). Routes can be further subdivided. For example, ingestion can be by direct food and water ingestion, plus by nondietary ingestion. The latter term includes exposures such as pica (e.g., children eating paint chips or dirt) and contaminants transferred by hand to mouth from surfaces.

The exposure term is not as direct as it may first appear. For example, the applied dose and the amount of the chemical that is present at the interface (e.g., the mouth, nose, or skin) may be different from the amount that actually finds its way into the body. This difference is reflected in an organism's (or person's) absorption fraction—the percentage of a substance that penetrates through the boundary into the body. Furthermore, the absorbed dose is usually larger than the internal and biologically effective doses and varies according to an organism's ability to metabolize and detoxify chemicals that cross the exposure boundary.

In addition to the metabolic factors affecting exposure, a person's behavior is extremely important, often dictating the amount, if any, of a person's exposure to a contaminant. Therefore, the exposure estimate must account for a population's activity pattern (i.e., information concerning the activities of various individuals, the length of time dedicated to performing various activities, the places where people spend their time, and the length of time that people spend within those various microenvironments, such as indoors at home, in the car, or in a restaurant). After documenting these activities, the engineer needs to attribute physical characteristics to each microenvironment, especially air exchange rates (or residence times for chemical agents) and interzonal airflows.

Indoor exposures result from air exchanges (which affect penetration or infiltration of chemicals from ambient air to indoors), interzonal airflows, and local circulation. The exchange rates are the combined rates of air leakage through windows and doors, as well as intakes and exhausts, including adventitious openings (i.e., cracks and seams) that constitute a building's leakage configuration. Air exchange includes the movement of air through the building envelope combined with both natural and mechanical ventilation. The interzonal airflows include doors, ductwork, and service chaseways that connect rooms and zones within a building structure, while the local circulation of the air moves by convection and advection within each zone. The airflow distribution across the building envelope can be calculated from the distribution of interior pressures in the building structure and the forces leading to the airflow, especially temperature differentials, mechanical ventilation (especially by heating and air-conditioning systems), and external meteorology (especially winds).[17] Residential volume is the volume of the structure in which an individual resides and where that individual may be exposed to a chemical agent, especially airborne contaminants.

A person's exposure may be estimated using either direct or indirect approaches. Direct approaches are actual measurements of the person's contact with the chemical in the various pathways over a defined time interval. Direct methods include personal exposure monitors (PEMs) that are worn by people continuously as they conduct activities in every microenvironment. Indirect approaches apply already existing information about the amount of a chemical or mixtures of chemicals in microenvironments, combining it with information about the time that a person spends in each microenvironment and the person's explicit activities. The indirect approach then applies mathematical models to estimate exposure.

Discussion: What Goes on in the Laboratory?

The quality of exposure estimates depends on the quality of sample collection, preparation, and analysis. Analytic chemistry is an essential part of exposure characterization. Every exposure equation begins with the concentration of the chemical and then considers other factors, such as surface area of contact, exposure duration, and absorption processes. So it is critical that reliable analytic chemistry methods be employed.

Before the samples arrive at the laboratory, a monitoring plan, including quality assurance provisions, must be in place. The plan describes the procedures to be employed to examine a particular site. It should include all environmental media (e.g., soil, air, water, and biota) that are needed to characterize the exposure. It should explicitly point out which methods will be used. For example, if dioxins are being monitored, the U.S. EPA specifies certain sampling and analysis methods.[18] The environmental monitoring plan usually includes the kinds of samples to be collected (e.g., "grab" or integrated soil sample of *x* mass or *y* volume), the number of samples needed (e.g., for statistical significance), the minimum acceptable quality as defined by the quality assurance (QA) plan and sampling standard operating procedures (SOPs), and sample handling after collection. Handling includes specifications on the temperature range needed to preserve the sample, the maximum amount of time the sample can be held before analysis, special storage provisions (e.g., some samples need to be stored in certain solvents), and chain-of-custody provisions (i.e., only certain authorized persons should possess samples after collection).

When the sample arrives at the laboratory, the next step may be extraction. Extraction is needed for two reasons: (1) the environmental sample may be in sediment or soil, where the chemicals of concern are sorbed to particles and must be freed for analysis to take place; and (2) the actual collection may have been by trapping the chemicals onto sorbents. So, to analyze the sample, the chemicals must first be freed from the sorbent matrix. Again, dioxins provide an example. Under environmental conditions, dioxins are fat soluble and have low vapor pressures, so they may be found on particles, in the gas phase, or in the water column suspended to colloids (and very small amounts dissolved in the water itself). Therefore, to collect the gas-phase dioxins, the standard method calls for trapping it on polyurethane foam (PUF). Thus to analyze dioxins in the air, the PUF and particle matter must first be extracted, and to analyze dioxins in soil and sediment, those particles must also be extracted.

Extraction uses physics and chemistry. For example, many compounds can be simply extracted with solvents, usually at elevated temperatures. A common solvent extraction is the Soxhlet extractor, named after the German food chemist, Franz Soxhlet (1848–1913). The Soxhlet extractor (the U.S. EPA Method 3540) removes sorbed chemicals by passing a boiling solvent through the media. The heated solvent is condensed by cooling water, and the extract is collected over an extended period, usually several hours. Other automated techniques apply some of the same principles as solvent extraction but allow for more precise and consistent extraction, especially when large volumes of samples are involved.

For example, supercritical fluid extraction (SFE) brings a solvent—usually carbon dioxide—to the pressure and temperature near the critical point of the solvent, where the solvent's properties are rapidly altered with very slight variations of pressure.[19] Solid-phase extraction (SPE) uses a solid and a liquid phase to isolate a chemical from a solution and is often used to clean up a sample before analysis. Combinations of various extraction methods can enhance the extraction efficiencies, depending on the chemical and the media in which it is found. Ultrasonic and microwave extractions may be used alone or in combination with solvent extraction.

For example, the U.S. EPA Method 3546 provides a procedure for extracting hydrophobic (i.e., not soluble in water) or slightly water-soluble organic compounds from particles such as soils, sediments, sludges, and solid wastes. In this method, microwave energy elevates the temperature and pressure conditions (i.e., 100–115°C and 50–175 psi) in a closed extraction vessel containing the sample and solvent(s). This combination can improve recoveries of chemical analytes and can reduce the time needed compared to the Soxhlet procedure alone.

Chromatography consists of separation and detection. Separation uses the chemicals' different affinities for certain surfaces under various temperature and pressure conditions. The first step, injection, introduces the extract to a column. The term *column* is derived from the time when columns were packed with sorbents of varying characteristics, sometimes meters in length, and the extract was poured down the packed column to separate the various analytes.

Today columns are of two major types: gas and liquid. Gas chromatography (GC) uses hollow tubes (columns) coated inside with compounds that hold organic chemicals. The columns are in an oven, so that after the extract is injected into the column, the temperature—as well as the pressure—is increased, and the various organic compounds in the extract are released from the column surface differentially, whereupon they are collected by a carrier gas (e.g., helium) and transported to the detector.

Generally, the more volatile compounds are released first (they have the shortest retention times), followed by the semivolatile organic compounds. This is not always the case, however, because other characteristics such as polarity can greatly influence a compound's resistance to be freed from the column surface. Therefore, numerous GC columns are available to the chromatographer (different coatings, interior diameters, and lengths). Rather than coated colums, liquid chromatography (LC) uses columns packed with sorbing materials with differing affinities for compounds. Also, instead of a carrier gas, LC uses a solvent or blend of solvents to carry the compounds to the detector. In the high-performance LC (HPLC), pressures are also varied.

Detection is the final step for quantifying the chemicals in a sample. The type of detector needed depends on the kinds of pollutants of interest. Detection gives the peaks that are used to identify compounds (Figure 2-2). For example, if hydrocarbons are of concern, GC with flame ionization detection (FID) may be used. GC-FID counts the number of carbons, so, for example, long chains can be distinguished from short chains. The short chains come off the column first and have peaks that appear before the long-chain peaks. If pesticides or other

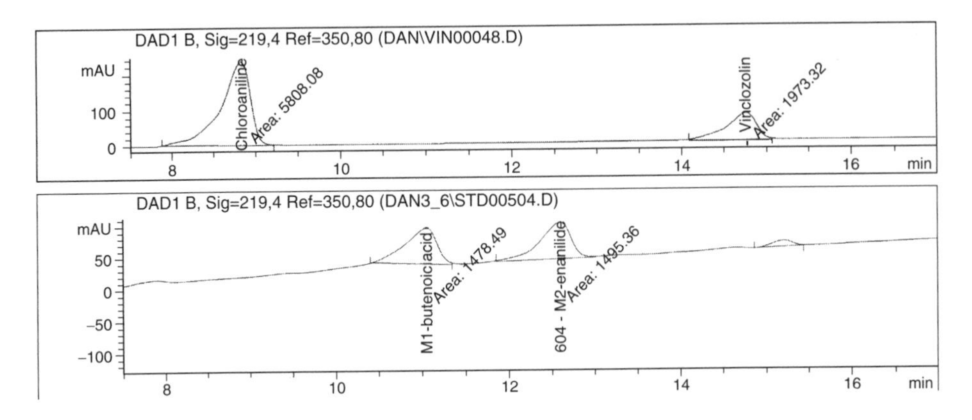

FIGURE 2-2. High-performance liquid chromatograph/ultraviolet detection peaks for standard acetonitrile solutions: 9 mg L^{-1} 3,5-dichloroaniline and 8 mg L^{-1} the fungicide vinclozolin (*top*); and two of vinclozolin's degradation products, 7 mg L^{-1} M1 and 9 mg L^{-1}M2 (*bottom*).

halogenated compounds are of concern, however, electron capture detection (ECD) is a better choice.

Several detection approaches are also available for LC; probably the most common is absorption. Chemical compounds absorb energy at various levels, depending on their size, shape, bonds, and other structural characteristics. Chemicals also vary in whether they will absorb light or how much light they can absorb depending on wavelength. Some absorb very well in the ultraviolet (UV) range, but others do not. Diode arrays help identify compounds by giving absorption ranges in the same scan. Some molecules can be excited and will fluoresce. The Beer-Lambert Law tells us that energy absorption is proportional to chemical concentration:

$$A = eb[C] \tag{2-3}$$

where, A is the absorbency of the molecule, e is the molar absorptivity (portionality constant for the molecule), b is the light's path length, and $[C]$ is the chemical concentration of the molecule. Thus, the concentration of the chemical can be ascertained by measuring the light absorbed.

One of the most popular detection methods is mass spectrometry (MS), which can be used with either GC or LC separation. The MS detection is highly sensitive for organic compounds and works by using a stream of electrons to consistently break compounds into fragments. The positive ions resulting from the fragmentation are separated according to their masses. This is referred to as the mass-to-charge ratio (m/z). No matter which detection device is used, software is used to decipher the peaks and to quantify the amount of each contaminant in the sample.

For inorganic substances and metals, the additional extraction step may not be necessary. The actual measured media (e.g., collected airborne particles) may be measured by surface techniques like atomic absorption (AA), X-ray fluorescence (XRF), inductively coupled plasma (ICP), or sputtering. As for organic compounds, the detection approaches can vary. For example, ICP may be used with absorption or MS. If all one needs to know is elemental information (e.g., to determine total lead or nickel in a sample), then AA or XRF may be sufficient; however, if speciation (i.e., knowing the various compounds of a metal) is necessary, then significant sample preparation is needed, including a process known as *derivatization*. A sample is derivatized by adding a chemical agent that transforms the compound in question into one that can be recognized by the detector. This is done for both organic and inorganic compounds (e.g., when the compound in question is too polar to be recognized by MS). The physical and chemical characteristics of the compounds being analyzed must be considered before visiting the field and throughout all the steps in the laboratory.

Exposure models[20] typically use one of two general approaches: (1) a time-series model to estimate microenvironmental exposures sequentially, following people's contact with chemicals through time; or (2) a time-averaged approach to estimate microenvironmental exposures using mean microenvironmental chemical concentrations for the total time that people spend in each microenvironment. Although the time-series approach to modeling personal exposures provides the appropriate structure to estimate personal exposures, the time-averaged approach usually is applied when the data needed to support a time-series model cannot be found.

So, exposure is a function of the concentration of the chemical or mixture of chemicals in a hazardous waste, and personal exposure can then be simply defined as:

$$E = \int_{t=t_1}^{t=t_2} C(t)\,dt \tag{2-4}$$

where, E is the personal exposure during the integrated time period from t_1 to t_2, and $C(t)$ is the concentration near the human interface (e.g., on the skin, in the mouth, or at the nose and mouth not impacted by exhaled air). This concentration is sometimes referred to as the potential dose (i.e., the chemical has not yet crossed the boundary into the body, but is present where it may enter the person).

Human exposures to chemicals associated with cancer and other chronic, long-term diseases are usually represented by estimates of lifetime average daily dose (LADD), which is a function of the concentration of the chemical, contact rate, contact fraction, and exposure duration per a person's body weight and life expectancy. For example, exposure from ingesting contaminated water can be calculated[21] as:

$$LADD = \frac{(C) \cdot (CR) \cdot (ED) \cdot (AF)}{(BW) \cdot (TL)} \tag{2-5}$$

where, LADD = lifetime average daily dose ($mg\ kg^{-1}\ d^{-1}$), C = concentration of the contaminant in the drinking water ($mg\ L^{-1}$), CR = rate of water consumption ($L\ d^{-1}$), ED = duration of exposure (d), AF = portion (fraction) of the ingested contaminant that is physiologically absorbed[22] (dimensionless), BW = body weight (kg), and TL = typical lifetime (d).

Some general human factors to be applied in such equations are provided in Table 2-6. Keep in mind that these factors are based on average conditions and can vary among populations and under various environmental conditions, such as differences in occupational versus general environmental exposures.

Discussion: Time Is of the Essence!

Why is the concept of time so important in risk? Time is a part of every formula and equation used to calculate risk. It is a factor in calculating dose, exposure, health outcomes, environmental effects, and risk.

Strictly speaking, dose is not time-dependent; it is simply the amount of a contaminant received by an organism. The actual dose is the mass of contaminant per mass of the organism (e.g., $mg \times kg^{-1}$ body weight). Time becomes important when determining exposure (e.g., $mg \times kg^{-1}\ day^{-1}$); however, in estimating effects and calculating dose-response curves, time does come into play. For example, newspaper editorial cartoons have frequently depicted the poor, bloated lab rat who has just consumed 20,000 cans of artificially sweetened cola. While this makes for interesting journalism, there is a credible reason for such research in comparative biology. The chronic diseases, especially cancer, are rare (thankfully!) in human populations in a statistical sense. They also require long periods before an effect occurs and symptoms are observable (the so-called latency period). How can such a rare, yet important disease be studied? One answer is the substitution of time for dose. The large dosages replace the decades of time and the huge population size of laboratory animals needed to test compounds. So, in a manner of speaking, time really is important for dose and health effects.

TABLE 2-7
Expressions of Chemical Analytical Limits

Type of Limit	*Description*
Limit of detection (LOD)	Lowest concentration or mass that can be differentiated from a blank with statistical confidence. This is a function of sample handling and preparation, sample extraction efficiencies, chemical separation efficiencies, and capacity and specifications of all analytical equipment being used (See IDL below).
Instrument detection limit (IDL)	The minimum signal greater than noise detectable by an instrument. The IDL is an expression of the piece of equipment, not the chemical of concern. It is expressed as a signal-to-noise (S:N) ratio. This is mainly important to the analytical chemists, but the engineer should be aware of the different IDL's for various instruments measuring the same compounds, so as to provide professional judgment in contracting or selecting laboratories and deciding on procuring for appropriate instrumentation all phases of remediation.
Limit of quantitation (LOQ)	The concentration or mass above which the amount can be quantified with statistical confidence. This is an important limit because it goes beyond the presence-absence of the LOD and allows for calculating chemical concentration or mass gradients in the environmental media (air, water, soil, sediment, and biota).
Practical quantitation limit (PQL)	The combination of LOQ and the precision and accuracy limits of a specific laboratory, as expressed in the laboratory's quality assurance/quality control (QA/QC) plans and standard operating procedures (SOPs) for routine runs. The PQL is the concentration or mass that the engineer can consistently expect to have reported reliably.

Time is critical for exposure to hazardous substances. In fact, it is included in both the numerator and the denominator of the exposure formulae. Actually, the numerator contains two time-relative terms. The exposure is increased when the *exposure duration* term is increased. Exposure duration is expressed in units of time (e.g., days). In other words, if Persons A and B are identical in every way except that Person A was exposed to Chemical X for 30 days in a lifetime and Person B was exposed for 3,000 days, Person A's exposure is 1% that of Person B. Also, because risk is the product of exposure and toxicity, Person A's risk is also 1% of the risk to Person B (because the two hypothetical persons were the same in every other manner except the exposure and their exposure was to the same chemical).

Also in the numerator is the time-related term, *exposure length*. This term is expressed in dimensionless (e.g., hours per day = time/time) units and is important because it distinguishes the person's activities. For example, an occupational exposure is usually about 8 to 10 hours per day, 5 days per week, whereas an indoor or personal exposure can be highly variable. If a person spends most of his or her time indoors (which most do, incidentally), then that person's exposure length would be much longer than an occupational exposure. If a person engages in an activity for only one hour per day where the exposure occurs, however, then the exposure and risk will be 1/24th of a shut-in person exposed to an indoor exposure. The time-relative term in the denominator of the exposure formulae is the typical lifetime, which is expressed in units of time (e.g., days). The longer the typical lifetime, the less the exposure and risk. From a purely mathematical perspective, the longer one lives, the more the exposure is diluted. So, if you live a long time, in your case "dilution is the solution to pollution."

It is not just time that is important, but also the timing of exposure. This point has recently been made by the increasing awareness of endocrine disruptors or hormonally active chemical agents. There are critical times during an organism's life when it is particularly sensitive to the effects of chemical compounds. For example, the usual time-relative terms in the exposure formulae that we just discussed are inadequate for compounds that affect reproduction, gestation, and tissue development in its prolific stages. If the current research linking exposures to some phthalates and pesticides to breast cancer holds up, the exposure duration term will need to be modified to address the times in one's life when one is especially vulnerable and sensitive to the exposures, such as in utero, neonatal, and puberty.

Some compounds have been shown to have effects on ova and sperm, so the exposures lead to reproductive effects. Even more complicated in terms of time are the *transgenerational* effects, where the person who gets the disease is not the person who is exposed, but the parent (or maybe the grandparent). The most infamous case is probably that of DES (diethylstilbestrol), a synthetic hormone that was prescribed to pregnant women from 1940 to 1971 as a treatment to prevent miscarriages. Unfortunately, it has been subsequently classified as a known carcinogen. The major concern was not with the treatment mothers, but with the in utero exposure that led to a high incidence of cervical cancers in the daughters of the treated mothers.

Time is also important to the mechanical and chemical processes that determine the movement and fate of chemical compounds in the environment. The concept of *persistence* is crucial in hazardous waste engineering. Chemists apply the concept of *half-life* ($t_{1/2}$) to a compound's environmental persistence. The $t_{1/2}$ is the amount of time it takes for half of the mass of the compound to be broken down into degradation products. So, a compound with $t_{1/2} = 2$ years has twice the persistence of a compound with $t_{1/2} = 1$ year. Half-life depends on both the inherent properties of the compound and the conditions of the environmental medium where the compound is found. For example, if Compound X's water $t_{1/2} = 10$ minutes, but its $t_{1/2} = 6$ months if it is sorbed to a clay particle, one would expect to find greater residues of Compound X in soil and sediment than in surface waters (assuming a steady-state source). This phenomenon explains much of the persistence of dioxins, which can have half-lives of months in the laboratory, but decades when sorbed to soil particles.

Where Does the Engineer Fit in the Risk-Assessment Paradigm?

A glance at Figure 2-1 shows that there are numerous places where engineers can help with risk assessment and management; however, particular steps in the figure define a flow diagram where the engineer's role is most prominent, which is highlighted in Figure 2-3. One basic flow diagram is:

- *Source characterization.* Much of what is done in the site investigation and environmental assessment phases of hazardous waste sites is conducted and led by engineers. The federal Hazard Ranking System[23] (HRS) defines a "source" as "any area where a hazardous substance has been deposited, stored, disposed, or placed, plus those soils that have become contaminated from migration of a hazardous substance." Thus the engineer plays a critical role in providing information for the preliminary assessment and site investigation of a hazardous waste site.
- *Exposure assessment.* The engineer conducts several tasks associated with assessing exposures from hazardous wastes. These include

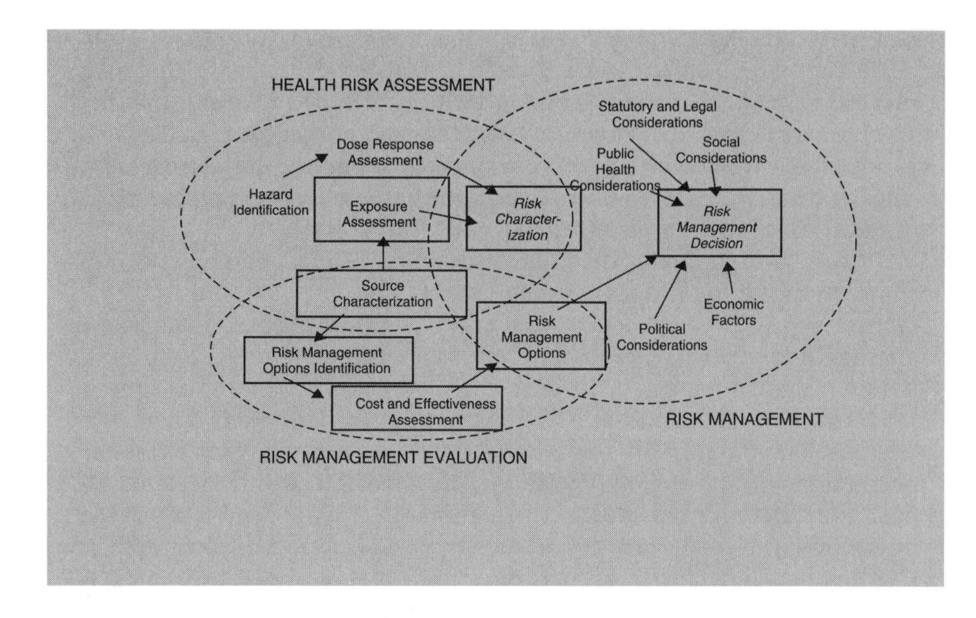

FIGURE 2-3. Steps in risk assessment and management during which the engineer has the most impact (boxed functions).

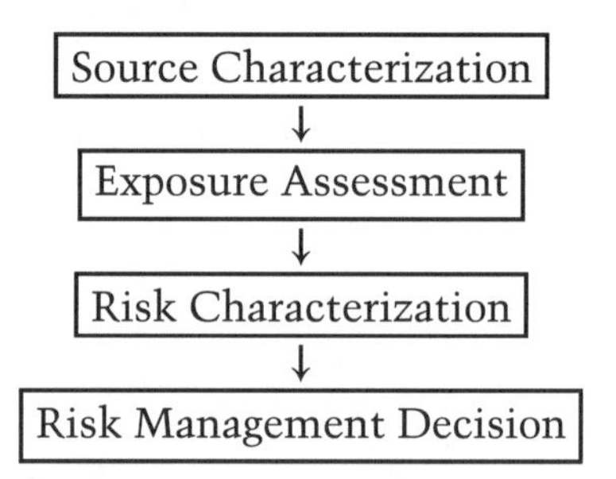

Another engineering risk flow diagram is:

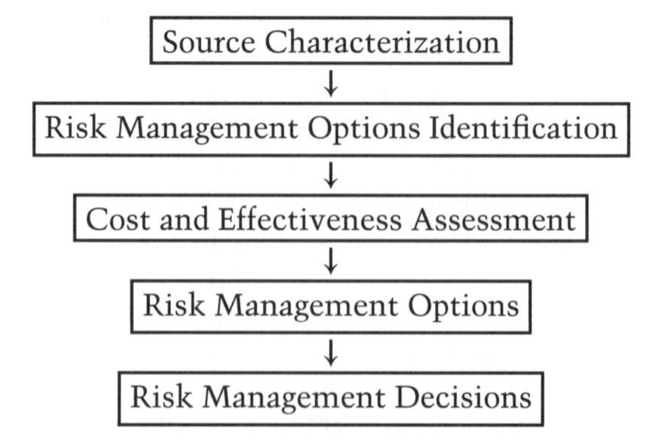

measuring concentrations of chemicals in all environmental media and modeling the movement and transformation of those chemicals in the environment. Once the environmental concentrations are measured and modeled, the engineer must estimate the concentrations that may come in contact with individuals from the various pathways. For example, concentrations of chemicals in groundwater must be extrapolated and modeled to estimate the amount that will enter drinking water supplies.

- *Risk characterization*. In this step, the engineer is called on to "pull everything together." This requires teamwork with chemists, biologists, toxicologists, and other experts to arrive at an estimate of risk from the chemicals.

Discussion: Ecologic Risk Assessment

This text is primarily concerned with human health risks; however, the engineer responsible for planning and implementing remedies to hazardous waste problems should be aware of other endpoints of concern. Some of these other endpoints are built into the hazard identification process discussed earlier in this chapter, such as the link between material damage and the corrosiveness of a chemical; however, others may not be addressed sufficiently. These include hazards to ecologic resources.

Ecologic risk assessment[24] is a process employed to determine the likelihood that adverse outcomes may occur in an ecosystem as a result of exposure to one or more stressors. The process systematically reduces and organizes data, gathers information, forms assumptions, and identifies areas of uncertainty to characterize the relationships between stressors and effects. As is the case for human health risk assessments, the stressors may be chemicals; however, ecological risk assessments must also address physical and biologic stressors. For example, the placement of a roadway or the changes wrought by bulldozers and earthmovers are considered to be physical stressors to habitats. The accidental or intentional introduction of invasive biota (e.g., grass carp [fauna] and kudzu [flora] in the Southern United States) are examples of biologic stressors. The identification of possible adverse outcomes is crucial. These outcomes alter essential structures or functions of an ecosystem. The severity of outcomes is characterized by their type, intensity, and scale of the effect, as well as the likelihood that an ecosystem will recover from the damage imposed by a single or multiple stressors.

The characterization of adverse ecologic outcomes can range from qualitative, expert judgments to statistical probabilities. Complete quantification of harm is never possible because of the many variables in the condition of an ecosystem. Ecologic risk assessments may be prospective or retrospective, but often are both. The Florida Everglades provides an example of an integrated risk approach. The population of panthers—a top terrestrial carnivore in Southern Florida—was found to contain elevated concentrations of mercury (Hg) in the 1990s. This was observed through retrospective ecoepidemiologic studies. The findings were also used as scientists recommended possible measures to reduce Hg concentrations in sediment and water in Florida. Prospective risk assessments can help estimate expected declines in Hg in panthers and other organisms in the food chain from a mass balance perspective. That is, as the Hg mass entering the environment through the air, water, and soil is reduced, how has the risk to sensitive species concomitantly been reduced? Integrated retrospective–prospective risk assessments are employed where ecosystems have a history of previous impacts and the potential for future effects exists from a wide range of stressors. This may be the case for hazardous waste sites.

The ecologic risk assessment process embodies two elements: characterizing the adverse outcomes and determining the exposures. From these elements, three steps are undertaken (Figure 2-4):

1. Problem formulation.
2. Analysis.
3. Risk characterization.

In problem formulation, the rationale for conducting the assessment is fully described, the specific problem or problems are defined, and the plan for analysis and risk characterization is laid out. Tasks include integrating available information about the potential sources, the description of all stressors, effects, and the characterization of the ecosystem and the receptors. Two basic products result from this stage of ecologic risk assessment: assessment endpoints and conceptual models.

The analysis phase consists of evaluating the available data to conduct an exposure assessment (i.e., exposure to stressors is likely to occur or to have occurred). From these exposure assessments, the next step is to determine the possible effects and how widespread and severe these outcomes will be. During analysis, the engineer should investigate the strengths and limitations of data on exposure, effects, and ecosystem and receptor characteristics. Using these data, the nature of potential or actual exposure and the ecologic changes under the circumstances defined in the conceptual model can be determined. The analysis phase provides an exposure profile and stressor-response profile, which together form the basis for risk characterization. Thus the ecologic risk assessment provides valuable

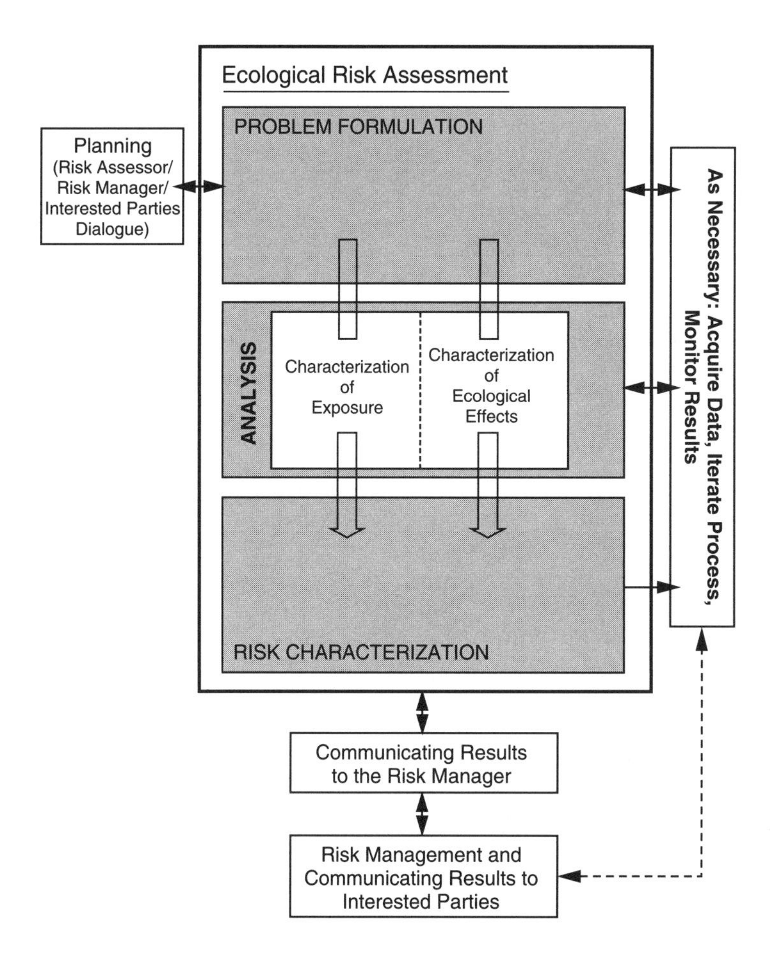

FIGURE 2-4. The ecologic risk-assessment framework. (*Source:* U.S. Environmental Protection Agency.)

information to the hazardous waste engineer through the following means:

- Provides information to complement the human health information, thereby improving environmental decision making.
- Expresses changes in ecologic effects as a function of changes in exposure to stressors, which is particularly useful to the decision maker, who must evaluate tradeoffs, examine different options, and determine the extent to which stressors must be reduced to achieve a given outcome.

- Characterizes uncertainty as a degree of confidence in the assessment, which aids the engineer's focus on those areas that will lead to the greatest reductions in uncertainty.
- Provides a basis for comparing, ranking, and prioritizing risks, as well as information to conduct cost-benefit and cost-effectiveness analyses of various remedial options.
- Considers management needs, goals, and objectives, in combination with engineering and scientific principles to develop assessment endpoints and conceptual models during problem formulation.

Risk characterization is the stage when the engineer summarizes the necessary assumptions, describes the scientific uncertainties, and determines the strengths and limitations of the analyses. The risks are articulated by integrating the analytic results, interpreting adverse outcomes, and describing the uncertainties and weights of evidence. Engineers are adept at applying diverse information to arrive at solutions to complicated problems. Chemical hazardous waste risk assessment is no exception.

Risk Roles for the Engineer

The environmental engineer is a pollution preventer and controller. The engineer is also a professional problem solver, or better yet, a problem foiler. Engineers control the risks associated with hazardous wastes. The professional engineer can participate at various points in the risk paradigm, but he or she mainly deals with those activities associated with preventing the risk in the first place and intervening to reduce or eliminate exposures of any chemical that could be a threat to human health and the environment.

The engineer prepares designs and plans for manufacturing processes that limit, reduce, or prevent pollution. The engineer looks for ways to reduce risks in operating systems. The engineer assesses sites and systems for possible human exposures to hazardous and toxic substances. The engineer then designs systems to reduce or eliminate these exposures. Finally, the engineer participates in the means to remedy the problem and intervenes to ameliorate health, environmental, and welfare damages. Specific technologies used by the engineer to address hazardous waste problems are covered in Chapter 4. At this point, however, it is useful to discuss these roles briefly.

Hazardous waste management is expensive, so it is most desirable for the engineer to work with the client to avoid generating the waste in the first place, while still efficiently producing the target product. Process engineering should identify the generation of hazardous wastes at any step in the value-added manufacturing chain. Pollution is generally a manifestation of inefficiency, so avoiding waste generation may improve operational efficiencies and reduce costs. It is certainly worth a comprehensive evaluation.

Discussion: Toxic Dyes and Pigments versus New Optics Paradigms: Thinking Outside the Light Box

The progress of science, engineering and technology is often not predictable and follows paths that can be quite surprising. Let us consider our colorful society. Few would tolerate a monochromatic lifestyle. Our products, our gardens, our artwork, our CD labels, our brochures, and our clothing create a tapestry of colors.

In fact, historians of science tell us that the seminal works of the Renaissance presaging modern science were Isaac Newton's short essay entitled "New Theory about Light and Colors" that was published in 1672, and his full-length treatise, 1704 *Opticks*. Thus, color was the subject of Newton's first written work, predating even his masterpiece *Philosophiae Naturalis Principia Mathematica* by 15 years. Newton's numerous breakthroughs in optics include light dispersion using glass prisms.[25]

Unfortunately, the way we have learned to achieve these kaleidoscopes in our products has for the most part depended on the use of large organic molecules and metallic dyes. A site where commercial dyes, paints or pigments are or have been produced or used can be contaminated with naphthalene, toluene, xylene, and other petroleum distillate solvents, as well as potassium dichromate and the colorful metals, such as cadmium (for the color red).

The challenges for the engineer to date have been how to remove these contaminants from the waste stream during coloration, pigmentation, and dyeing operations, or cleaning up waste sites left behind by these operations. In addition, the coloring-related contaminants have found their way to sanitary landfills and waste combustors, allowing them to re-enter the environment as parent compounds or degradation products.

But, does it have to be this way? What we really need is a sustainable approach to colors. Waste minimization and pollution prevention can help reduce or even eliminate the use of many of the chemicals used for coloration.[26] We need new manufacturing processes. Some of the answer can come by emulating nature.

Consider the brightly colored birds, and how their plumage can provide intense colorations that are important for mating and other biological functions such as camouflage.

Now consider how we humans, with our existing technologies, would color the blue jay (*Cyanocitta cristata*). We may choose to use an aniline dye, a heavy metal or even some cyanide-laden pigment source to obtain the blue plumage. But pull a feather from the blue jay and look at it under a microscope. You will find that the feather lacks any pigment, and some feathers are even transparent! In fact, you could pluck the poor bird bald and look at each "blue" feather and find no color blue. What gives?

Why should nature use heavy elements and molecules that will dramatically increase the bird's weight, when another means of obtaining colors is available? So, the bird's plumage does not rely on differential absorption and reflection (i.e., the cadmium or red dye absorbs all other light wavelengths, but reflects in the red part of the spectrum). And, the blue jay is not the only bird that uses microstructures to be blue, purple, or even green.

Unlike humans, most birds can see well into the ultraviolet range of the light spectrum, so they possess efficient means of producing colors at the violet to UV range, without pigmentation. These birds' colors come from the feathers' individual "microstructures" made of keratin and air bubbles that create diffraction or scattering (Tyndall effect or Raleigh scattering) of light or by interferences[27] of light by microscopic contours and shapes in the feather. The blue feathers result from differences in the distances traveled by light waves that are reflected. The diffracted light returns the blue color to the mate's (and our) eyes.

The birds may be telling us something. Can we apply Newton's seminal discussion of optics to discover breakthroughs in coloring technology? It could even pave the way for lucrative and more environmentally friendly businesses. Who knows? We may all be wearing transparent, microstructured clothing, products and driving cars with microstructure coatings in the near future. We may not have pigments, but we will be as colorful as the blue jay. And, when we finally wear out that old shirt, or throw out the box for our new High Definition TV, or junk the old car, we won't have to worry about toxic dyes in the environment.

Please accept our thanks in advance, Sir Isaac.

Paper and pulp bleaching provides an example of how processes can be changed in the planning stages to prevent waste generation. In the late 1970s, recycled paper had first become available. It was readily identifiable by its tan color and grainy texture. During the 1980s, the public seemed to increase its tolerance for the darker paper because of its earth friendliness, but this acceptance seemed to fall with time. The decrease may have been associated with the greater frequency of paper jams in copiers and printers or the increasing darkness of copies made from the tan originals. Whatever the reason, the consumer demand for white paper has increased.

Until recently, almost all bleaching of pulp and paper was accomplished with chlorine compounds, such as HOCl. Under the manufacturing conditions, the heat and makeup water would generate chemical reactions between the lignin and other pulp substances with the chlorine compounds and ions. As a result, halogenated contaminants, including chlorinated dioxins, were released with the effluent to surface and groundwater. The consumer was demanding white paper; regulatory agencies were requiring the dioxins to be eliminated; and lawsuits were pending. The paper companies were faced with hard choices. The companies could stop making the bleached products and lose market share; they could lose the battle of public opinion and continue the status quo; or they could reassess their whole manufacturing process and look for environmentally acceptable methods for producing white paper. The companies ended up choosing the last alternative, completely eliminating the use of halogenated bleaches and instead using nonchlorine oxidation bleaching processes.

The engineer is a controller. To control the generation and release of hazardous chemicals, the engineer must understand every aspect of the mass balance, or in the words of waste management, "from cradle to grave." That means that the engineer must be involved in the planning stages regarding decisions about whether to produce or use certain chemicals. The engineer must also be involved—even lead—the design of manufacturing that includes hazardous chemicals. The engineer must oversee the transport of hazardous chemicals and wastes. Finally, when problems occur or when a previously unknown waste is uncovered (either literally or figuratively), the engineer is called on to identify options for remediation and cleanup. This is a huge mandate, but who better to respond to it than the well-trained engineer?

Case Study: U.S. Army's Site Level Waste Reduction

The best time to deal with hazardous waste is before it is generated. Some simple approaches have been used to prevent creating hazardous wastes in the first place. One recent example is the U.S. Army's Hazardous Material Management Program (HMMP). The HMMP is designed to reduce operational costs and waste generation by centralizing its material handling and tracking these materials through an automated data system.[28] The system is a comprehensive management program at each Army installation, incorporating approaches for reporting and regulation, maintenance, and occupational safety and industrial hygiene at these facilities. Each Army base or other installation is allowed to use the HMMPs to improve operational efficiencies and to save costs. The HMMP has been implemented at Pine Bluff Arsenal, Arkansas; at Fort Campbell and Fort Knox, Kentucky; and at Schofield Barracks, Hawaii.

Each installation has a Hazardous Material Management Control Center that serves as the single point of contact for hazardous waste management. The center establishes stock criteria, coordinates resource management, and authorizes a list of approved materials. The center manager also directs the Hazardous Material Management Team, which consists of supply managers, technical inspectors, and experts in safety, industrial hygiene, and contracting.

A Microsoft Windows-based program is used to capture materials and use-related data. Training courses and hands-on demonstrations share lessons learned from how things worked well in other installations. The pollution prevention program has even improved defense readiness by enhancing personnel health and safety, increasing compliance, improving shelf-life management, and reducing procurement and disposal costs. This software application has led to a 20% reduction in the volume of hazardous material stock items entering the system.

The so-called pharmacy, or HAZMART, has been particularly successful. It is a centralized system for inventory and management of hazardous materials. Military installations have indicated that 20% to 80% of hazardous waste streams is required because of the expiration of chemicals beyond their usable dates. The HAZMART system reduces this waste and is similar in concept to the materials management system that is widely used in manufacturing. The program has been successful—environmentally and financially—and is being adopted more widely by the Army.

CHAPTER 3

The Fate, Transformation, and Transport of Hazardous Chemicals

How Hazardous Compounds Move and Change in the Environment

The chemistry and physics of hazardous substances are complex. Nearly 100,000 chemicals are commercially available throughout the world. Their rate of production greatly increased in the second half of the 20th century. The chemicals and their degradation products can be found in the atmosphere, in surface and groundwater, in sediments, in the tissue of fish and wildlife, in ecosystems, and in humans.[1]

The properties of a chemical and its behavior in the environment determine how best to control the fate and possible exposures to hazardous substances. The essential properties addressed by hazardous waste chemistry, biochemistry, and physics are described in Table 3-1. Unlike the fields of analytic or physical chemistry, engineering requires an understanding of chemicals in their unpure states. That is, the table provides a description of the physical and chemical properties of the compounds themselves within the context of where they are found in the environment. Although a compound may behave a certain way in a pure solution, its behavior in the environment is dictated not only by its theoretical attributes, but also by the presence of other factors in the environment. Later in this chapter, several of these properties are discussed relative to example compounds of concern to the field of hazardous waste engineering.

There are insufficient toxicity data available for most of these substances to assess the risks posed to human health and the environment. The so-called hazardous compounds include many different chemical and physical classifications, but for the purposes of our discussions, they can be

TABLE 3-1
Chemical and Physical Properties Affecting the Fate and Transport of Hazardous Substances in the Environment

Property of Substance or Environment	*Importance to Fate and Chemical Degradation*	*Importance to Movement Within or Among Environmental Compartments*
Molecular Weight	Larger molecules are broken down at more vulnerable functional groups first.	Heavier molecules have lower vapor pressures, so are less likely to exist in gas phase. Heavier molecules are more likely to remain sorbed to soil and sediment particles.
Chemical Bonding	Chemical bonds determine the persistence to degradation. Ring structures are generally more stable than chains. Double and triple bonds add persistence to molecules compared to single-bonded molecules.	Large, aromatic compounds have affinity for lipids in soil and sediment. Solubility in water is enhanced by the presence of polar groups in structure. Sorption is affected by presence of functional groups and ionization potential.
Stereochemistry	Stereochemistry is the spatial configuration or shape of a molecule. Neutral molecules with cross-sectional dimensions > 9.5 Angstroms (A) have been considered to be sterically hindered in their ability to penetrate the polar surfaces of the cell membranes. A number of fate and transport properties of chemicals are determined, at least in part by a molecule's stereochemistry.	Lipophilicity of neutral molecules generally increases with molecular mass, volume, or surface area. Solubility and transport across biologic membranes are affected by a molecule's size and shape. Molecules that are planar, such as polycyclic aromatic hydrocarbons, dioxins, or certain forms of polychlorinated biphenyls, are generally more lipophilic than are globular molecules of similar molecular weight. The restricted rate of bioaccumulation of octachlorodibenzo-p-dioxin (9.8 A) and decabromobiphenyl (9.6 A) has been associated with these compounds' steric hindrance.
Solubility	Lipophilic compounds may be very difficult to remove from particles and may require highly destructive (e.g., combustion) remediation	Hydrophilic compounds are more likely to exist in surface water and in solution in interstices of pore water of soil, vadose zone, and aquifers underground. Lipophilic compounds

TABLE 3-1 *(continued)*

Property of Substance or Environment	*Importance to Fate and Chemical Degradation*	*Importance to Movement Within or Among Environmental Compartments*
	techniques. Insoluble forms (e.g., certain valence states) may precipitate out of water column or be sorbed to particles.	are more likely to exist in organic matter of soil and sediment.
Vapor Pressure or Volatility	Volatile organic compounds (VOCs) readily move to the gas phase since their vapor pressures in the environment are usually greater than 10^{-2} kilopascals, while semivolatile organic compounds (SVOCs) have vapor pressures between 10^{-2} and 10^{-5} kilopascals, and nonvolatile organic compounds (NVOCs) have vapor pressures $<10^{-5}$ kilopascals.	Volatility is a major factor in where a compound is likely to be found in the environment. Higher vapor pressures mean larger fluxes from the soil and water to the atmosphere. Lower vapor pressures, conversely, cause chemicals to have a greater affinity for the aerosol phase.
Fugacity	Expressed as Henry's Law Constant (K_H) (i.e., the vapor pressure of the chemical divided by its solubility of water). Thus, high fugacity compounds are likely candidates for remediation using the air (e.g., pump-and-treat and air-stripping).	Compounds with high fugacity have a greater affinity for the gas phase and are more likely to be transported in the atmosphere than those with low fugacity. Care must be taken not to allow these compounds to escape prior to treatment.
Octanol-Water Coefficient (K_{ow})	Substances with high K_{ow} are more likely to be found in the organic phase of soil and sediment complexes than in the aqueous phase. They may also be more likely to accumulate in organic tissue.	Transport of substances with higher K_{ow} values is more likely to be on particles (aerosols in the atmosphere and sorbed to fugitive soil and sediment particles in water), rather than in water solutions.
Partitioning and Sorption	Adsorption (onto surfaces) dominates in soils and sediments low in organic carbon (solutes precipitate	Partitioning determines which environmental media will dominate. Strong sorption constants indicate that soil

(continued)

TABLE 3-1 *(continued)*

Property of Substance or Environment	*Importance to Fate and Chemical Degradation*	*Importance to Movement Within or Among Environmental Compartments*
	onto soil surface). Absorption is a major sorption process in soils and sediments high in organic carbon (partitioning into organic phase/aqueous phase matrix surrounding mineral particles).	and sediment may need to be treated in place. Phase distributions favoring the gas phase indicate that contaminants may be off-gassed and treated in their vapor phase. This is particularly important for semivolatile compounds, which exist under typical environmental conditions in both the gas and solid phase.
Dissociation	Molecules break down by a number of types of dissociation.	Hydrolysis involves the dissociation of compounds via acid–base equilibria among hydroxyl ions and protons and weak and strong acids and bases. Dissociation may also occur by photolysis directly by the molecules absorbing light energy and indirectly by energy or electrons transferred from another molecule that has been broken down photolytically. Other dissociation processes include dissociation of complexes and nucleophilic substitution.
Substitution, Addition, and Elimination	These processes are important for treatment and remediation. For example, dehalogenation (e.g., removal of chlorine atoms) of organic compounds often renders them much less toxic. Adding or substituting a functional group can make the compound more or less toxic. Hydrolysis is an important substitution mechanism where a water molecule or hydroxide ion substitutes for an atom or group on a molecule. Phase 1 metabolism by organisms also uses hydrolysis and redox reactions (discussed below) to break down complex molecules at the cellular level.	These processes can change the physical phase of a compound (e.g., dechlorination can change an organic compound from a liquid to a gas), can change their affinity to or from one medium (e.g., air, soil, and water) to another. That is properties such as fugacity, solubility, and sorption will change and may allow for more efficient treatment and disposal. New species produced by hydrolysis are more polar and, thus, more hydrophilic than their parent compounds, so they are more likely to be found in the water column.

TABLE 3-1 *(continued)*

Property of Substance or Environment	*Importance to Fate and Chemical Degradation*	*Importance to Movement Within or Among Environmental Compartments*
Reduction-Oxidation	Reduction is the chemical process where at least one electron is transferred to another compound. Oxidation is the companion reaction where an electron is transferred from a molecule. These reactions are important in hazardous waste remediation. Often, toxic organic compounds can be broken down ultimately to CO_2 and H_2O by oxidation processes, including the reagents ozone, hydrogen peroxide, and molecular oxygen (i.e., aeration). Reduction is also used in treatment processes. For example, hexavalent chromium is reduced to the less toxic trivalent form in the presence of ferrous sulfate: $2CrO_3 + 6FeSO_4 \rightarrow 3Fe_2(SO_4)_3 + Cr_2(SO_4)_3 + 6H_2O$ The trivalent form is removed by the addition of lime, where it precipitates as $Cr(OH)_3$.	Reductions and oxidations are paired into so-called redox reactions. Such reactions occur in the environment, leading to chemical speciation of parent compounds into more or less mobile species. For example elemental or divalent mercury is reduced to the toxic species, mono- and di-methylmercury in sediment and soil low in free oxygen. The methylated metal species have greater affinity than the inorganic species for animal tissue.
Diffusion	Diffusion is the mass flux of a chemical species across a unit surface area. It is a function of the concentration gradient of the substance. A compound may move by diffusion from one compartment to another (e.g., from the water to the soil particle).	The concentration gradients within soil, underground water, and air determine to some degree the direction and rate that the contaminant will move. This is a very slow process in most environmental systems; however, in rather quiescent systems* ($<2.5 \times 10^{-4}$ cm s^{-1}), such as aquifers and deep sediments, the process can be very important.
Isomerization	A congener is any chemical compound that is a member of a chemical family, the members of which have different molecular weights and various substitutions (e.g., there are 75 congeners of chlorinated dibenzo-p-dioxins). Isomers are chemical species	The fate and transport of chemicals can vary significantly depending on the isomeric form. For example, the rates of degradation of left-handed chiral compounds (mirror images) are often more

(continued)

TABLE 3-1 *(continued)*

Property of Substance or Environment	*Importance to Fate and Chemical Degradation*	*Importance to Movement Within or Among Environmental Compartments*
	with identical molecular formulae but that differ in atomic connectivity (including bond multiplicity) or spatial arrangement. An enantiomer is one of a pair of molecular species that are nonsuperimposable mirror images of each other.	rapid than for right-handed compounds (possibly because left-handed chirals are more commonly found in nature, and microbes have acclimated their metabolic processes to break them down). Isomeric forms also vary in their fate (see the discussions of hexachlorocyclohexane isomers later in this chapter).
Biological Transformation	Many of the processes discussed in this table can occur in or be catalyzed by microbes. These are biologically mediated processes. Reactions that may require long periods to occur can be sped up by biological catalysts (i.e., enzymes). Many fungi and bacteria reduce compounds to simpler species to obtain energy. Biodegradation is possible for almost any organic compound, although it is more difficult in very large molecules, insoluble species, and completely halogenated compounds.	Microbial processes will transform parent compounds into species that have their own transport properties. Under aerobic conditions, the compounds can become more water soluble and are transported more readily than their parent compounds in surface and groundwater. The fungicide example given later in this chapter is an example of how biologic processes change the flux from soil to the atmosphere.
Potential to Bioaccumulate	Bioaccumulation is the process by which an organism takes up and stores chemicals from its environment through all environmental media. This includes bioconcentration (i.e., the direct uptake of chemicals from an environmental medium alone), and is	Numerous physical, biologic, and chemical factors affect the rates of bioaccumulation needed to conduct environmental risk assessment. For chemicals to bioaccumulate, they must be sufficiently stable, conservative, and resistant to chemical degradation. Elements, especially metals, are inherently

TABLE 3-1 *(continued)*

Property of Substance or Environment	*Importance to Fate and Chemical Degradation*	*Importance to Movement Within or Among Environmental Compartments*
	distinguished from biomagnification (i.e., the increase in chemical residues in organisms that have been taken up through two or more levels of a food chain).	conservative, and are taken up by organisms either as ions in solution or via organometallic complexes, such as chelates. Complexation of metals may facilitate bioaccumulation by taking forms of higher bioavailability, such as methylated forms. The organisms will metabolize by hydrolysis that allows the free metal ion to bond ionically or covalently with functional groups in the cell, like sulfhydryl, amino, purine, and other reactive groups. Organic compounds with structures that shield them from enzymatic actions or from nonenzymatic hydrolysis have a propensity to bioaccumulate; however, readily hydrolyzed and eliminated compounds are less likely to bioaccumulate (e.g., phosphate ester pesticides like parathion and malathion). Substitution of hydrogen atoms by electron–withdrawing groups tends to stabilize organic compounds like the polycyclic aromatic hydrocarbons (PAHs). For example, the chlorine atoms are large and highly electronegative, so chlorine substitution shields the PAH molecule against chemical attack. Highly chlorinated organic compounds such as the polychlorinated biphenyls (PCBs) bioaccumulate to high levels since they possess properties that allow them to be easily taken up, but do not allow easy metabolic breakdown and elimination.

*W. Tucker and L. Nelken, 1982, "Diffusion Coefficients in Air and Water," in *Handbook of Chemical Property Estimation Methods* (New York: McGraw-Hill, 1982).

classified into two basic types: organic and inorganic. These classifications are not sufficient for particular actions, such as remediation and storage, but some general characteristics can be helpful as a first step in assessing the potential hazards, exposure, and risks posed by chemicals in the environment.[2]

An understanding of the degradation and flux of hazardous chemicals is essential to the protection of public health and the environment. Engineers and scientists need high-quality data to predict exposures and risks to human populations. Protection of sensitive ecosystems also requires better information regarding the fate and transport of hazardous chemicals and their degradation products.

Physicochemical Properties of Chemicals

Several chemicals, especially organic compounds, are synthesized with active organochlorine and organometallic functional groups, frequently causing them to be persistent in the environment and to partition between physical phases under environmental conditions; that is, they are semivolatile substances (vapor pressures = 10^{-2} to 10^{-8} kilopascals). In addition to the persistent groups, these compounds often contain more reactive groups (e.g., pesticides possess reactive groups that provide some or all of the biocidal qualities of the molecule).

For example, the dicarboximide fungicide vinclozolin, 3-(3,5-dichlorophenyl)-5-methyl-5-vinyl-oxzoli-dine-2,4-dione (Figure 3-1) is a dichlorobenzene coupled to a heterocyclic carbamate-like ring. The fungicide's molecular weight is 286.11, its boiling point is 131°C, its melting point is 108°C, and its density is 1.51. At 20°C, vinclozolin's solubility is 3.5 mg L^{-1} in water, 17,700 mg L^{-1} in ethanol, and 550,000 mg L^{-1} in acetone (as reported in Merck Index Number 9890). The halogenated aromatic ring in the structure engenders persistence to the molecule, while the carbamate-like group provides reactivity, and thus is the less stable part of the molecule. Two principal degradation products result from opening the carbamate-like ring: (1) a butenoic acid (2-[(3,5-dichlorophenyl)-carbamoyl]oxy-2-methyl-3-butenoic acid), referred to as M1; and (2) an enanilide (3′,5′-dichloro-2-hydroxy-2-methylbut-3-enanilide), called M2. Both degradation products have been isolated in plants and soils.[3] The M1 reaction is reversible, wherein the butenoic acid closes into a carbamate ring to return to the vinclozolin structure. The M2 pathway gives a nonreversible degradation product. Both reactions will continue to cleave the right segment of the respective molecule, to form the third degradate, 3,5-dichloroaniline.

Degradation Mechanisms in the Environment

Hazardous chemicals may undergo degradation by abiotic chemical transformation, as well as by photochemical and biochemical processes.

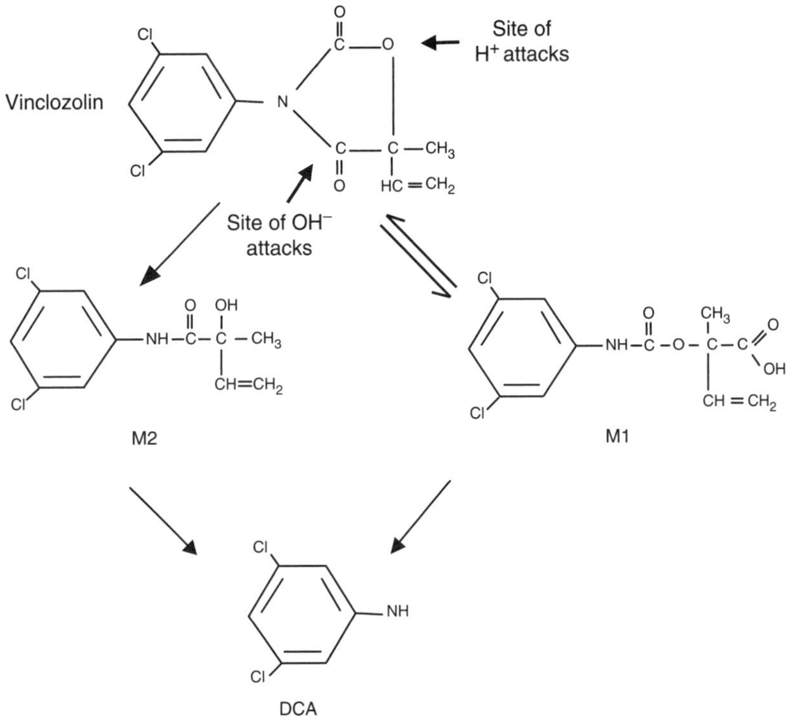

FIGURE 3-1. Structural formulae and degradation pathways of a hazardous chemical and its principal degradation products. (*Source:* After S.Y. Szeto, N.E. Burlinson, J.E. Rahe, and P.C. Oloffs, "Kinetics of Hydrolysis of the Dicarboximide Fungicide Vinclozolin," *Journal of Agricultural and Food Chemistry*, 37: 523–529, 1989.)

Abiotic chemical transformation processes are those reactions that occur in the dark and without being mediated by microbes. Abiotic reactions for hazardous chemicals include hydrolytic reactions of acid derivatives and surface-mediated reactions. Photochemical transformation results from the molecule's absorption of light, usually at the visible and ultraviolet wavelengths. Biochemical transformations are those that are mediated by microorganisms and occur along a pathway where the organic compound becomes increasingly mineralized by the microbes (i.e., being degraded into more simple inorganic molecules, such as carbon dioxide and water). These mechanisms occur simultaneously in the environmental substrate, but for clarity should be considered separately.

In the vinclozolin example, hydrolysis to M1 appears to take place at the amide in the fifth position of the carbamate ring (the right side of the vinclozolin molecule shown in Figure 3-1). The hydrolysis takes place at elevated pH levels (i.e., greater concentrations of hydroxyl ions), where hydroxyl ions attack the carbonyl group and break the single bond between the N and the vinyl group, leaving an acid moiety (butenoic acid). At lower pH, protons attack the molecule and reduction appears to occur in the degradation of vinclozolin to M2. The site of protonation is at the third position of the carbamate group (Figure 3-1). The degradation continues with cleavage of the remaining structure until the dichloroaniline molecule is formed with an amine (NH) moiety at the first position of the aromatic ring.

Abiotic Hydrolysis in Solution

The reaction pathway and kinetics of hazardous chemicals are partly determined by solution characteristics, especially the hydroxyl and hydrogen ion content. Hydrolysis can be the principal pathway for the degradation of many compounds and may be a necessary first step before biodegradation can occur. Hazardous chemicals can be degraded by hydrolytic reactions of acid derivatives, wherein a nucleophile reacts with a doubly bonded carbon atom. In our vinclozolin example, this attack takes place at the carboxylic acid ester position of the molecule and depends highly on solution pH. Acid-catalyzed hydrolysis occurs when the molecule's carboxylic acid ester is attacked via protonation (i.e., substitution by H^+ ions). Base-catalyzed hydrolysis takes place when an ester group is attacked by hydroxide ions.

The pH of pore water in soil and groundwater, or the pH of surface water can influence or determine the principal degradation pathway taken by hazardous chemicals. Some chemicals are less persistent in the field than in laboratory studies, possibly the result of increased photodegradation. Moisture in the soil column would also affect protonation and polarity over time. This point underscores the importance of characterizing both soil surface characteristics and solution acidity in a hazardous chemical's persistence. Acidity and ionic strength of soils can also greatly influence the degradation pathways and the rate at which the secondary degradation products may form. Just because a high or low pH rain falls on soil does not mean the pore water in the soil will have high or low pH, respectively. The ionic strength of the soil can provide significant buffering capacity. This phenomenon is demonstrated in Figure 3-2.

Surface-Mediated Hydrolysis

Surface characteristics have been found to mediate hydrolysis for certain organic compounds. Soil provides absorbents for the retention of organic compounds, with most sorption surfaces in clay and organic matter. Cation exchange capacity, charge, and the nature of cations on the exchange complexes will determine the amount of sorption of organic compounds. Double-layer phenomena often control sorption at the soil particle solid–liquid interface. These thin films around soil particles have thicknesses of 1–10 nm and are composed of ions of opposite charge to the particle.

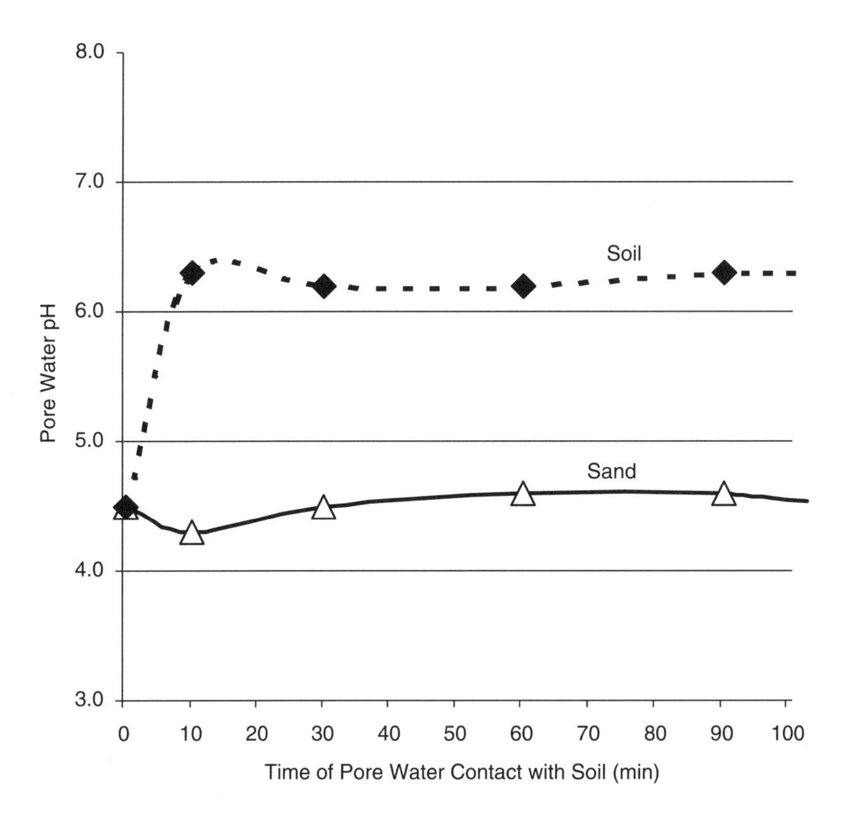

FIGURE 3-2. Resultant pore fluid pH in a North Carolina Piedmont aquic hapludult soil and 20–30 mesh Ottawa sand following challenges with pH 4.5 Piedmont rainwater prepared from 30 microequivalents (μeq) H_2SO_4 and 20 μeq NH_4Cl in de-ionized water. Note that within 10 minutes in the soil the low pH rain reached the soil water pH level of 6.3. This is the result of the high ionic strength and buffering capacity of the soil.

The hydrolysis can be acid or base catalyzed, depending on the pH of the soil water and other soil conditions. For example, by adding titanium dioxide particles or dissolved natural organic matter particles to a solution, surface catalyzed, neutral hydrolysis can dominate across a large pH range, even when acid or base catalysis had been observed. Such a relationship is important for engineers trying to control the degradation of toxic compounds. Additives like surfactants present in soil micellar solutions can also affect hydrolysis and even modify the kinetics of the chemical/soil/water system.

Photolysis

Hazardous chemicals can be broken down by photolysis. As has been found in other chemical degradation, photolysis may be inhibited or catalyzed by surfaces. For example, hazardous chemicals' photolysis will decrease with increasing particle concentrations in water, probably because of light attenuation. The photodegradation of hazardous chemicals can be enhanced in the presence of organic matter in pore fluid. Soil-bound hazardous chemicals are photodegraded in ultraviolet light more rapidly in the presence of fulvic acid than in the presence of humic acid. Photochemistry can also be catalyzed by metal oxides in the soil complex. If an organic compound is in solution, the type of solvent also affects the rate of photodegradation. This includes photoaddition and dehalogenation (e.g., higher alcohol content can bring about higher rates of photodegradation than in pure water).

Microbially Mediated Hydrolysis

Microbes play an important role in degrading organic compounds. For example, the reversible chemical reaction pathway in Figure 3-1 can be mediated by bacteria. *Corynebacterim* spp., *Pseudomonas* spp., *Nitrosomonas*, *Xanthomomonas* spp., and other bacterial spp., can enzymatically mediate the hydrolysis of numerous aliphatic and aromatic compounds, including halogenated compounds. Exposure to similarly structured pesticides may improve acclimation by microbes, as in the instance of soils that are pretreated with one compound that also elicit slightly faster degradation of other compounds in that chemical class (e.g., PCB or dicarboximide fungicides).

As is true for wastewater treatment processes, biologic degradation of hazardous wastes depends on a combination of abiotic and biotic processes. For example, the rate of microbial mediation of hydrolysis is affected by abiotic factors, such as pore fluid pH. As is always the case in environmental engineering, the fate of hazardous chemicals after their release is complex and depends on a host of variables. The engineer is well advised to know as much about the chemistry of these wastes as possible to assess the problem and devise remedial actions.

Physical Transport Mechanisms

Turning to physical processes, diffusion and mass flow mechanisms are highly variable among compounds, following several equilibrium and non-equilibrium sorption isotherms. Semivolatile, lipophilic compounds may be transported by at least five processes: (1) sorption, exudation, and retention in plant material; (2) dissolved or suspended runoff; (3) sorption to soil organic matter, clays, and mineral surface; (4) vapor-phase diffusion; and (5) advection and dispersion in the aqueous phase. The characteristics of the environmental medium and the compound largely determine which of these processes will dominate or be a rate-limiting step. In our chamber experiments, for example, the most important transport mechanisms were phase partitioning, diffusion, and mass flow.[4]

If a hazardous waste is applied to or buried in soil, the soil contains pockets of high and low concentrations of chemical compounds and diffusion will play a major role in transport. In a single phase, the chemical mass flows from the higher to lower concentrations until after some time the soil chemical concentration becomes uniform. In multiple-phase systems like those in soils, however, mass flow is more complex, and phase distributions among particles, solution, and air come into play.

The flux of a compound and its degradation products through the air–soil interface to the troposphere depends on desorption and transport mechanisms. These processes are not well understood for all hazardous chemicals. Semivolatile organic compounds (SVOCs), for example, do not readily dissipate and have greater affinity for solid and liquid phases than do the more volatile compounds. Semivolatile compounds' flux rates from soil are predominantly dictated by concentration gradients in soil and pore fluid and are proportional to vapor pressure; however, SVOCs may also be transported in the absence of partitioning, especially when they become sorbed to soil particles and are emitted as aerosols. Upon reaching the troposphere, these pollutants may remain in the vapor phase or be sorbed to particles that can travel hundreds of kilometers, subsequently be deposited, and ultimately become concentrated in various environmental media compartments. From these compartments, the compounds may enter the food chain and be bioaccumulated in the tissues of humans and wildlife.

Depending on their persistence, chemicals can be transported and transformed before reaching their ultimate fate. Within soil and sediment, sorption and degradation processes control the fate of organic chemicals. Once sorbed, the vapor pressure of the chemicals decreases.

The rate of water evaporation from the soil affects a compound's transport. Physicochemical properties and soil conditions also control plant uptake of chemicals and affect the movement of these compounds between soils and ground and surface water. A chemical's stability can be affected by ionic strength; for example, a compound can be quite unstable in extremely alkaline soils. As we saw in our earlier discussion, factors such as soil water content, physical and chemical properties, compound concentrations, and soil properties, especially soil organic matter, are important in controlling sorption and other partitioning rates.

The soil organic matter (SOM) content plays an important role in the mobility of a chemical. Sorption can increase in proportion to the available SOM dissolved in the soil water. The humic components of the SOM appear to be more important to sorption than the fulvic components. In addition, the SOM plays a role in partitioning hydrophobic compounds between aqueous and solid phases. The polarity of SOM has been identified as a factor in the differences between organic carbon-normalized sorption coefficients (K_{OC}) of soils from those in sediment. In addition to the role that surfactants play in surface-mediated degradation, surfactant-derived organic matter from organoclays can enhance the sorption of organic compounds above that from natural SOM. So surfactants can play a key role in soil and groundwater remediation.

Inorganic soil components also affect the transport of organic compounds. The role of clay content in soil is an important consideration in the mobility and degradation of hazardous wastes. Pore fluid ionic strength and pH affect the transport of organic compounds in addition to their roles in degradation. Organic compounds may move more rapidly through soils under low-ionic-strength soil conditions than under more lime conditions.

Particles have also been found to facilitate the flow of organic compounds through soil. The particle-facilitation process depends on the ability of particles and colloids to sorb the compounds and later be transported through the porous media. In addition, the mechanism of preferential flow allows the organic compounds to be transported, but in a more confined region of the soil column. Both particle-facilitated transport and preferential flow are involved in organic compound movement through the soil and are interrelated in soil macropores, where colloidal transport velocities are relatively high and preferential flow plays a larger role.

Discussion: Pollutant Transport: The Four Ds

Experts who model the movement of chemicals in the environment usually think about physical processes. Pollutant transfer systems are almost always concerned with transport by fluid. A fluid is any substance that will change its shape continuously when shear stress is applied. Any amount of shear stress will deform a fluid. The most common fluids in the environmental arena are air and water. Fluid mechanics abides by five basic laws:

1. Conservation of mass.
2. Newton's second law of motion (Force = mass × acceleration, $F = mA$).
3. Conservation of momentum.
4. First law of thermodynamics (Conservation of energy).
5. Second law of thermodynamics (Efficiency).

In a static fluid system, the volume and mass of fluid are constant, but the shape changes. In a dynamic fluid system, the volume and mass of the fluid also remain constant, but the area changes. Obeying the conservation law, mass is conserved within the domain. The fluid is moving, so identifying the actual amount of mass can be somewhat difficult. Before developing a model on transport, it is advisable that one become familiar with the principles of fluid dynamics. For this discussion, we are concerned with the movement of a contaminant in a fluid across an area. The flux of the contaminant is expressed as mass per area per time (units may be mg cm^{-2} s^{-1}).

The properties of the fluid are also important in fluid dynamics. A Newtonian fluid is one that has a constant viscosity at all shear rates at a constant temperature and pressure. Water and most solvents are Newtonian fluids. However, environmental engineers are confronted with Non-Newtonian fluids, i.e., those with viscosities not constant at all shear rates. Sites contaminated with drilling fluids and oils have large quantities of Non-Newtonian fluids onsite. Another consideration is whether the fluid is compressible. So, when using software programs, it is important to ensure that the defaults represent the fluids being investigated.

Dynamics combines the properties of the fluid and the means by which it moves. This means that the continuum fluid mechanics vary by whether the fluid is viscous or inviscid, compressible or incompressible, and whether flow is laminar or turbulent. For example the difference between a plume in a ground water aquifer and an air mass in the troposphere would be:

	Ground Water Plume	**Air Mass Plume**
General Flow Type	Laminar	Turbulent
Compressibility	Incompressible	Compressible
Viscosity	Low viscosity (1×10^{-3} kg m^{-1} s^{-1} @ 288 K)	Very low viscosity (1.781×10^{-5} kg m^{-1} s^{-1} @ 288 K)

The flux of a contaminant is equal to the mass flux plus the dispersion flux, diffusion flux, as well as source and sink terms. Sources can be the result of a one-time or continuous release of a chemical from a reservoir or result from desorption of the chemical along the way. Sinks can be the result of sorption and surface processes. This means that even though the source contribution is known, there will be sorption occurring in soil, sediment and biota that will either remove the chemical from the fluid, or under other environmental conditions, the chemical will be desorbed from the soil, sediment or biota. Thus, these interim sources and sinks must be considered in addition to the initial source and final sinks (i.e., the media of the chemical's ultimate fate).

Perhaps the easiest process to deal with is advection ($J_{Advection}$), the transport of dissolved chemicals with the water or airflow. The one-dimensional mass flux equation for advection can be stated as:

$$J_{Advection} = \bar{v}\eta_e[c] \qquad (3\text{-}1)$$

where,

$\bar{\nu}$ = average linear velocity (m s^{-1})
η_e = effective porosity (percent, unitless)
$[c]$ = chemical concentration of the solute (kg m^{-3})

But how do the chemicals find their way to the water body, aquifer or air mass in the first place? Generally, the pollutants follow a four-step trek from the source to the receiving reservoir: desorption; diffusion; dispersion; and dilution.[5]

Desorption

Sorption processes include four general processes:

1. *Adsorption* is the process wherein the chemical in solution attaches to a solid surface, which is a common sorption process in clay and organic constituents in soils.
2. *Absorption* is the process that often occurs in porous materials so that the solute can diffuse (we will discuss diffusion later in this section) into the particle and be sorbed onto the inside surfaces of the particle.
3. *Chemisorption* is the process of integrating a chemical into porous materials' surface via chemical reaction.
4. *Ion exchange* is the process by which positively charged ions (cations)[6] are attracted to negatively charged particle surfaces or negatively charged ions (anions) are attracted to positively charged particle surfaces, causing ions on the particle surfaces to be displaced. Particles undergoing ion exchange can include soils, sediment, airborne particulate matter, or even biota, such as pollen particles.

The mass of sorbed solute is:

$$M_{Sorb} = V \cdot (1 - \eta) \cdot \rho_{Soil}[c_{Sorb}] \tag{3-2}$$

where,

M_{Sorb} = mass of solute sorbed per dry unit weight of solid (kg kg^{-1})
V = volume of solid (m^3)
η = porosity (percent, unitless)
ρ_{Soil} = density of soil (kg m^{-3})
$[c_{Sorb}]$ = concentration of sorbed chemical (kg kg^{-1})

Thus, since the first step of freeing up a pollutant is that it is "desorbed" from its parent media, which can be defined as the reciprocal of the sorption equation.

Diffusion

Once the chemical is desorbed from its parent medium, it can move through an environmental compartment as result of a concentration gradient. The concentration gradient (i_c) is the change in concentration (in units of $kg\,m^{-3}$) with length (in meters), so the units of i_c are $kg\,m^{-4}$. Diffusion is therefore analogous to the physical potential field theories (that is, flow is from the direction of high potential to low potential, such as from high pressure to low pressure). This gradient is observed in all phases of matter, solid, liquid, or gas. Diffusion is really only a major factor of transport in porous media, such as soil or sediment, and can be ignored if other processes, such as advection, lead to a flow greater than $2 \times 10^{-5}\,m\,s^{-1}$ (See note 7).[7] However, it can be an important process for source characterization, since it may be the principal means by which a contaminant becomes mixed in a quiescent container (such as a drum, a buried sediment, or a covered pile) or at the boundaries near clay or artificial liners in landfill systems. Thus, diffusion is proportional to the concentration gradient, and is expressed by Fick's first law:

$$J_{Diffusion} = -d_o i_c \quad (3\text{-}3)$$

And,

$$i_c = \frac{\partial[c]}{\partial x} \quad (3\text{-}4)$$

where,

x = distance between the points of $[c]$ measurements.

Dispersion

Dispersion is the mixing of the pollutant within the fluid body (e.g., aquifer, surface water or atmosphere). A basic question is in order. Is it better to calculate the dispersion from physical principles, using a deterministic approach, than to estimate the dispersion using statistics, actually probabilities? The eulerian model bases the mass balance around a differential volume. A Lagrangian model applies statistical theory of turbulence, assuming that turbulent dispersion is a random process described by a distribution function. The Lagrangian model follows the individual random movements of molecules released into the plume, using statistical properties of random motions that are characterized mathematically. Thus, this mathematical approach estimates the movement of a volume of chemical (particle) from one point in the

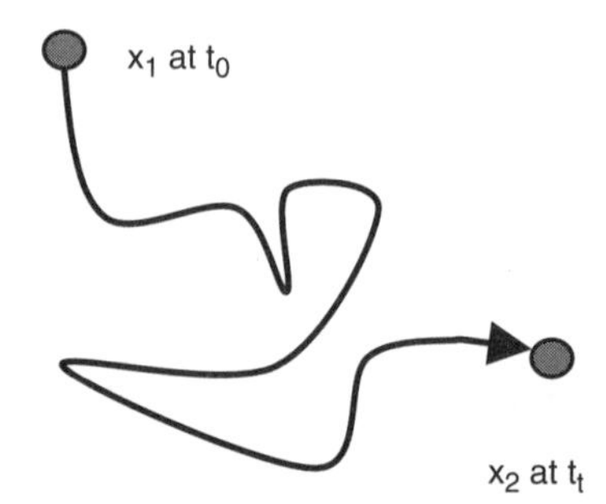

FIGURE 3-3. Hypothetical path of particle moving during time interval $(t_0 - t_t)$, as used in a Lagrangian model.

plume to another distinct point during a unit time.[8] Thus, the path (See Figure 3-3) each particle takes during this time, that is, an ensemble mean field relates to the particle displacement probabilities:

$$\overline{[c]}(x, y, z, t) = M_{Total} P(Dx, t) \qquad (3\text{-}5)$$

where,

Dx $= x_2 - x_1 =$ particle displacement (see Figure 3-3)
$P(Dx_2, t) =$ probability that the point x_2 will be immersed in the dispersing media at time t
M_{Total} = total mass of particles released at x_1
$\overline{[c]}$ = mean concentration of all released particles
= mass of particles the plume $dx{\cdot}dy{\cdot}dz$ around x_2

Gaussian dispersion models assume a normal distribution of the plume.

In a deterministic approach, the dispersion includes mixing at all scales. At the microscopic scale, the model accounts for frictional effects as the fluid moves through pore spaces, the path length around unconsolidated material (the tortuosity), and the size of the pores. At the larger scales, characteristics of strata and variability in the permeability of the layers must be described. So, a deterministic dispersion flux would be:

$$J_{Dispersion} = \underline{D} \cdot grad[c] \qquad (3\text{-}6)$$

$J_{Dispersion}$ = mass flux of solute due to dispersion ($kg\,m^{-2}\,s^{-1}$)
$\underline{D}$ = dispersion tensor ($m\,s^{-1}$)
$[c]$ = concentration of chemical contaminant ($kg\,m^{-3}$)

The $\underline{D}$ includes coefficients for each direction of dispersion, i.e., longitudinally, horizontally and vertically (D_{xx}, D_{xy}, D_{xz}, D_{yy}, D_{yz}, D_{zz}).

Comprehensive Effect of the Fluxes, Sinks, and Sources

These are the principal *physical* processes that determine transport, but chemical degradation processes are also at play in determining environmental fate of contaminants. Recall that one of five laws dictating fluid dynamics mentioned at the beginning of this discussion included conservation of mass. This is true, of course, but the molecular structure of the chemical may very well change. The change depends upon chemical characteristics of the compound (e.g., solubility, vapor pressure, reactivity, and oxidation state) and those of the environment (e.g., presences of microbes, redox potential, ionic strength, and pH). The chemical degradation can be as simple as a first order decay process:

$$\frac{\partial[c]}{\partial t} = -\lambda c$$

These lambda terms are applied to each chemical. In fact, the new degradation products must also be accounted for in the model (being true to conservation of mass), so an iterative approach to the transport and fate of each degradation product can be described. This is essential if the degradates are toxic. In the fungicide example described in this chapter, two of the degradation products are actually more toxic (antiandrogenic) than the parent compound!

So the expected total flux representing the fate (J_{Fate}) of the contaminant will look something like:

$$J_{Fate} = J_{Desorption} + J_{Diffusion} + J_{Dilution} + J_{Dispersion} + J_{Advection} - J_{Sorption} - \lambda[c] \qquad (3\text{-}7)$$

This describes the general components and relationships of pollutant transport, and should help the engineer to select the appropriate model for the chemical and environmental needs dictated by each project.

Chemical Sorption Kinetics[9]

Equilibrium between sorbed phases and solutions of chemical compounds is the point at which the rate of a forward reaction equals the reverse reaction rate. For a more thorough discussion of chemical equilibria, see Appendix 5. The sorption of a chemical compound in soil can be expressed by the Freundlich equation:

$$[S] = K_d[C]^\eta \tag{3-8}$$

where:

$[S]$ = concentration of the chemical sorbed to the soil (mg L^{-1})
K_d = equilibrium distribution coefficient (dimensionless)
$[C]$ = concentration of the chemical in aqueous phase (mg kg^{-1})
η = empirical order of sorption reaction (dimensionless).

Several mechanisms can affect sorption and the time it takes to reach sorptive equilibrium expressed in the Freundlich equation:

- Diffusion through aqueous solution.
- Diffusion through micropores in the soil matrix.
- Diffusion through organic solids.
- Overcoming molecular-level energy barriers to phase transfer.

One-dimensional, isotropic diffusion through aqueous solution is governed by Fick's first law:

$$J^n = K_d \frac{\delta [C]^n}{\delta x} \tag{3-9}$$

where:

J^n = flux of diffusing compound n (mg m^{-2} s^{-1})
$[C]^n$ = concentration of chemical compound n (mg L^{-1})
x = space coordinate measured normal to a section of matrix (dimensionless).

Thus, sorption is controlled by the diffusion of a chemical from higher concentrations to lower concentrations in the aqueous phase.

The micropores in the soil matrix provide another control over sorption processes. Pore sizes even in well-sieved soils can span large ranges, and soil aggregate structures provide larger complexes with different pore spaces than if only soil texture (i.e., size of individual soil particles) are considered. The transport mechanism active in the pore space is predominantly advection (i.e., flowing pore fluids). Advection decreases with decreasing pore space size.

Diffusion through organic solids, such as those in soils (e.g., humic and fulvic acids, humins, polysaccharides, and partially decomposed detritus) is generally slower than diffusion through an aqueous solution. The diffusion K_d of a compound in organic matter generally decreases with molecular size and weight, but the solution K_d (i.e., the distribution coefficient) of the compound increases with molecular weight and size. This means that the most soluble molecules are expected to diffuse at the slowest rate. Thus, even though a hazardous chemical is slightly soluble in water, its transport in the soil may be enhanced by diffusion.

The kinetics of sorption are tied to the activation energies at the molecular level. These include Van der Waals–London interactions, such as the energy of interaction between permanent dipole and induced dipoles from neighboring molecules. Energies also include hydrophobic bonding created from the entropy gain resulting from breaking up water molecules around nonpolar regions of the pesticide molecule, charge transfer from the orbital overlap of electron-rich functional groups, ligand exchange with metal ions in the soil matrix, and ion exchange.

Solute advection and diffusion are the principal processes that determine the movement and emission of numerous organic compounds from soil matrices to the atmosphere. The flux is driven by the volumetric water flux to some extent. This flux is assumed to be proportional to the hydraulic potential gradient in the soil, and the matric potential is assumed to be primarily a function of soil moisture.

Organic Chemistry Discussion: Why Are Carbon Compounds Called Organic?

Many of the so-called organic compounds are anything but natural, so one may wonder why they possess a name that is associated with living things.

Paracelsus, the 16th century alchemist and medical doctor, was among the first to ascribe chemical principles to human bodily functions and disease.[10] In fact, he frequently diagnosed sick people as having a deficiency in what he called *elements*. Beginning in the 18th century, certain chemical compounds were being associated with systems in living organisms, especially that these chemicals were thought to emanate from a vital force that controlled the process of life. The term *organic* was coined by Jons Jacob Berzelius early in the 19th century, after he was able to isolate certain compounds from living creatures. At this time, however, any such substance extracted from living things was considered impossible to synthesize in the laboratory and could only be derived from organisms; however, serendipitously in 1928, Fredrich Wöhler synthesized urea by heating ammonium cyanate (chemists had been able to isolate urea from the urine of dogs).[11] So, the first organic synthesis was:

$$NH_4^+NCO \xrightarrow{heat} H_2N - CO - NH_2 \qquad (3\text{-}10)$$

Following this first foray from vitalism, theory led to the synthesis of thousands of organic compounds. Unfortunately, several have been found to be toxic and hazardous to those life processes that science had first thought were the only sources of organic compounds.

What Kinds of Hazardous Chemicals Are There?

Chemists have employed numerous ways of classifying compounds. The broadest distinctions have been between organic and inorganic compounds. We will use these two broad classes to describe the compounds that are of most importance to hazardous waste engineering and risk assessment.

Organic Compounds

Simply stated, organic chemistry is interested in compounds that contain carbon. More correctly, organic chemistry is the chemistry of compounds possessing carbon-to-carbon and carbon-to-hydrogen bonds.[12] Most of the hazardous compounds are organic. Generally, most organic compounds are more lipophilic (fat soluble) and less hydrophilic (water soluble) than most inorganic compounds; however, there are large ranges (Figure 3-4) of solubility for organic compounds, depending on the presence of polar groups in their structure. For example, adding the alcohol group to n-butane to produce 1-butanol increases the solubility several orders of magnitude. Organic compounds can be further classified into two basic groups: aliphatics and aromatics.

Aliphatic compounds are classified into a few chemical families. The alkanes contain a single bond between each carbon atom and include the simplest organic compound, methane (CH_4), and its derivative chains such as ethane (C_2H_6) and butane (C_4H_{10}). Alkenes contain at least one double bond between carbon atoms. For example, 1,3-butadiene's structure is $CH_2{=}CH{-}CH{=}CH_2$. The numbers 1 and 3 indicate the position of the double bonds. The alkynes contain triple bonds between carbon atoms, the simplest being ethyne, $CH{\equiv}CH$, which is commonly known as acetylene (the gas used by welders).

The aromatics are all based on the six-carbon "ring" configuration of benzene (C_6H_6). The carbon–carbon bond in this configuration shares more than one electron, so that benzene's structure (Figure 3-5) allows for resonance among the double and single bonds (i.e., the actual benzene bonds flip locations). Benzene is the average of two equally contributing resonance structures. To show this average resonance, the benzene molecule is often depicted as it is shown in Figure 3-6 (usually with the H atoms understood).

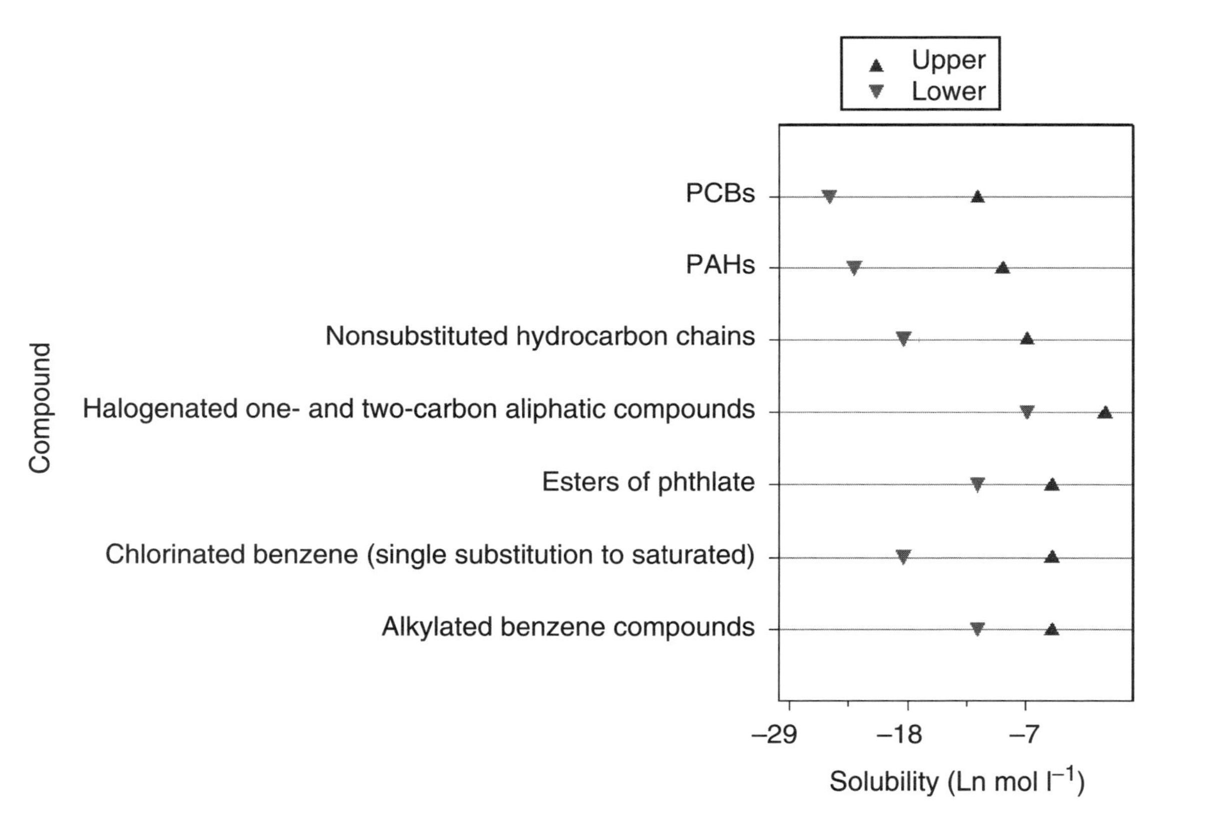

FIGURE 3-4. Natural logarithms of solubilities of classes of toxic organic compounds in water (mol l^{-1}at 25°C). (*Source:* From data in R.P. Schwarzenbach, P.M. Gschwend, and D.M. Imboden, *Environmental Organic Chemistry* [New York: John Wiley & Sons, 1993].)

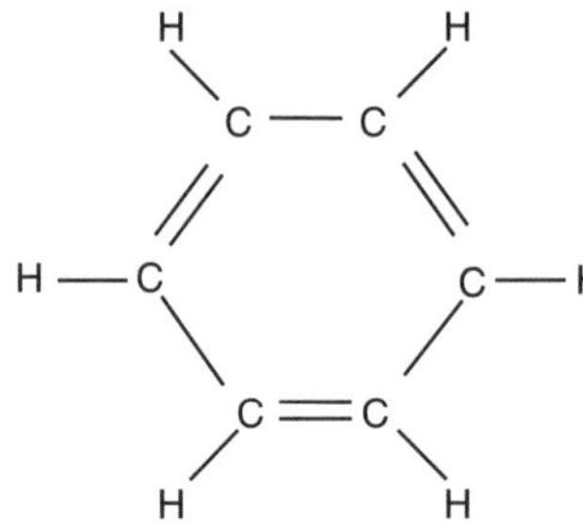

FIGURE 3-5. Structure of the benzene ring.

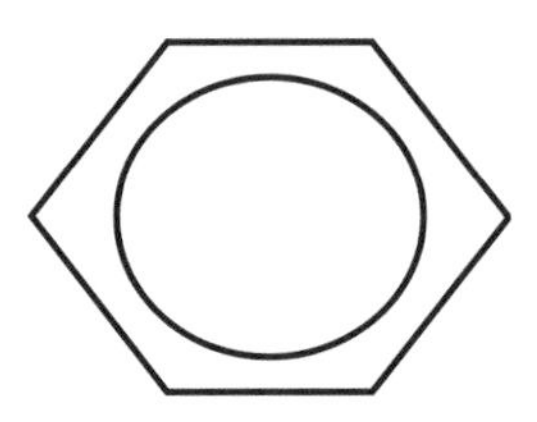

FIGURE 3-6. The benzene ring structure, accounting for resonance.

The term *aromatic* comes from the observation that many compounds derived from benzene were highly fragrant, such as vanilla, wintergreen oil, and sassafras. Aromatic compounds, thus, contain one or more benzene rings. The rings are planar (i.e., they remain in the same geometric plane as a unit); however, in compounds with more than one ring, such as the highly toxic polychlorinated biphenyls (PCBs), each ring is planar, but the rings may be bound together and may or may not be planar. This is actually an important property for toxic compounds. Some planar aromatic compounds are more toxic than their nonplanar counterparts, possibly because living cells may be more likely to allow planar compounds to bind to them and to produce nucleopeptides that lead to biochemical reactions associated with cellular dysfunctions, such as cancer or endocrine disruption.

Both the aliphatic and aromatic compounds can undergo substitutions of the hydrogen atoms. These substitutions render new properties to the compounds, including changes in solubility, vapor pressure, and toxicity. For example, halogenation (substitution of a hydrogen atom with a halogen) often makes an organic compound much more toxic.[13] Trichloroethane is a highly carcinogenic liquid that has been found in drinking water supplies, whereas nonsubstituted ethane is a gas with relatively low toxicity. Substitution chemistry is also a means for treating the large number of waste sites contaminated with chlorinated hydrocarbons and aromatic compounds by using dehalogenation techniques.

Persistent Organic Pollutants

Many of the synthetic chemicals over the past half century have left society with a burden of large quantities of very persistent compounds that resist breakdown in the environment. We should discuss some of the principal types—polychlorinated biphenyls (PCBs) and dioxins. However, several other groups of organic compounds are also persistent.

Polychlorinated Biphenyls. The PCBs and dioxins provide examples of the complexity of dealing with persistent hazardous organic compounds.[14] About 700,000 tons (6.4×10^{10} g) of pure PCBs were produced in the United States from 1929 to 1977. Most of these PCBs were used as dielectric or insulating fluids in electrical devices, especially transformers, but the compounds were also used in resins, paints, and coatings, as well as in hydraulic and heat-transfer fluids. None of these items are currently manufactured with PCBs, but products manufactured before PCBs were banned are still in service.

The fate of PCBs expected when their regulation began is shown in Figure 3-7. The ubiquitous use of PCBs has meant that the compounds

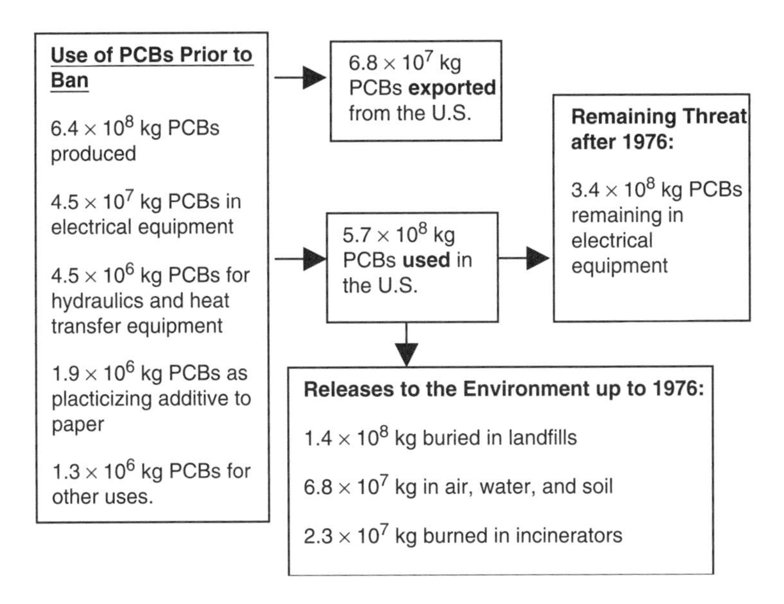

FIGURE 3-7. The mass balance of PCBs in the United States as of 1976. (*Source:* Committee for Environmental Cooperation, North America Free Trade Agreement.)

may be found anywhere. A recent inventory of PCBs remaining in service in the United States was conducted for the Electric Power Research Institute (EPRI).[15] The data were collected from the 100 largest U.S. utilities, augmented by nonutility data. Large-volume PCB wastes, such as dredged materials, contaminated sediments, sewage treatment sludges, and demolition wastes, must be disposed of in federally permitted, commercial PCB disposal facilities. As of 1991, PCBs constituted the predominant waste of 20% of Superfund's largest facilities (the so-called National Priority Listing sites) and 7% of all Superfund sites. PCBs accounted for about 34,070,000 cubic yards of material at the National Priority Listing sites.[16]

Anyone storing or disposing of PCBs commercially in the United States must have approval from the U.S. EPA.

Securing this approval entails preparing closure plans and obtaining disposal permits allowing for the maximum available capacity for storing and disposing PCBs. Commercial storage facilities are often also a permitted PCB disposal facility. Figure 3-7 shows the principal places where PCBs were found in the second half of the 20th century. Figures 3-8 (numbers of items) and 3-9 (volumes of PCBs) show the trends in storage from 1990 to 1993. About 26,000,000 kg of PCB wastes were being stored in 1993. PCB disposal is being conducted at four federally permitted incinerators in Kansas, Utah, and Texas. One mobile incinerator is also in operation. Seven waste landfills (Figure 3-10) are dedicated to PCB disposal, including physical separation, dechlorination, transformer

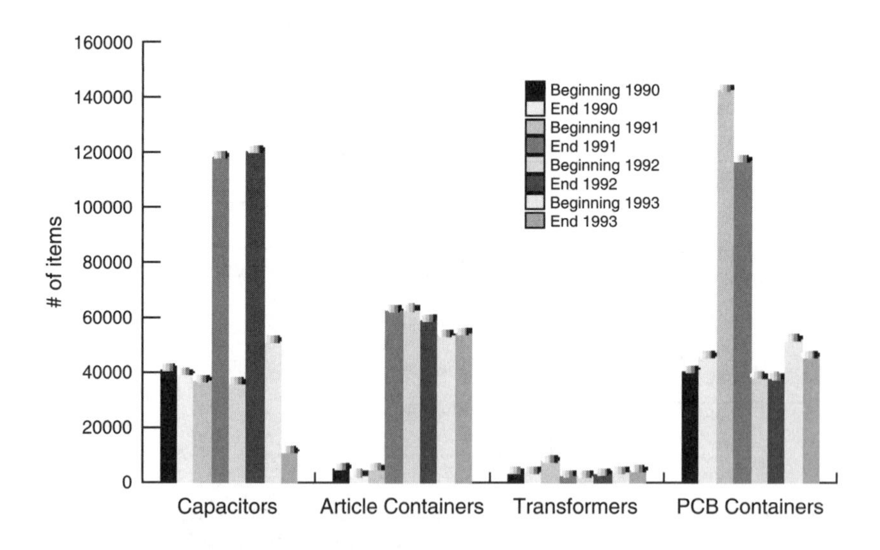

FIGURE 3-8. Numbers of PCB-laden items in storage, 1990–1993.

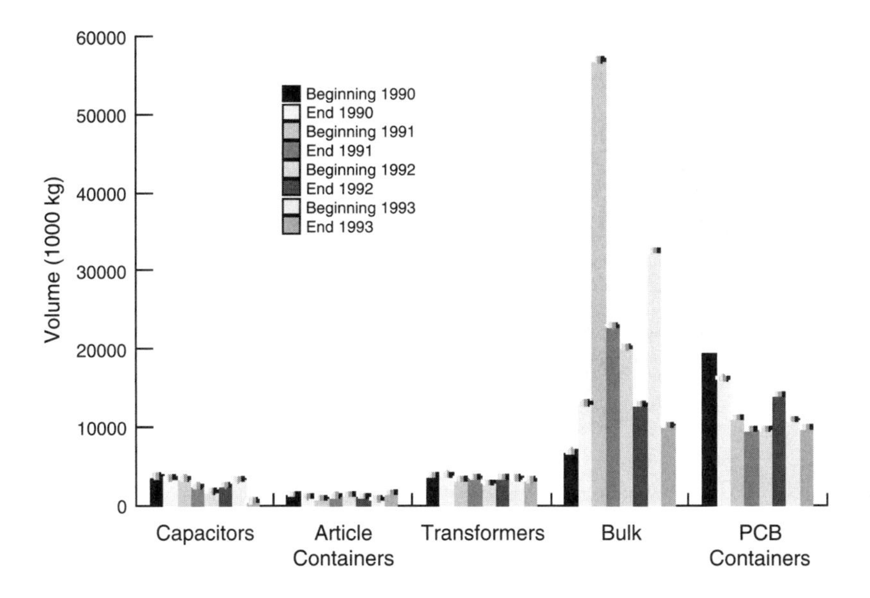

FIGURE 3-9. Volume of PCBs in storage, 1990–1993.

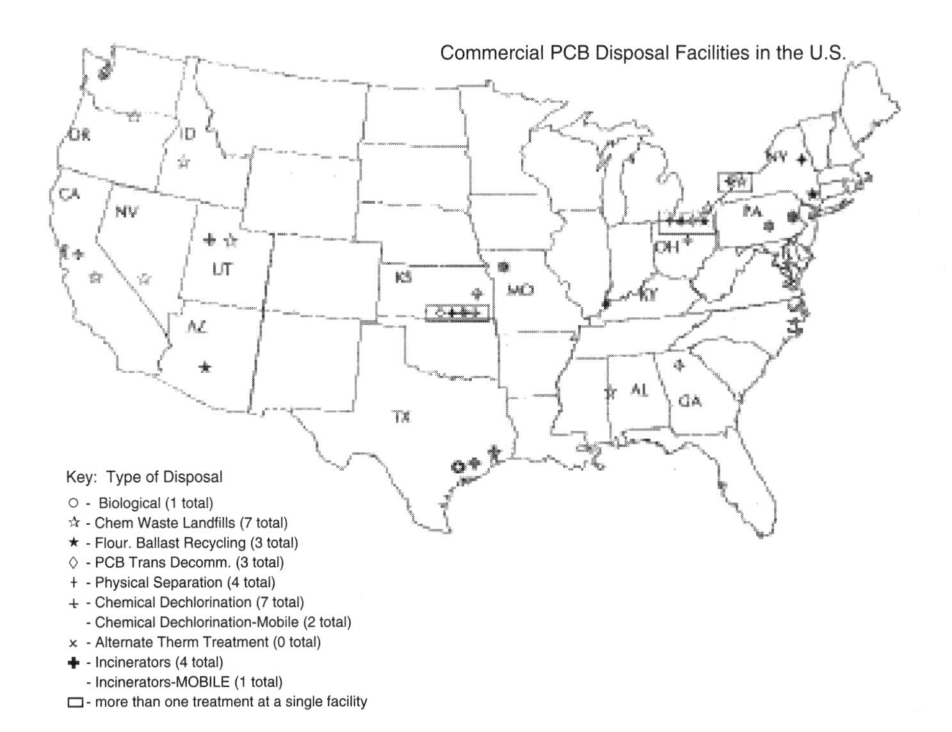

FIGURE 3-10. PCB disposal sites in the United States.

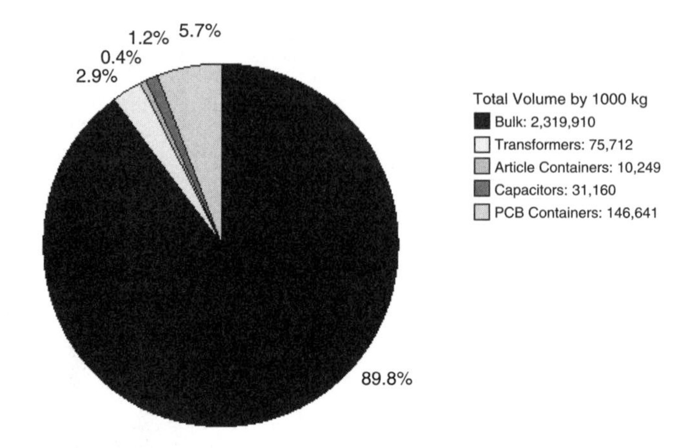

FIGURE 3-11. Total PCB disposal volumes, 1990–1993.

decommissioning, and fluorescent light ballast recycling (see pie chart in Figure 3-11).

Dioxins. *Dioxin* is a general term that describes a group of hundreds of persistent, aromatic compounds. The most toxic form, 2,3,7,8-tetrachlorodibenzo-p-dioxin (TCDD), is produced when Cl atoms substitute for H atoms at the 2, 3, 7, and 8 positions. Unlike PCBs, which were intentionally manufactured, the dioxins and furans (Figure 3-12) are produced accidentally by other processes, including waste incineration, chemical and pesticide manufacturing, and pulp and paper bleaching. Dioxin was in the defoliant Agent Orange and was the major contaminant that led to the famous evacuation and remediation efforts at Love Canal, New York; Times Beach, Missouri; and Seveso, Italy.

How Are Dioxins Formed?

Incinerators of chlorinated wastes are the most common environmental sources of dioxins, accounting for about 95% of the volume. The emission of dioxins and furans from combustion processes can follow three general pathways. The first pathway occurs when the feed material going to the incinerator contains dioxins and/or furans and a fraction of these compounds survives thermal breakdown mechanisms and passes through to be emitted from vents or stacks. This process is not considered to account for a large volume of dioxin released to the environment, but it may account for the production of dioxin-like, coplanar PCBs.

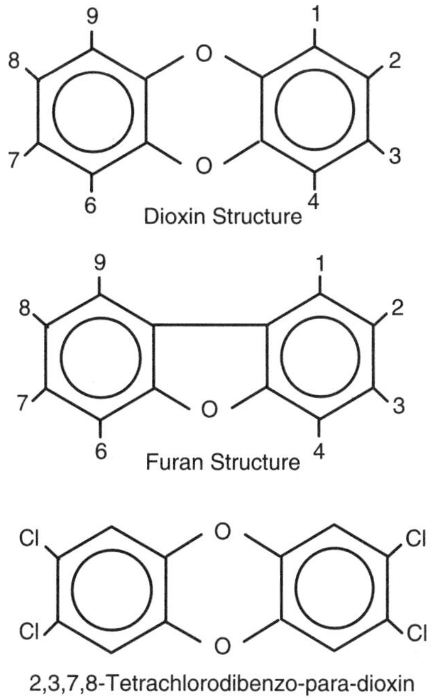

Dioxin Structure

Furan Structure

2,3,7,8-Tetrachlorodibenzo-para-dioxin

FIGURE 3-12. Molecular structures of dioxins and furans.

The second process is the formation of dioxins and furans in the thermal breakdown and molecular rearrangement of precursor compounds, such as the chlorinated benzenes, chlorinated phenols (such as pentachlorophenol, PCP), and PCBs, which are chlorinated aromatic compounds with structural resemblances to the chlorinated dioxin and furan molecules. Dioxins appear to form after the precursor has condensed and adsorbed onto the surface of particles, such as fly ash. This is a heterogeneous process, where the active sorption sites on the particles allow for the chemical reactions, which are catalyzed by the presence of inorganic chloride compounds and ions sorbed to the particle surface. The process occurs within the temperature range of 250–450°C, so most of the dioxin formation under the precursor mechanism occurs away from the high-temperature zone in the incinerator, where the gases and smoke derived from combustion of the organic materials have cooled during conduction through flue ducts, heat exchanger and boiler tubes, air pollution control equipment, or the vents and the stack.

The third means of synthesizing dioxins is *de novo* within the so-called cool zone of the incinerator, wherein dioxins are formed from moieties different from those of the molecular structure of dioxins, furans, or precursor compounds. Generally, these can include a wide range of both halogenated compounds like polyvinylchloride (PVC) and non-halogenated organic compounds like petroleum products, nonchlorinated plastics (polystyrene), cellulose, lignin, coke, coal, and inorganic compounds like particulate carbon and hydrogen chloride gas. No matter which *de novo* compounds are involved, however, the process needs a chlorine donor (i.e., a molecule that donates a chlorine atom to the precursor molecule). This leads to the formation and chlorination of a chemical intermediate that is a precursor. The reaction steps after this precursor is formed can be identical to the precursor mechanism discussed in the previous paragraph.

Other processes generate dioxin pollution. A source that has been greatly reduced in the last decade is the paper production process, which formerly used chlorine bleaching. This process has been dramatically changed, so that most paper mills no longer use the chlorine bleaching process. Dioxin is also produced in the making of PVC plastics, which may follow chemical and physical mechanisms similar to the second and third processes discussed previously.

Because dioxin and dioxin-like compounds are lipophilic and persistent, they accumulate in soils, sediments, and organic matter and can persist in solid and hazardous waste disposal sites.[17] These compounds are semivolatile, so they may migrate away from these sites and be transported in the atmosphere either as aerosols (solid and liquid phase) or as gases (the portion of the compound that volatilizes). Therefore, the engineer must take great care in removal and remediation efforts not to unwittingly cause releases from soil and sediments via volatilization or perturbations, such as landfill and dredging operations.

Inorganic Compounds

By definition, inorganic compounds include all those that are not organic. Thus any reaction that does not involve a molecule with a carbon-to-carbon bond is considered inorganic. For engineers who are interested in controlling hazardous wastes, several areas of inorganic chemistry are important.

Engineers are often concerned with mass balances. No mass balance is possible without an understanding of the inorganic chemical processes that occur in the environment. This knowledge is particularly important to the hazardous waste engineer. The engineer must know the form and amount of a toxic substance from its generation to its commercial movement to its use to its ultimate disposal (the cradle-to-grave concept). Likewise, once a contaminant is released to the environment, the way that it moves and changes in the environment must be known. In the environment, inorganic processes dictate where the hazardous compounds are likely to be found. Inorganic processes, such as dissolution and precipitation, are used to collect and treat contaminated environmental resources.

Toxic metals and metalloids (such as arsenic) and their compounds are foremost among the inorganic substances of concern. Several metals have received much attention for their role in environmental contamination. The neurotoxic metals, especially lead and mercury, and more recently manganese, are well-known for major contamination events and exposures to large numbers of susceptible populations, such as small children. Mercury is particularly difficult to address because its mobility in the environment and its toxicity to humans and animals is determined by its chemical form. For example, dimethylmercury is highly toxic, accumulates in the food chain, and has high affinity for organic tissues, but elemental mercury is much less toxic and is slower to bioaccumulate.

Metals exist in the environment in several oxidation or valence states. Each of these forms has its own toxicity and dictates its fate in the environment. Chromium, for example, in its trivalent (Cr^{+3}) form is an essential form of the metal. Although toxic at higher levels, Cr^{+3} is much less toxic than the hexavalent Cr^{+6}, which is highly toxic to aquatic fauna and is a suspected human carcinogen. Thus, in a reduced environment (such as a bog or wetland), the more reduced forms of metals are formed. In fact, the presence of ferrous iron (Fe^{+2}) is a method for determining the extent of reduction of environments. For example, an environment with a higher ferrous to ferric (Fe^{+3}) iron ratio than another environment is an indication that the former is more reduced.[18] Toxicity, persistence, and fate are also determined by the metal's equilibrium chemistry, especially the amount in ionic forms and the amount that forms salts with nonmetals.

Discussion: Sources, Movement, and Fate of Semivolatile Organic Compounds in the Environment

Environmental engineers monitor the release of toxic chemicals and are concerned with the fate of these compounds in the environment. Engineers must understand the factors that lead to the release and degradation of these compounds. Engineers must identify approaches to abate the transport of chemicals in order to reduce or prevent human and environmental exposures. Determining the movement and chemical transformation of toxic substances in the environment is crucial to understanding potential exposures to humans and wildlife.

The transport of a compound in the environment can be complex. For example, organic pesticide migration increases with larger rain

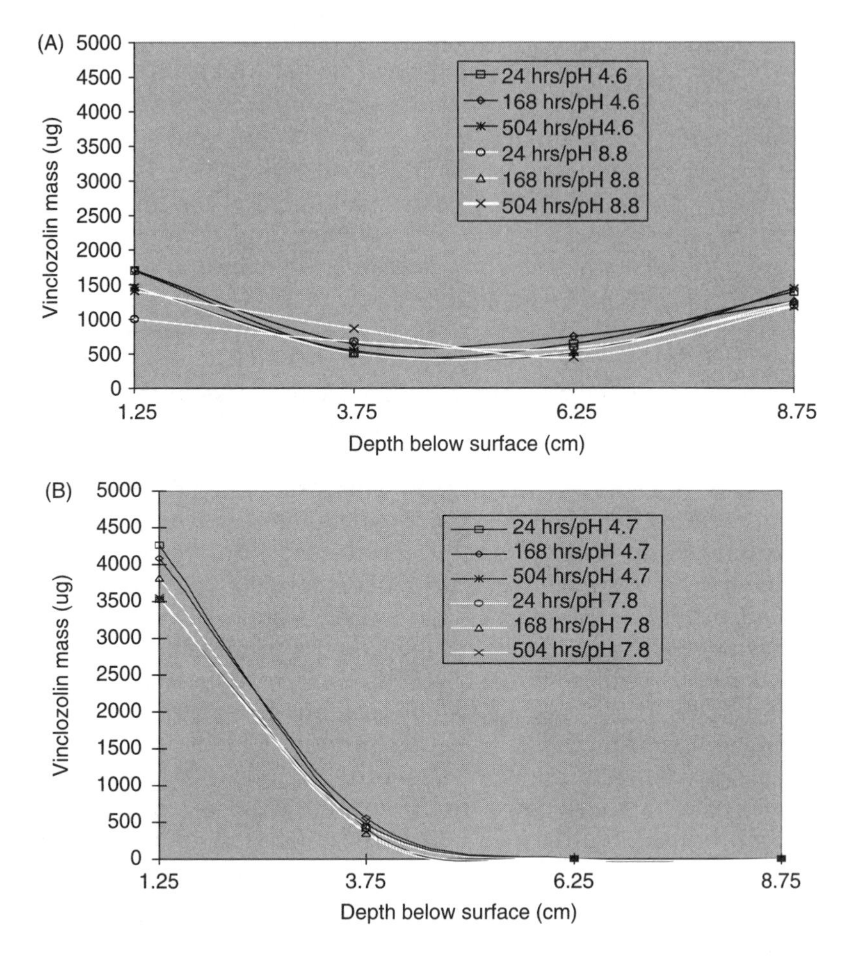

FIGURE 3-13. Mass distribution trends of an organic fungicide in (A) 20–30 mesh sand at pH 4.6 and 8.8; and (B) an aquic hapludult soil at pH 4.7 and 7.8 following an application of 5,000 mg vinclozolin and 3.1 mm rain event effects.

events and varies by soil type (Figure 3-13). Soil characteristics drive the adsorption mechanisms that are a principal determinant of pesticide downward migration in laboratory soil columns.[19]

Advection, dispersion, and gravimetric forces assist in water flow through soil and unconsolidated material. Compounds can exist in solid, liquid, and gas phases. Solid-phase (particle) movement within soil is governed by the amount of water available to transport the particle downward through the soil column and by the soil porosity. Soil contains a complex matrix of macropores and micropores, which limit

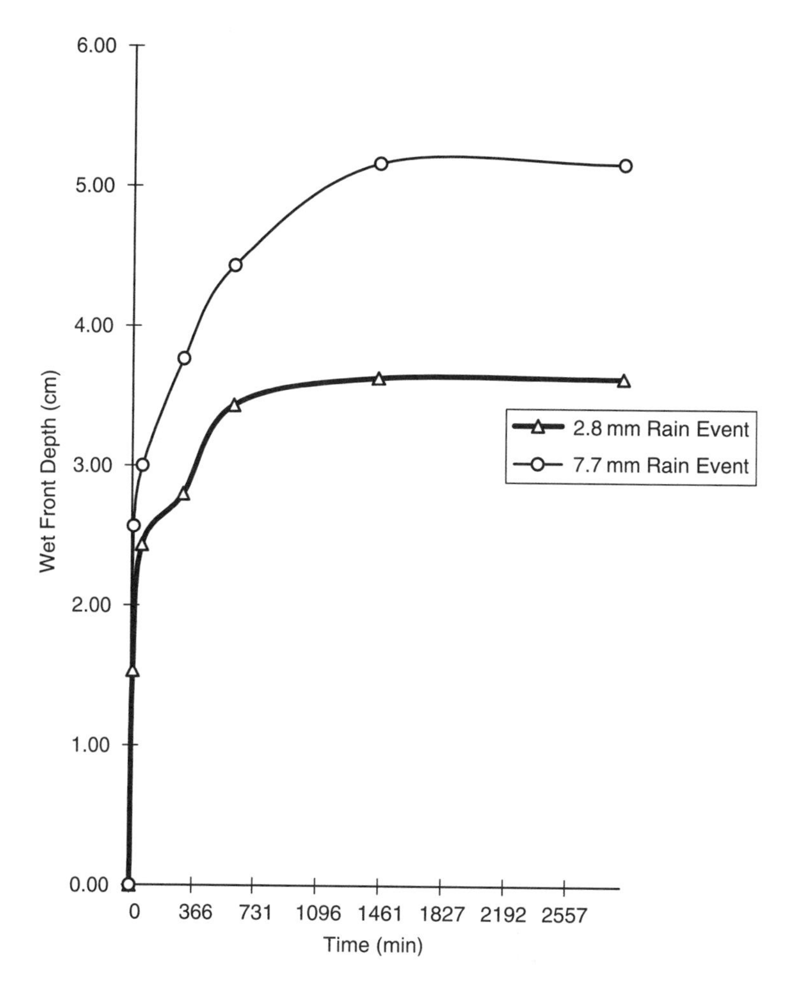

FIGURE 3-14. Measured wetting front movement in a North Carolina Piedmont aquic hapludult soil for 2.8 mm and 7.7 mm rain events as functions of time.

the soil fluid migration through the soil column. Particles that form microcoagulates of the chemical remain adhered to soil, while the fluid may continue its downward path, but the particles are filtered as they flow through smaller pore spaces. The size exclusion may explain the retention of some organic particles in the top layers of soil, even with a strong flushing event like a large storm (Figure 3-14). Conversely, larger-grained sands may filter particles, like the rapid sand filter in water treatment facilities. The size and characteristics of particles are

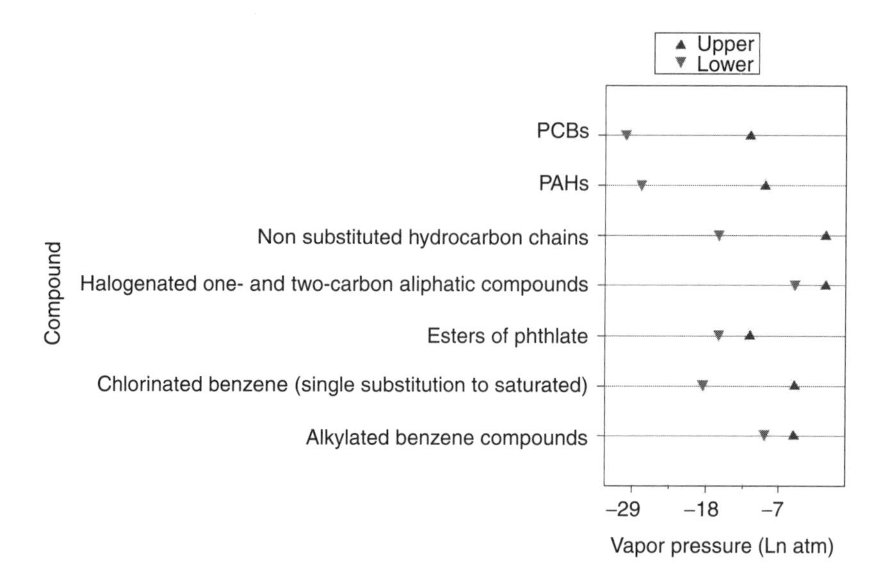

FIGURE 3-15. Ranges (natural logarithms) of vapor pressure (P^0) at 25°C for classes of organic compounds found in the environment. (*Source:* From data in R.P. Schwarzenbach, P.M. Gschwend, and D.M. Imboden, *Environmental Organic Chemistry* [New York: John Wiley & Sons, 1993].)

important in the solid particle filtration process as solid particles are filtered from water by way of solid–solid interception.[20] As solid organic particles flow past stationary sand particles, an attachment by inertia can occur when the chemical particles travel off streamline. High-impact velocities at the surface could expedite the inertia attachment of the organic particles on sand and soil particles.

Movement from soil to air is driven by several factors, especially the vapor pressure (Figure 3-15), which generally depends on the molecule size (Figure 3-16). Compounds that have vapor pressures between 10^{-5} and 10^{-2} kilopascals are referred to as semivolatile because they may be transported from soil in the gas phase or as aerosols.[21]

The water solubility of the compound (see Figure 3-4) and the characteristics of the soil are also factors in how easily a compound will be transported. The combination of the chemical's propensity to remain dissolved in water and to resist volatilization is inversely related to the *fugacity* of the chemical in the environment. Fugacity is the tendency or ease with which a compound is partitioned from one medium to another. A common measure of fugacity is that of an equilibrium partitioning known as the Henry's law constant, K_H (Figure 3-17).

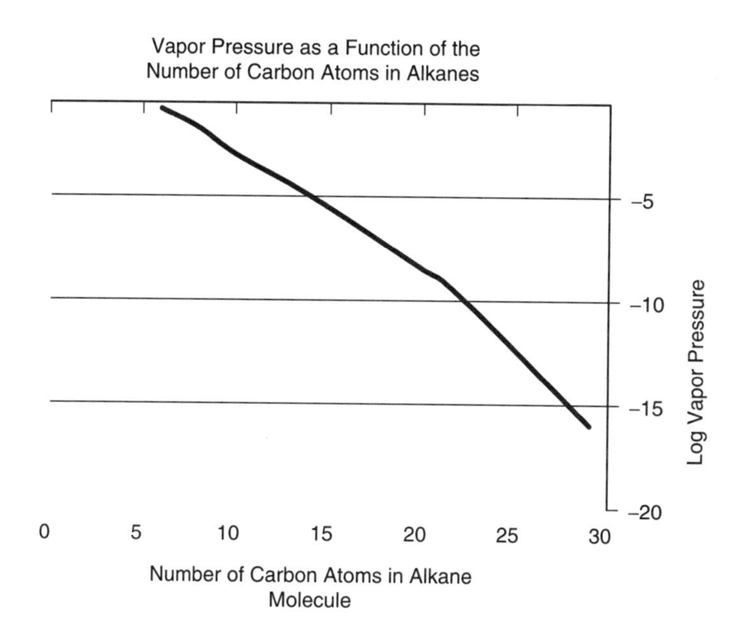

FIGURE 3-16. Change in vapor pressure (P^0) at 25°C in alkanes as a function of molecular size. (*Source:* From data in R.P. Schwarzenbach, P.M. Gschwend, and D.M. Imboden, *Environmental Organic Chemistry* [New York: John Wiley & Sons, 1993].)

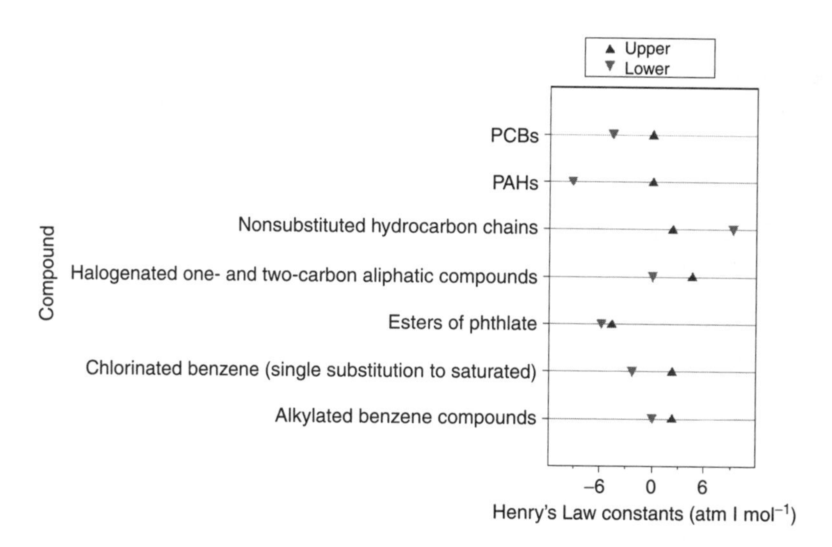

FIGURE 3-17. Natural logarithms of Henry's law constants (K_H) at 25°C for classes of organic compounds found in the environment. (*Source:* From data in R.P. Schwarzenbach, P.M. Gschwend, and D.M. Imboden, *Environmental Organic Chemistry* [New York: John Wiley & Sons, 1993].)

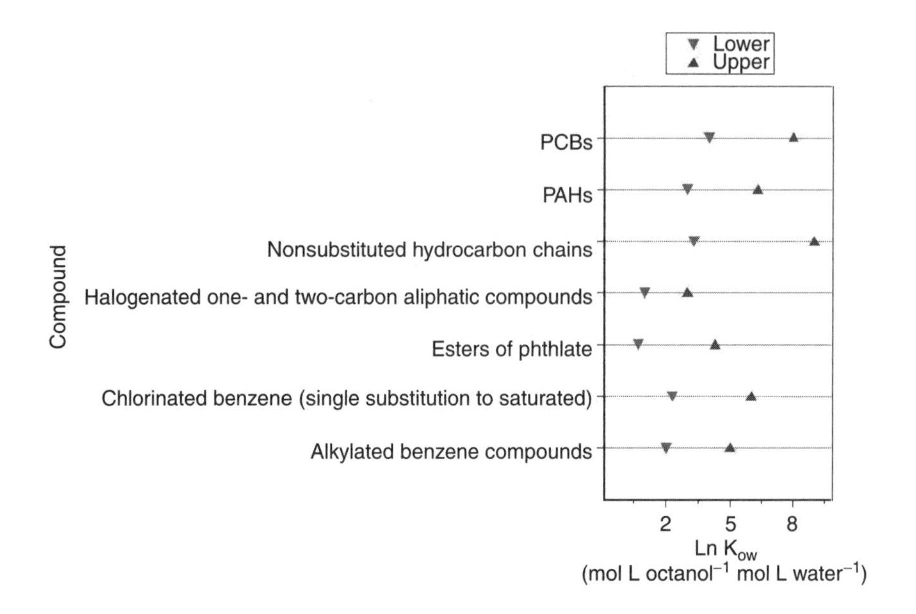

FIGURE 3-18. Natural logarithms of octanol-water partition constants (K_{OW}) at 25°C for classes of organic compounds found in the environment. (*Source:* R.P. Schwarzenbach, P.M. Gschwend, and D.M. Imboden, *Environmental Organic Chemistry* [New York: John Wiley & Sons, 1993].)

Simply stated, K_H is the ratio of a compound's mass in the gas phase to its mass in the aqueous phase. Although the ranges cover several orders of magnitude, in general the larger, halogenated ringed compounds, such as PCBs and dioxins, have less affinity to the aqueous phase than to the short, nonsubstitued chains.

Another important measure of partitioning compounds is the octanol-water partition constant, K_{ow} (Figure 3-18), which reflects a compound's affinity for an organic medium (less polar) versus its affinity for the aqueous medium (more polar). The larger the K_{ow}, the more likely the compound will be found in organic matter. So, when the engineer confronts a compound with a very high K_{ow} value, this may mean it will be difficult to remove the compound from any medium, such as soil, that contains organic matter.

The chemical breakdown and reaction pathways and kinetics of organic compounds are partly determined by the pH of the soil pore fluid. Bacteria can also mediate the degradation rates.

The mass balance for semivolatile organic compounds (SVOCs) can be expressed as an equation of advection, dispersion, and reaction.[22]

The equation can be stated as:

$$\frac{d[C]^{SVOC}}{dt} = \left[\frac{Q[C]_0^{SVOC}}{V_A} + \frac{J^{SVOC}A}{V_A}\right] - \left[\frac{\Psi^{SVOC}A'}{V_A} + \frac{Q[C]_f^{SVOC}}{V_A}\right] + R_A - D_A \quad (3\text{-}11)$$

where:

$[C]^{SVOC}$ = SVOC concentration in the chamber ($\mu g\ L^{-1}$)
$[C]_0^{SVOC}$ = SVOC concentration entering the site ($\mu g\ L^{-1}$)
$[C]_f^{SVOC}$ = SVOC concentration exiting the site ($\mu g\ L^{-1}$)
J^{SVOC} = SVOC emission from the soil, flux per unit area ($\mu g\ cm^{-2}\ s^{-1}$)
A = Soil–air interface (cm^2)
V_A = Air volume above the soil (L)
ψ^{SVOC} = SVOC deposition of SVOC to A'_A ($\mu g\ cm^{-2}\ s^{-1}$)
A' = Surfaces in contact with air volume, except A (cm^2)
Q = Airflow ($L\ s^{-1}$)
R_A = Chemical production rate for SVOC in headspace ($\mu g\ L^{-1} s^{-1}$)
D_A = Chemical destruction rate for SVOC in headspace ($\mu g\ L^{-1} s^{-1}$)

Under steady-state conditions, soil-to-air flux can be simplified to be:

$$J\,h^{-1} = (QV^{-1} + \Psi h^{-1})[C]^{SVOC} \quad (3\text{-}12)$$

where, h^{-1} = Inverse height of the atmospheric mixing zone (cm).

Soil characteristics, meteorologic conditions, and agricultural practices can be highly variable even within the same test area in the field. Chemical degradation rates are affected by soil moisture, partly because soil microbes have optimal ranges for growth. Biologic activation in the soil column depends on available oxygen and moisture. Thus biotic degradation depends on the contact of microbes to constituents essential to their growth and metabolism: water, nutrients (i.e., the organic compounds), and air.

Treatment of the soil can also have profound effects on the chemical breakdown of hazardous substances. For example, raising the soil alkalinity by liming and increasing soil moisture by irrigation may increase the atmospheric transport and degradation rates of otherwise persistent organic compounds in soil and groundwater.[23]

Orphan Pesticides: The Complicated Example of Lindane[24]

Heavy use of organochlorine insecticides, especially in the past, has led to the dispersal of these pollutants throughout the global environment. One compound of major concern is 1,2,3,4,5,6-hexachlorocyclohexane (HCH). HCH can persist in the environment and be transported long distances from the areas of application. While the bulk of HCH used today is in the form of the γ-HCH isomer of the compound, there is concern that this isomer can be transformed into other isomers that exhibit greater persistence and have potentially more deleterious effects on humans and wildlife.

Environmental contamination caused by lindane and the other HCH isomers is an international concern. For example, under the North American Free Trade Agreement (NAFTA), Canada, Mexico, and the United States have agreed to manage certain persistent, bioavailable, and toxic compounds such as HCH that warrant special attention because of the risks they pose to human health and the environment.

Although the main form of HCH presently used in North America is γ-HCH, high concentrations of other isomers, particularly α-HCH, in the Arctic suggest that γ-HCH is transformed into other isomers in the environment. To accurately assess and mitigate the environmental impacts of γ-HCH, it is necessary to know the degree to which it is transformed into the more harmful HCH isomers.

There are eight geometric isomers of HCH. The isomers differ in the axial and equatorial positions of the chlorine atoms. One of the isomers, α-HCH, exists in two enantiomeric forms. HCH is commercially produced by photochemical chlorination of benzene. The product, technical-grade HCH, consists principally of five isomers: α-HCH (60–70%), β-HCH (5–12%), γ-HCH (10–15%), δ-HCH (6–10%), and ε-HCH (3–4%).[25] This mixture is marketed as an inexpensive insecticide, but because γ-HCH is the only isomer that exhibits strong insecticidal properties, it has been common to refine it from the technical HCH and market it under the name lindane; however, all commercially produced lindane contains trace amounts of other HCH isomers.

All of the HCH isomers are acutely toxic to mammals, and chronic exposure has been linked to a range of health effects in humans. Of the different isomers, α-HCH exhibits the most carcinogenic activity and has been classified along with technical-grade HCH as a Group B2 probable human carcinogen by the U.S. EPA.[26] As the most metabolically stable isomer, β-HCH is the predominant isomer accumulating in human tissues. All isomers of HCH have high water solubilities compared to other aromatics. They also have moderately high vapor pressures when compared to other organochlorine pesticides. So, HCH is usually present in the environment as a gas or dissolved in water, with only a small percentage adsorbed onto particles.

The compounds have fairly long lifetimes in air and can be transported long distances.

The physical and chemical properties of HCH vary among the isomers. For example, the vapor pressure of α-HCH is somewhat less than that of γ-HCH. α-HCH has also been shown to be slightly more lipophilic than γ-HCH (log K_{OW} 3.8 versus 3.6). The Henry's law constant for α-HCH is about twice that of γ-HCH, so α-HCH is more likely to partition to the air. Another important difference between the isomers is the persistence of the β-isomer. β-HCH is resistant to environmental degradation. It is also more lipophilic than the other isomers. These properties may result from its significantly smaller molecular volume. Because β-HCH's bonds between H, C, and Cl at all six positions are equatorial (i.e., within the plane of the ring), the molecule is denser and small enough to be stored in the interstices of lipids in animal tissues. Thus even though the isomers possess identical chemical composition, the difference in their molecular arrangement leads to very different fate and transport properties.

Production and Use of HCH Worldwide

Presently, agricultural uses of lindane in the United States are limited to seed treatment. A medical formulation of lindane is also used on the skin to control head lice and scabies. Commercial production of technical-grade HCH began in 1943. Its extremely low cost led to its wide use, particularly in some developing countries; however, its strong odor can be taken on by crops, so the odorless lindane has been more widely used in developed countries. Total global production and use of the different HCH isomers is difficult to determine, and estimates vary considerably, but the usage of technical-grade HCH between 1948 and 1997 was recently estimated to be around 10 million tons.

Environmental and human health concerns led to the banning of technical-grade HCH in many countries during the 1970s. China, India, and the former Soviet Union remained the largest producers and users of HCH in the early 1980s. China, whose total production was estimated at 4.5 million tons, banned production in 1983, although residual stocks may have been used until 1985. In 1990, production of technical-grade HCH was also prohibited in the former Soviet Union, and use of residual stocks was restricted to public health and specific crop uses. Of the 90,000 tons applied in 1990, 51,000 tons were used in India.

In North America, the use of technical-grade HCH was banned in Canada and the United States in the 1970s but continued in Mexico until 1993 (see Table 3-2). The use of lindane, however, continues in all three countries.

In addition to HCH isomers produced and applied, considerable unused stockpiles of both technical-grade HCH and lindane exist. The 1998 Food and Agriculture Organization Inventory of Obsolete, Unwanted and/or Banned Pesticides found a total of 2,785 tons of technical-grade HCH, 304 tons of

TABLE 3-2
Estimated Annual Usage of α-HCH and γ-HCH in 1980 and 1990 for the United States, Canada, and Mexico

	α-HCH use (t/yr)		*γ-HCH use (t/yr)*	
	1980	1990	1980	1990
United States	0	0	268	114
Canada	0	0	200	284
Mexico	105	1218	23	261

(*Source:* B.G.E. De March, C.A. de Wit, and D.C.G Muir, AMAP Assessment Report: Arctic Pollution Issues [Oslo, Norway: Arctic Monitoring and Assessment Program, 1998], 183–373.)

lindane, and 45 tons of unspecified HCH material scattered in dump sites in Africa and the Near East. Some of the containers have deteriorated and are leaking, creating a serious threat to humans and wildlife. Stockpiles associated with earlier manufacturing of technical-grade HCH may also be causing problems in Eastern Europe.

Presence of HCH Isomers in the Environment

When HCH (either lindane or technical grade) is applied to the soil or when it is left in an abandoned site, it can either persist there sorbed to soil particles or be removed through several processes. The primary process for removing HCH from soil is volatilization into the air, although microbial and chemical degradation and uptake by crop plants can also occur. HCH can also enter the air adsorbed onto resuspended particulate matter, but this process does not appear to contribute as much as volatilization to the movement of HCH isomers. Once HCH isomers enter the environment, they can be found in air, surface water, soil, and living organisms.

The most common isomers found in the environment are α-HCH, β-HCH, and γ-HCH. α-HCH is typically predominant in ambient air as well as in ocean water. β-HCH is the predominant isomer in soils and animal tissues and fluids, including human because its all equatorial (eeeeee) configuration favors storage in biologic media and affords it greater resistance to hydrolysis and enzymatic degradation. The detection of HCH isomers in the Arctic and Antarctic, where lindane and technical-grade HCH have not been used, is evidence of long-range transport.

Evidence for Isomerization of Lindane

Several hypotheses have been suggested to explain why so much of the HCH residue found in the environment is in the form of the α isomer.

Photoisomerization in Air. γ-HCH may be transformed by sunlight into α-HCH, which is more photostable, during long-range transport; however, other factors such as the different rates of atmospheric volatilization and deposition of the isomers, as well as seasonal use of lindane or technical-grade HCH, could also explain the fluctuations in the isomerization.

Bioisomerization in Soil and Sediments. Laboratory evidence shows that γ-HCH can be transformed into other isomers in soil or sediments through biologic degradation. The orientation of chlorine atoms on the γ isomer makes irreversible transformation into α- or β-HCH the most likely form of isomerization.

Other Explanations for the Abundance of α-HCH in the Environment

Differences in the Henry's law constants of the α and γ isomers could affect their movement in the environment. The lower the value of the coefficient, the more likely a compound will dissolve into the water. At 20°C in freshwater, the constant is $0.524\,Pa\,m^{-3}/mol$ for α-HCH and 0.257 for γ-HCH. Therefore, during long-range transport in air over oceans, γ-HCH is more likely to be removed either by direct partitioning into water or through washout in rain, leaving proportionately more α-HCH in the air.

The case of lindane and the HCH isomers indicates the complexity of the fate and transport of hazardous compounds. It also underlines the need for excellent data regarding the physical, chemical, biologic, and environmental properties of substances during all phases of hazardous waste management.

Using Physical Movement and Chemical Changes to Estimate Possible Chemical Risks

The engineer is often asked to solve a groundwater contamination problem. The movement of a hazardous chemical toward groundwater resources is a major threat that must be eliminated. Let us begin by considering how water moves through porous media.

Following precipitation, water can flow across land surfaces and reach surface waters. In porous soil, the water can seep into the ground by infiltration. The infiltrated water sorbs to soil particles and may be taken up by plants via capillarity in the root hairs. Plants release water vapor to the atmosphere. Excess soil moisture may migrate by gravity downward into underlying soil and porous media, such as sand and gravel. At a certain depth, water fills the pore spaces of the unconsolidated material. This is the zone of saturation, and the top of this zone is known as the *water table*. The water beneath the water table is called *groundwater*.

The flow rate of groundwater is determined by the slope of the impervious rock layer underlying the groundwater, as well as by the permeability of the unconsolidated material in the zone of saturation. Water-filled pore space is an expression of the amount of water a material can store, whereas permeability is an expression of the ease with which water moves through the material. If the zone consists of less permeable material (e.g., fine-grained material like clay), then the flow rate is slower than with a larger-grained material like sand. Porosity and permeability are related, but distinct terms. Clays are highly porous materials, but the amount of surface area of each clay particle is large, so that water is held tightly. Thus the permeability of clays is much lower than that of silts, sands, and gravels. The groundwater flow rate can range from several meters per day to less than a meter per century.

An aquifer is a geologic stratum (or groups of strata) that holds groundwater.[27] Aquifers are usually limited to such strata that contain sufficient amounts of water to be pumped to the surface and used by humans. Aquifer thickness varies from a few meters to thousands of meters, and in spatial terms are from less than a square kilometer to thousands of kilometers. The amount of water an aquifer holds depends on the volume of the soil and rock in the saturated zone, the size and number of the pores and fractures that can be saturated, and the permeability of the unconsolidated material. Bedrock usually has low water-filled pore space, as well as low permeability. Aquifers are often threatened by mishandled hazardous wastes.

Let us employ a case study[28] to consider the physical and chemical processes that determine the fate of a chemical compound after it is released into the environment. Both porous flow and transport models can be used to estimate flow through a fractured rock aquifer; however, this is at best an approximation or abstraction of the actual physical system. The alternative—and more accurate—approach is the exact modeling of flow and chemical transport along the fractures. This approach requires the detailed mapping of all the fractures, which is usually not feasible.

The Duke Forest Gate 11 waste site is the site of a previously operational landfill on property owned by Duke University in the Duke Forest, Durham, North Carolina. The landfill had been approved for disposal of low-level radioactive wastes, including animal carcasses, laboratory wastes, scintillation vials, and fluids. Wastes were buried at the site from 1961 to 1970. Monitoring wells were installed in the early 1980s. Figure 3-19 shows the topography, the location of monitoring wells, and water table height measurements as of March 2000.

Slope trends in all directions away from the site, but more steeply to the north and the south. The burial site boundary is demarcated in Figure 3-18 by a square-shaped clearing at the crest of a knoll. A clay cap was installed in 1985 to reduce infiltration from rainwater, but pollutants appear to have continued to migrate into the site because the cap was inadequately engineered. Monitoring beyond the site allows for calculations of hydraulic head and water quality. Hydraulic conductivity, electrical resistivity, other hydrogeologic data, and the location of buried containers were determined using pumping tests, video logging, and extensive geophysical surveys.

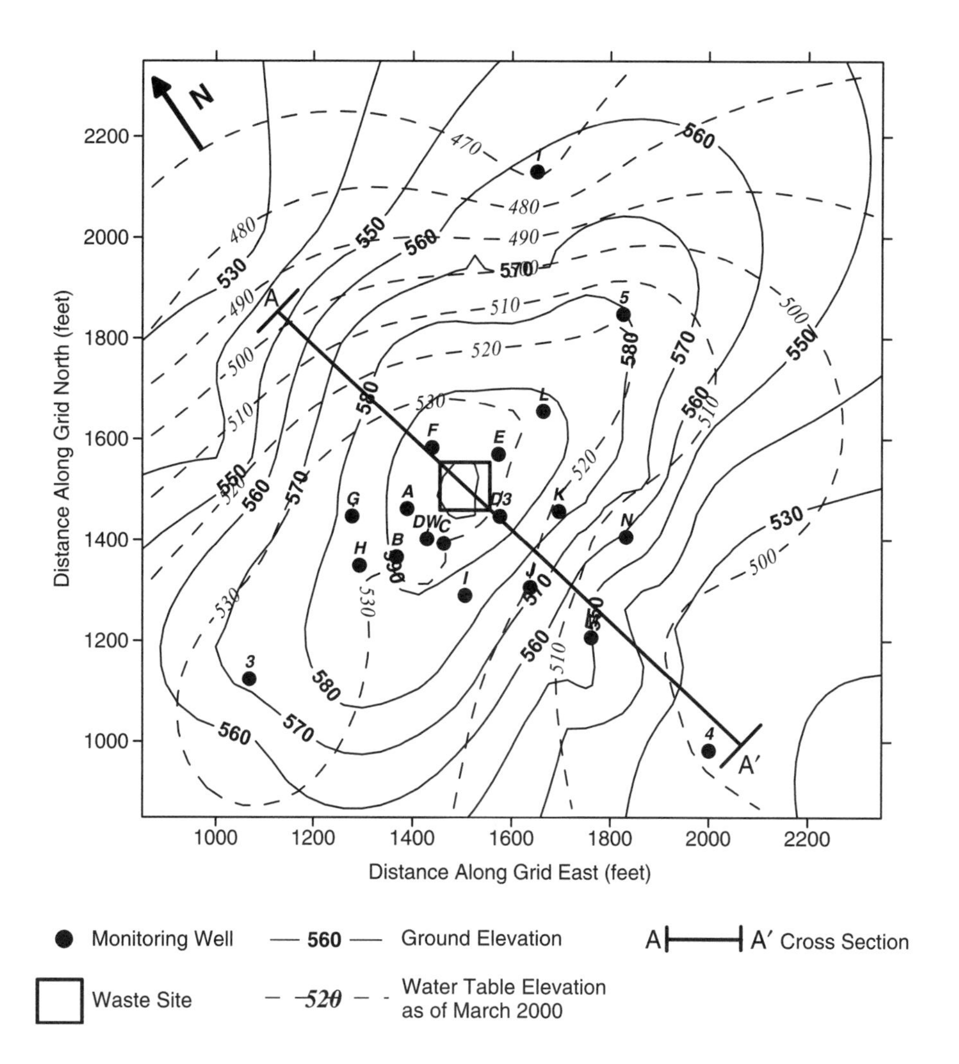

FIGURE 3-19. Topography, monitoring wells, and water table of Duke Forest Gate 11 waste site. (*Source:* From M.A. Medina, Jr., W. Thomann, J.P. Holland, and Y-C. Lin, "Integrating Parameter Estimation, Optimization and Subsurface Solute Transport," *Hydrological Science and Technology*, 17: 259–282, 2001. Used with permission from the author.)

What Is the Hydrogeology of the Site?

The fractured bedrock has low storage capacity, decreasing to nearly zero at 122 meters below the ground surface. Topographic depressions have the shallowest depth to the water table and the largest saturated thickness of regolith if they are underlain by heavily fractured bedrock, and drainage linears are a good indication of fracture control. Figure 3-20 shows

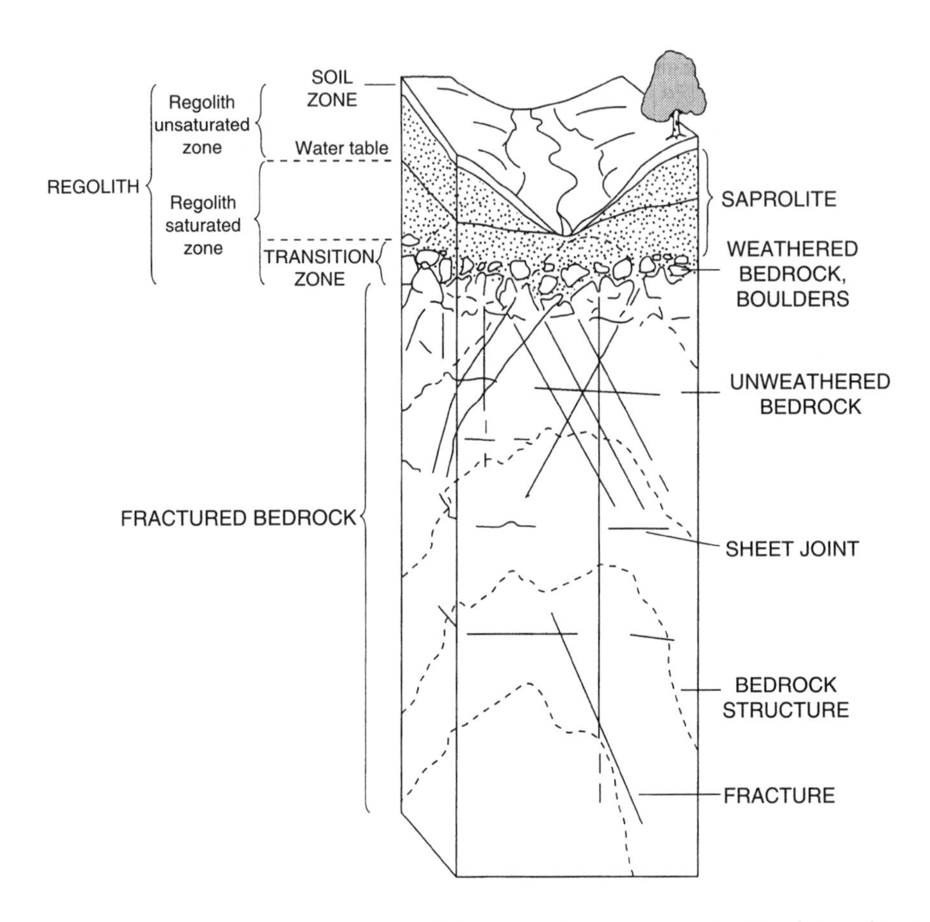

FIGURE 3-20. Principal components of the groundwater system in North Carolina's Piedmont region. (*Source:* D.A. Harned and C.C. Daniel III, "The Transition Zone between Bedrock and Regolith: Conduit for Contamination?" in C.C. Daniel III, R.K. White, and P.A. Stone, eds., *Ground Water in the Piedmont*, Proceedings of a Conference on Ground Water in the Piedmont of the Eastern United States, Charlotte, NC, Oct. 16–19, 1989 [Clemson, SC: Clemson University, 1992], 336–348.)

that the foliated crystalline rocks below the site are mantled by thick regolith. This layering provides a possible conduit for subsurface chemical contamination.

The soils underlying the site are a surface layer of silt loam with 10% to 15% fine gravel. The subsoil consists of silty clay and silty clay loam. The fraction of gravel increases (60% to 90%) with depth to bedrock. A saturated fractured section, and finally the lower, less fractured bedrock, are depicted in Figure 3-21 for cross-section A-A′ from Figure 3-17. The water table is within the fractured bedrock.

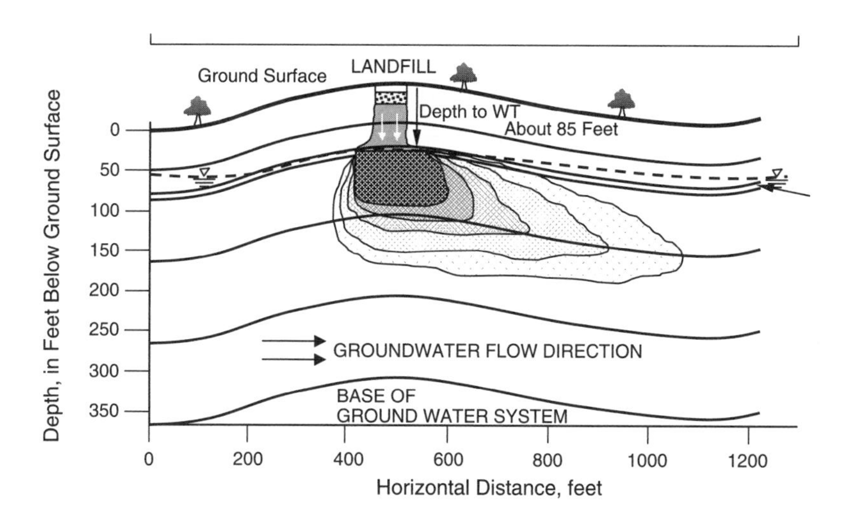

FIGURE 3-21. Profile along Section A-A′ of Figure 3-17, showing subsurface layers. (*Source:* From M.A. Medina, Jr., W. Thomann, J.P. Holland, and Y-C. Lin, "Integrating Parameter Estimation, Optimization and Subsurface Solute Transport," *Hydrological Science and Technology*, 17: 259–282, 2001. Used with permission from the author.)

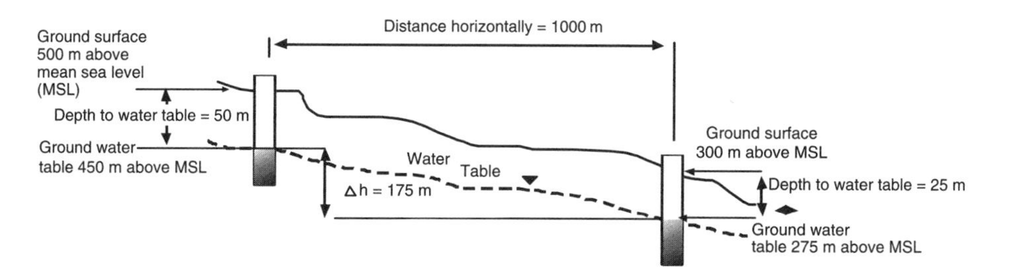

FIGURE 3-22. Calculation of hydraulic gradient (K), which is the change in hydraulic head (h) over a unit distance. In this case, the horizontal distance is 1,000 m. The Δh is the difference between the upper h (450 m) and the lower h (275 m), thus $\Delta h = 175$ m. So, $K = 175\text{ m}/1{,}000\text{ m}$, or 0.175 (dimensionless).

Groundwater generally moves by laminar flow.[29] Darcy's law states that groundwater moves through the porous media, such as sand, gravel, or silt, at a velocity that is proportional to the hydraulic gradient (Figure 3-22), and is stated as:

$$V = Ki \tag{3-13}$$

where,

V = velocity of water flow (m sec^{-1})
K = Hydraulic conductivity (dimensionless coefficient of permeability)
i = Hydraulic gradient

Hydraulic conductivity is expressed as:

$$K = \frac{Nd^2 \rho_f g}{\eta} \tag{3-14}$$

where,

ρ_f = fluid density (g L^{-1})
g = gravitational acceleration (m sec^{-2})
η = the fluid's (dynamic) viscosity (g m^{-1} sec^{-1})
d = mean grain diameter (m)
N = factor depending on the shape of pore spaces (dimensionless)

The hydraulic conductivity (K) of unconsolidated materials ranges from the rapid K values of gravels and course sands (as high as 3×10^{-4} m s^{-1}) to the very low values for unweathered clays (as low as 8×10^{-13} m s^{-1}).[30] The K values measured in the aquifer at each Duke waste site's monitoring wells ranged from <1 m yr^{-1} to >700 m yr^{-1}. Recharge from the surface drives the hydraulic gradient. Flow radiates from the waste site.

How Is Groundwater Contamination Characterized at the Site?

Groundwater contaminants have been detected in monitoring wells at the Duke Forest site. The major contaminant of concern is the compound paradioxane (CAS No. 123-91-1). The source of the paradioxane contamination has been attributed to the landfill. Before remediation, paradioxane concentrations were routinely greater than 1,000 $\mu g\,l^{-1}$, and at times reached levels as high as 2,800 $\mu g\,l^{-1}$. Paradioxane is a synthetic organic compound used as a solvent. It is classified as a probable human carcinogen[31] and a priority drinking water pollutant in North Carolina, with a maximum concentration level (MCL) = 7 $\mu g\,l^{-1}$. Paradioxane is completely miscible in water and has a very low K_{ow}. Paradioxane's miscibility and estimated Henry's law constant (4.88×10^{-6} atm m^3 $mole^{-1}$) would be expected to inhibit volatilization. Its K_{oc} of 1.2 suggests that paradioxane does not easily adsorb onto soil surfaces.

The wastes from the original disposal site were excavated in 1997 and 1998. This resulted in substantial reductions in contaminant concentrations of paradioxane in all of the monitoring wells adjacent to the site. Paradioxane concentrations are presented in Figure 3-23. The paradioxane concentrations of Wells I and N are plotted against the smaller-scale secondary axis (right

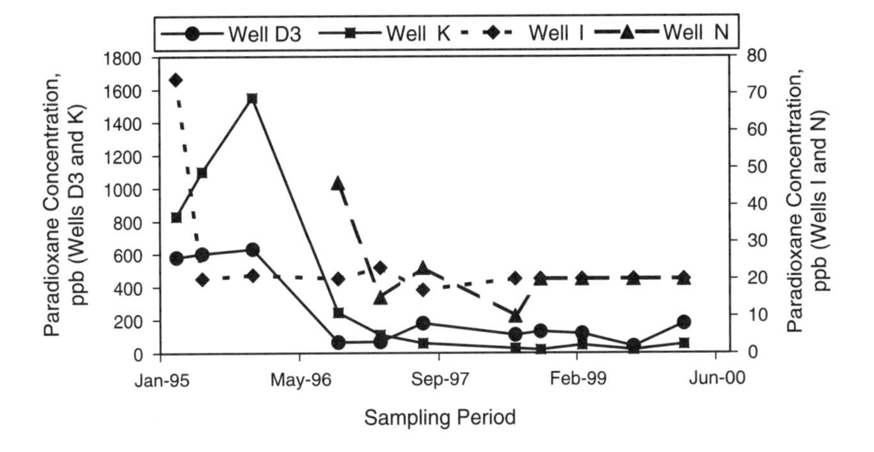

FIGURE 3-23. Paradioxane concentrations measured at four monitoring wells. Left scale (larger concentrations) applies to Wells D3 and K, and right scale applies to Wells I and N. The removal of contaminated soils occurred between October 1997 and February 1998, followed by a concomitant decline in paradioxane levels in groundwater. (*Source:* From M.A. Medina, Jr., W. Thomann, J.P. Holland, and Y-C. Lin, "Integrating Parameter Estimation, Optimization and Subsurface Solute Transport," *Hydrological Science and Technology*, 17: 259–282, 2001. Used with permission from the author.)

ordinate) so as to view their historical profiles on the same graph with Wells D3 and K.

How Can Contaminant Transport Models Be Applied to Remediation?

Models can be used to simulate three-dimensional solute transport. These models can be employed by the engineer to evaluate various intervention options. One type of model mathematically derives the changes in concentration of a dissolved chemical over time, based on advection, hydrodynamic dispersion (both mechanical dispersion and diffusion), mixing from fluid sources, and chemical reactions. Chemical transformation is estimated from linear sorption, which is calculated from a retardation factor, and decay rates. The solute transport model is then combined with a groundwater flow model to solve the transient flow equation.

If the principal axes of the second-order hydraulic conductivity tensor K_{ij} are aligned with the x-y-z coordinate axes ($K_{ij} = 0$ when $i \neq j$), the groundwater flow equation may be written to include explicitly all hydraulic conductivity terms as:

$$\frac{\partial}{\partial x}\left(K_{xx}\frac{\partial h}{\partial x}\right) + \frac{\partial}{\partial y}\left(K_{yy}\frac{\partial h}{\partial y}\right) + \frac{\partial}{\partial z}\left(K_{zz}\frac{\partial h}{\partial z}\right) - W = S_s\frac{\partial h}{\partial t} \qquad (3\text{-}15)$$

where,

> h = hydraulic head
> S_s = specific storage of the aquifer
> W = volumetric flux per unit volume
> K_{xx}, K_{yy}, and K_{zz} = hydraulic conductivities along the x, y, and z axes, respectively.

Darcy's law provides the mean groundwater velocity:

$$V_i = -\frac{K_{ij}}{\varepsilon}\frac{\partial h}{\partial x_i} \tag{3-16}$$

where ε is the effective porosity of the porous medium. This velocity can represent the advective flux in the solute transport equation:

$$\frac{\partial(\varepsilon C)}{\partial t} + \frac{\partial(\rho_b \bar{C})}{\partial t} + \frac{\partial(\varepsilon C V_i)}{\partial x_i} - \frac{\partial}{\partial x_i}\left(\varepsilon D_{ij}\frac{\partial C}{\partial x_j}\right) - \sum C' W + \lambda\left(\varepsilon C + \rho_b \bar{C}\right) = 0 \tag{3-17}$$

where, C is the concentration of the chemical, ρ_b is the bulk density of the aquifer media, $\bar{C}$ is the mass concentration of solute sorbed on or contained within the solid aquifer material, D_{ij} is a second-rank tensor of dispersion coefficients, C′ is the volumetric concentration in the sink/source fluid, and ∂ is the decay rate.

Three remediation scenarios were modeled:

1. Natural attenuation, without human intervention.
2. Intermittent pumping using extraction wells and based on optimization, but without clean water recharge wells.
3. Modified optimal pumping strategy, complemented with clean water recharge wells in addition to the extraction wells.

Three-dimensional models uncovered a severe limitation in applying the pumping strategy without recharge wells (i.e., the wells would become dry quickly, requiring an artificially higher precipitation rate than was required for the historical rate to maintain convergence in the numerical simulations).

What Would Happen without Intervention?

Initial conditions were prescribed by the model to be the March 2000 data (paradioxane detection limit = 20 ppb). Existing wells A, C, D2, K, and E (see Figure 3-22) withdrew water with the highest concentrations of paradioxane, so these wells were selected to represent remediation pumping wells. Wells G, H, J, L, and I were outside the perimeter of the contaminant plume, so they were used to monitor changes in groundwater concentrations of paradioxane.

The infiltration rate was estimated to be 4.6×10^{-9} to $8.2 \times 10^{-9}\,\mathrm{m\,s^{-1}}$), based on annual precipitation of 130–260 mm. The mean retardation factor for paradioxane was estimated to be 1.2.

Simulations predicted the paradioxane plume migration 50 years into the future. The plume confined within the 50 parts per billion (ppb) concentration isoline would be expected to remain essentially unchanged for the first 15 years; however, the highest concentrations in the center of the plume would decrease from 210 ppb at 2 years after the start of simulation to more than 90 ppb at 15 years after. Twenty years after the initial conditions, the boundary defined by the 50 ppb concentration isoline would start to retreat. After 50 years, the peak concentration in the center of the plume falls to about 45 ppb (Figure 3-24).

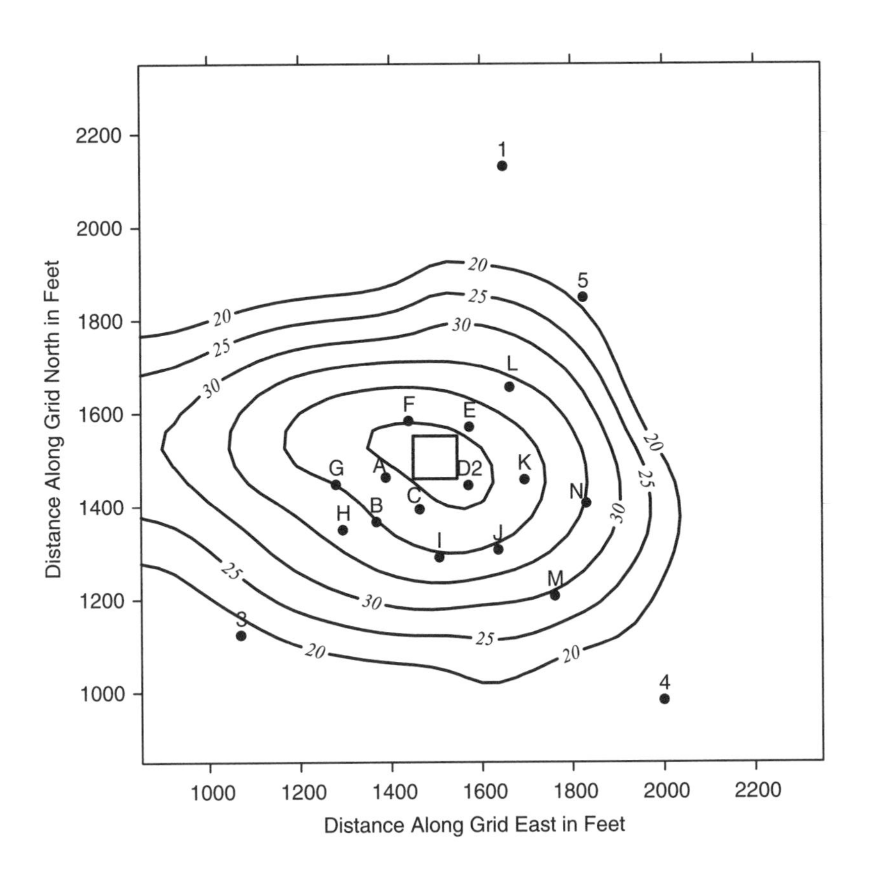

FIGURE 3-24. Modeled paradioxane plume after 50 years of natural attenuation. (*Source:* From M.A. Medina, Jr., W. Thomann, J.P. Holland, and Y-C. Lin, "Integrating Parameter Estimation, Optimization and Subsurface Solute Transport," *Hydrological Science and Technology*, 17: 259–282, 2001. Used with permission from the author.)

How Does This Compare to Pumping with Recharge?

The predicted plume expected from pumping is shown in Figure 3-25. Each pumping cycle stage lasts 36 hours. The pumps are then shut off for 120 hours, before resuming the next 36-hour pumping period. This alternation between pumping and recess continues for the predicted duration of the remediation. Eight monitoring wells on the plume's perimeter were selected to be recharging wells (wells G, F, H, L, J, M, N, and I). Five years after the initial conditions, the peak paradioxane concentration would be reduced to 75 ppb, a 55% removal rate. After 10 years, the peak concentration would be reduced to 65 ppb, about twice the expected effectiveness of natural remediation.

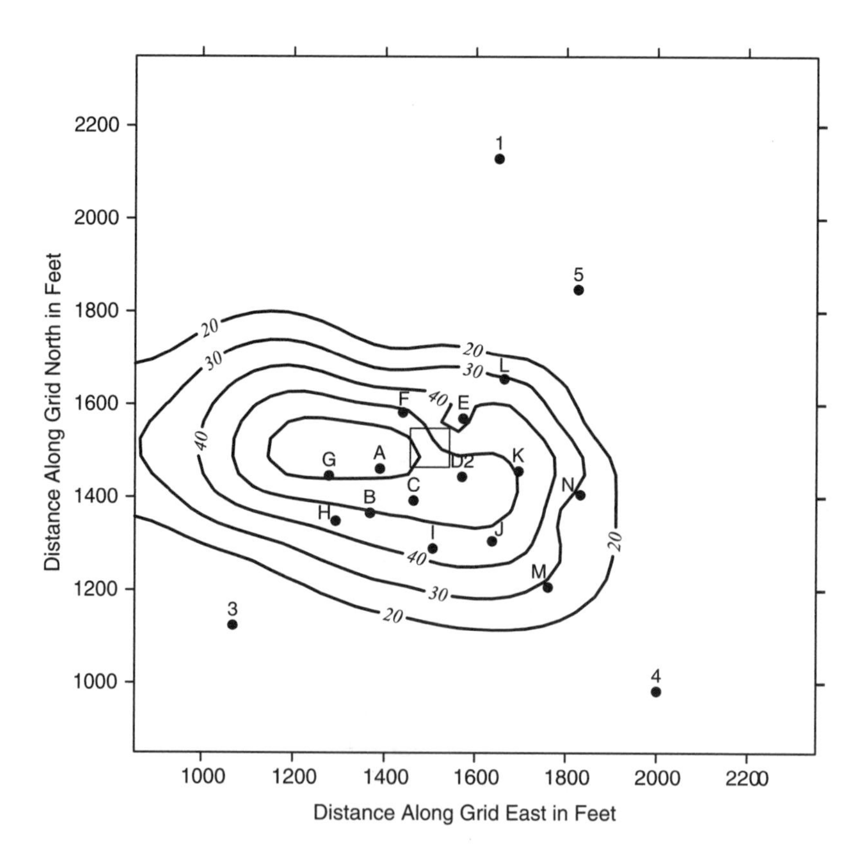

FIGURE 3-25. Paradioxane plume modeled after 10 years of pump recharge remediation. (*Source:* From M.A. Medina, Jr., W. Thomann, J.P. Holland, and Y-C. Lin, "Integrating Parameter Estimation, Optimization and Subsurface Solute Transport," *Hydrological Science and Technology*, 17: 259–282, 2001. Used with permission from the author.)

In addition to replenishing the aquifer, the artificial recharge helps flush the groundwater with clean water, helping to strip paradioxane that is sorbed to soil and unconsolidated material. Risk managers must now decide whether the removal rate without human intervention is sufficiently different from that expected from a pump-and-treat approach. Natural attenuation is cheaper and easier, but are the higher potential exposures to paradioxane worth pursuing engineering controls? Generally, with a polar, hydrophilic chemical that is not strongly sorbed to soil particles, pumping the maximum amount of contaminated groundwater from the zone of contamination in the aquifer can be expected to provide effective remediation. In this instance, paradioxane is both miscible and weakly sorbed; however, the paradioxane plume seems to be contained within the site, so further intervention may not be necessary.

Case Study: Mixed Inorganic and Organic Hazardous Wastes: The Double Eagle Refinery, Oklahoma City, Oklahoma[32]

A 12-acre facility that refined used motor oil and other petroleum products from 1929 until 1980 is now listed as a hazardous waste site. The sediments, soil, and surface water on the site have been contaminated with both organics and metals, including cadmium, chromium, and lead. The groundwater is threatened with contamination. The site was placed on the National Priority Listing in 1989.

A remedial investigation and feasibility study (RI/FS) has been conducted for the on-site wastes. The primary contaminants of concern are lead and an oily sludge. The initial remedy selected in the record of decision (ROD) for the waste sources was on-site treatment of waste to a nonhazardous state and then off-site disposal of the material at a commercial facility.

The remedial design (RD) for the waste sources began in 1993, but after the remedy was selected, the EPA found evidence that wastes listed by the Resource Conservation and Recovery Act were accepted at the refinery. One hazardous waste manifest for trichloroethylene (TCE) was found, indicating that Double Eagle accepted and may have disposed of this waste on-site. Although the U.S. EPA was considering amending the ROD to include the stabilization of oil sludge wastes and subsequent disposal in an on-site landfill, it withdrew this demand because concentrations of TCE in the on-site wastes are below the health-based cleanup levels normally used. The EPA decided to evaluate the feasibility of delisting the waste and began formal procedures to do so. An amendment to the ROD was found to be needed, and the waste may have to be transported off-site after treatment as originally planned. The EPA determined that the waste will not be delisted. The State of Oklahoma and the federal government are evaluating the next actions needed.

Because the compounds and hydrology at this site are so complex, a separate groundwater RI/FS was completed in 1993. The remedial action proposed by the EPA includes installing additional groundwater monitoring wells and setting up and operating a groundwater monitoring program. The plan also includes a five-year review to determine if additional remedial actions are required to identify and eliminate health risks related to groundwater use.

The first phase of the remedial action began with the installation of 19 piezometer wells to provide data to map groundwater flow. Five holes were bored to log geophysical data and to define potential regions of slow groundwater movement (aquitards) within 200 feet beneath the site's surface. Also, a water production well was logged before plugging for background subsurface information.

Phase 2 will include a specified number, location, and depth of groundwater monitoring wells. Existing wells not used as part of the five-year monitoring program will be closed and plugged. Quarterly sampling of the monitoring wells will begin as part of the operation and maintenance of the remedy. All samples will be analyzed for both organic and inorganic contaminants.

Because metals such as lead and cadmium are of health concerns to young children, the State of Oklahoma, jointly with the Childhood Lead Poisoning Prevention Program, has been presenting programs at local elementary schools and Head Start centers on lead poisoning prevention techniques. Although the site is fenced to restrict public access, a creek flows by the site and appears to be contaminated, possibly from the runoff of the lagoons, so children are being advised to stay away from the water adjacent to the site. A recent site visit showed some evidence of transient people living east of the lagoons, with indications that the water was used at least for bathing. Efforts to verify this situation and place warning signs are being made by the East Oklahoma City community task force.

Do you believe that this is a sufficient response? How does an understanding of fate and chemistry of the contaminant help in efforts to prevent and reduce exposures to the Double Eagle refinery site? (Hint: Should someone be asking about the metal speciation and the various species' solubilities?)

Discussion: Use Rules of Thumb with Caution

The discussions in this chapter are aimed at giving the engineer a background in the major physical and chemical principles that drive the movement and change of chemicals after they have been released into the environment; however, these rules of thumb are fraught with exceptions.

Let us consider, for example, the nonaqueous phase liquids (NAPLs),[33] which are subdivided into the dense nonaqueous phase liquids (DNAPLs) and the light nonaqueous phase liquids (LNAPLs). The general rule of thumb is that the DNAPLs are heavier than water (i.e., their specific gravity is greater than 1), so they would be expected to sink in groundwater, whereas the LNAPLs, which are lighter than water, would be expected to float at the top of the saturated zone. This would be true if not for contravening factors such as those discussed in this chapter, especially the surface characteristics of the particles, which keep chemicals from sinking or rising in the ground water column (Figure 3-26). In fact, however, these physical characteristics of the NAPLs will combine with the chemical characteristics, such as solubility in water, and cause the compounds to become partitioned

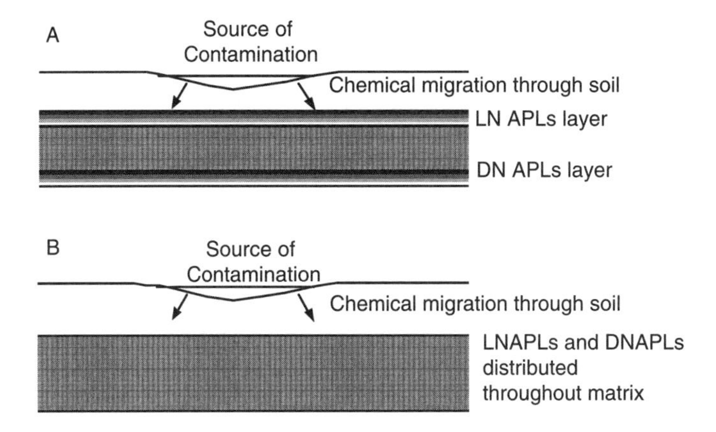

FIGURE 3-26. Effect of sorption and other physical processes on location of dense nonaqueous phase liquids (DNAPLs) and light nonaqueous phase liquids (LNAPLs) in an aquifer. Where particles have little effect (A), the DNAPLs sink and the LNAPLs float, but when particle effects are prominent (B), both classes of compounds may be distributed more evenly throughout the aquifer.

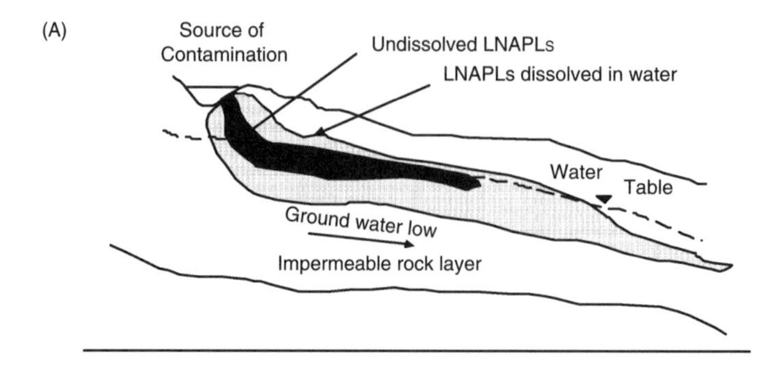

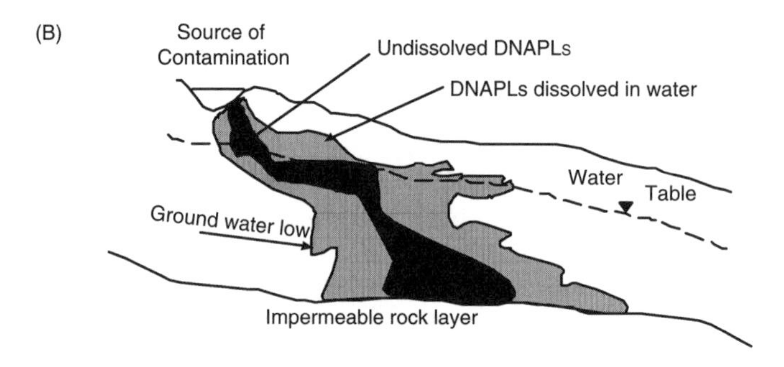

FIGURE 3-27. Displacement of plume for substances lighter than water (A) and denser than water (B), accounting for solubility.

between dissolved and pure forms in the vadose and saturated zones underground. Further, the movement of the groundwater will displace the plume so that there is a differential transport of the NAPLs (Figure 3-27).

Therefore, monitoring and remediation plans must account for the likelihood that chemicals will behave differently in the environment than they do under ideal and highly controlled conditions in the laboratory. The physical and chemical properties of hazardous substances that are published in engineering handbooks and manuals are often observed under such controlled conditions and are not usually found in the ambient environment.

How Can We Put the Physical and Chemical Properties to Work as We Deal with Hazardous Wastes?

Although much of this chapter has been concerned with the problems caused by the properties of hazardous chemicals, the engineer can also

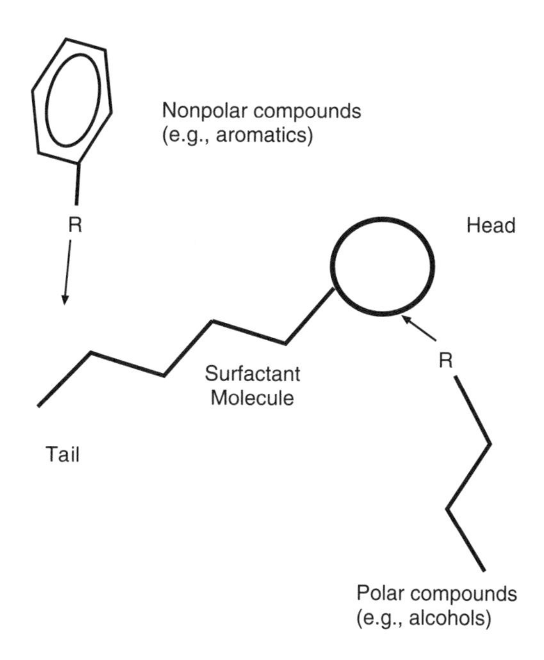

FIGURE 3-28. Structure of a surfactant prototype. The head is relatively polar and the tail is relatively nonpolar, so in the aqueous phase the more hydrophilic compounds will be dissolved by the surfactant molecule at the head position, whereas the more lipophilic compounds will be dissolved at the tail position. The result is that a greater amount of the organic compounds will be in solution when surfactants are present.

use these properties to the benefit of site remediation and cleanup. For example, many detergents are commercially available for use in environmental engineering. Detergents are in fact surfactants (Figure 3-28), which have the unique property of being both polar and nonpolar, so they are both water soluble (the head in Figure 3-28) and fat soluble (the tail in Figure 3-28).[34] This means that in a water column, the addition of a surfactant will increase the dissolved quantity of the dense and light NAPLs in the aquifer. Applying this increased hydrophilicity to Figure 3-26, the plume would be more widespread (Figure 3-29; however, the application of engineering controls, such as a slurry wall and a pump, along with the addition of surfactants, could allow for more efficient removal of the NAPLs from the groundwater and protection of water supplies [Figure 3-30]).

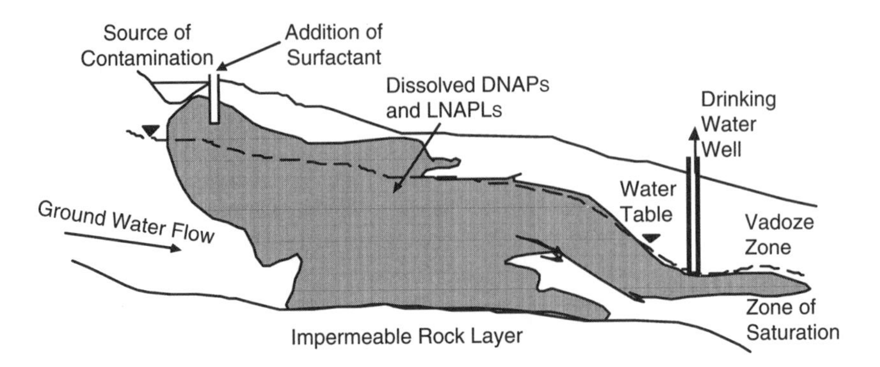

FIGURE 3-29. Effect on plume of nonaqueous phase liquids (NAPLs) following the addition of surfactants. In this instance the addition allowed a greater quantity of NAPLs to be dissolved compared to the simple displacement shown in Figure 3-25. This allowed more of the NAPLs to migrate downgradient and to contaminate a drinking water well.

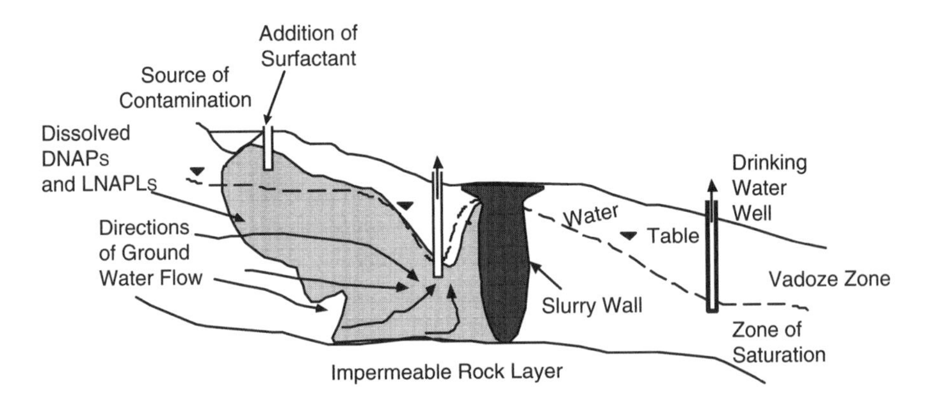

FIGURE 3-30. Effect on plume of nonaqueous phase liquids (NAPLs) with the installation of a barrier and pump. The slurry wall is used both to protect the water supply and to direct the flow of the plume to allow the NAPLs to be treated.

What Roles Can Models Play in Extending Our Understanding beyond Basic Rules of Thumb?

The various physical and chemical conditions and factors interact in complex ways. Thus scientists are continuously devising new models to predict the movement and change of compounds in the environment. Models provide a means of representing a real system in an understandable way.[35] They take many forms, beginning with conceptual models that explain how a system works, such as a delineation of all the factors and parameters of how a particle moves in the

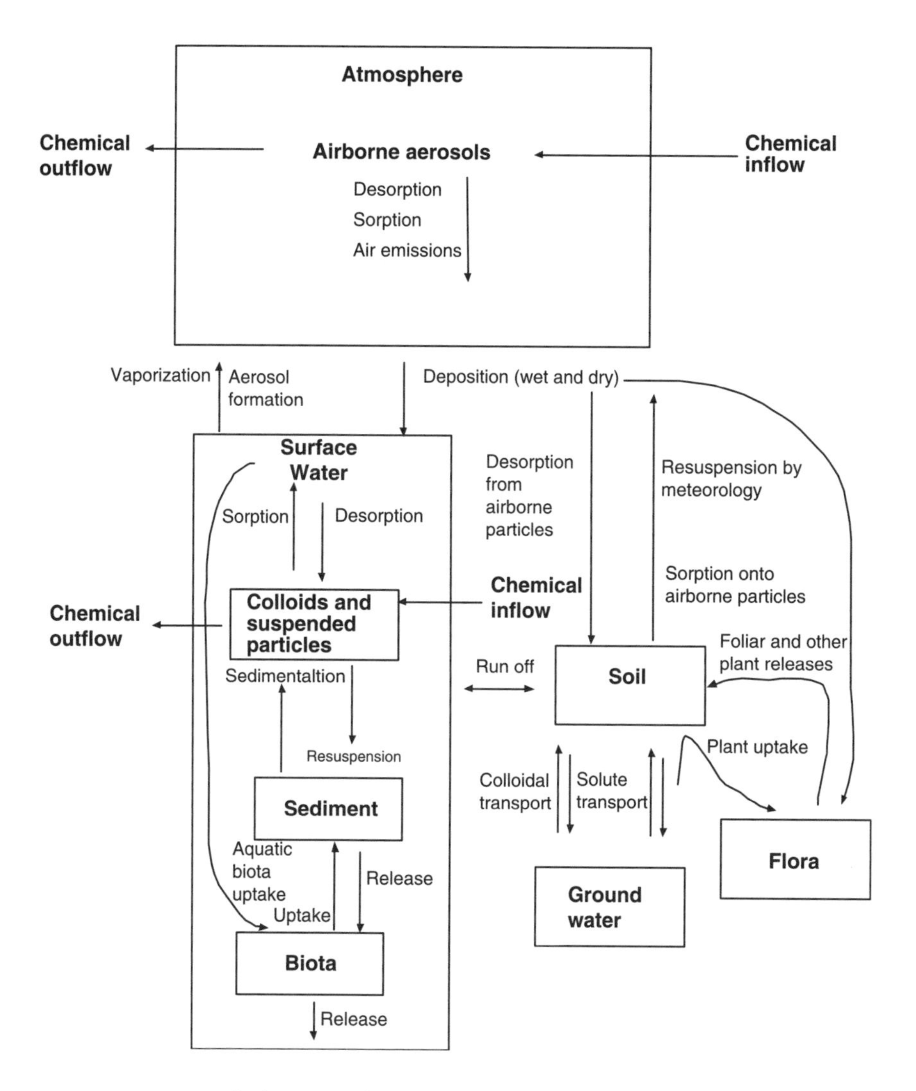

FIGURE 3-31. Example framework and flow of a multimedia, compartmental chemical transport and transformation model. Algorithms and quantities must be provided for each box. The partitioning coefficients, reaction rates, and other kinetics must be developed for each arrow.

atmosphere after its release from a power plant. These models help the engineer examine all the major influences on where a chemical is likely to be found in the environment and need to be developed to help target sources of data needed to assess an environmental problem.

Research scientists often develop physical or dynamic models to see how a chemical would be expected to move under controlled conditions, but at a much smaller scale. For example, the U.S. EPA's wind tunnel facility in Research Triangle Park, North Carolina, is sometimes used when mathematic models have too many data gaps or when terrain and other complex conditions render the models practically useless.

Recently, for example, the wind tunnel team built a scaled model of the town of East Liverpool, Ohio, and its surrounding terrain to estimate the movement of the plume from an incinerator. The plume could be observed under varying conditions, including wind direction and height of release. Only a few such facilities exist, however, so most hazardous waste sites have to use more virtual tools, such as computer simulations and geographic information systems (GIS). Like all models, the dynamic model's accuracy is dictated by the degree to which the actual conditions can be simulated and the quality of the information that is used.

Like exposure models, transport and fate models can be statistical and/or deterministic. Statistical models include the pollutant dispersion models, such as the Lagrangian models, that assume a Gaussian distribution of pollutants after their release (i.e., the pollutant concentrations are normally distributed in both the vertical and horizontal directions from the source). The Lagrangian approach is common for atmospheric releases. Stochastic models are statistical models that assume the events affecting the behavior of a chemical in the environment are random, so such models are based on probabilities.

Deterministic models are used when the physical, chemical, and other processes are sufficiently understood so as to be incorporated into a system that describes the movement and fate of chemicals. These models are difficult to develop because each process must be represented by a set of algorithms in the model. The relationship between and among the systems, such as the kinetics and mass balances, must also be represented. Thus the modeler must establish parameters for every important event following a chemical's release to the environment. Hybrid models using both statistical and deterministic approaches are often used (e.g., when one part of a system tends to be more random, whereas another has a strong basis in physical principles).

Numerous models are available to address the movement of chemicals through a single environmental media, but environmental scientists and engineers have increasingly begun to develop multimedia models, such as compartmental models that help predict the behavior and changes to chemicals as they move within and among the soil, water, air, sediment, and biota (see Figure 3-31).[36] Such models will likely see increased use in all environmental engineering, including hazardous waste management, in the future.

CHAPTER 4

Opportunities for Hazardous Waste Intervention by Engineers

Intervention to Prevent and Control the Risks Associated with Hazardous Wastes

Environmental engineers working in the public and private sectors have opportunities to control the risks associated with hazardous waste by relying on basic scientific principles, proven and developing engineering designs and processes, and acceptable technologies to address one or more of the six steps necessary and as a group sufficient to define that a hazardous waste polluting event has taken place. The discussions of the six steps that define a polluting event are followed by an introduction to the science, engineering, and technologies available to the engineer.

Six steps in sequence are necessary and sufficient to define an event that results in environmental pollution of the water, air, or soil from hazardous waste. These six steps—individually, and as a group—offer opportunities for engineers to intervene and control the risks associated with hazardous waste pollution and thus protect public health and the environment. The six steps that are necessary and sufficient for such pollution to occur from a hazardous generator, abandoned site, or accidental spill include the following:

SOURCE → RELEASE → TRANSPORT → RECEPTOR
→ DOSE → RESPONSE

For the engineer to have a problem with a particular hazardous waste, a source of the material must be identifiable. A hazardous component of the waste must be released from the source; be transported through the water, air, or soil environment; reach a human, animal, or plant receptor in a measurable dose; and have a quantifiable detrimental response in the form of death or illness. All six steps must exist for a hazardous waste problem to exist. No hazardous waste problem? No need for the engineer to intervene!

If a problem does exist, the engineer can intervene at any one of these six steps to control the risks to public health and the environment. Any intervention scheme and subsequent control by the engineer must be justified by the engineer as well as the public or private client in terms of scientific evidence, sound engineering design, technologic practicality, economic realities, ethical considerations, and the laws of local, state, and national governments.

Intervention at the Source of Hazardous Waste

A source of hazardous waste must be identifiable, either in the form of an industrial facility that generates waste byproducts, a hazardous waste processing facility, a surface or subsurface land storage/disposal facility, or an accidental spill into a water, air, or soil receiving location. The engineer can intervene to minimize or eliminate the risks to public health and the environment by utilizing technologies at this source that are economically acceptable and based on applicable scientific principles and sound engineering designs.

In the case of an industrial facility producing hazardous waste as a necessary by-product of a profitable item, as considered here for example, the engineer can take advantage of the growing body of knowledge that has become known as *life cycle analysis*.[1] In the case of a hazardous waste storage facility or a spill, the engineer must take the source as a given and search for possibilities for intervention at a later step in the sequence of six steps discussed as follows.

Under the life cycle analysis method of intervention, the engineer considers the environmental impacts that could incur during the entire life cycle of (1) all of the resources that go into the product, (2) all of the materials that are in the product during its use, and (3) all of the materials that are available to exit from the product once it or its storage containers are no longer economically useful to society. Few simple examples exist that describe how life cycle analysis is conducted, but consider for now any one of several household cleaning products. Consider that a particular cleaning product—a solvent of some sort—must be fabricated from one of several basic natural resources. Assume for now that this cleaning product currently is petroleum based. The engineer could intervene at this initial step in the life cycle of this product, as the natural resource is being selected, and consequently the engineer could preclude the formation of a source of hazardous waste by suggesting instead the production of a water-based solvent.

The engineer similarly could intervene in the production phase of this product's life cycle and suggest fabrication techniques that preclude the formation of a source of some hazardous waste. In this case, the recycling of spent petroleum materials could provide for more household cleaning product with less or zero hazardous waste generation, thus controlling the risks to public health and the environment.

In life cycle analysis, the engineer also must consider and attempt to control hazardous wastes associated with using the product under consideration. For example, this particular household cleaning product may result in unintended human exposure to buckets of solvent mixtures that fumigate the air in a home's kitchen or pollute the town's sewers as the bucket's liquid is flushed down a drain.

Under the plan of life cycle analysis, the engineer can also intervene and act to control the long-term risks associated with disposal of this solvent's containers. The challenge to engineers employing life cycle analysis is that every potential and actual environmental impact of a product's fabrication, use, and ultimate disposal must be considered.

Intervention at the Point of Release of the Hazardous Waste

Once a source of hazardous waste is identified, the engineer can intervene at the point at which the waste is released into the water, air, or soil environment. This point of release could be at the end of a pipe running from the source of pollution to a receiving water body like a stream, from the top of a stack running from the source of pollution to a receiving air shed, or from the bottom-most layer of a clay liner in a hazardous waste landfill connected to surrounding soil material. Similarly, this point of release could be a series of points as hazardous waste is released along a shoreline from a plot of land into a river or through a plane of soil underlying a storage facility. Physical, chemical, and microbiologic processes that are available to prevent the release of hazardous waste into the environment we discussed in Chapter 3. We will offer suggestions on how an engineer might organize an intervention strategy at the points at which hazardous wastes are released into the environment.

Intervention As the Hazardous Waste Is Transported in the Environment

Although the engineer cannot fight nature—because water will always run downhill and hot air will always rise—the engineer does have the opportunity to intervene and control the transport of hazardous waste through the environment with judicious site selection of facilities that generate, process, and store hazardous waste. For example, in the water environment, the distance from a source to a receiving body of water is of critical concern to the engineer who is interested in controlling the quantity and characteristics of waste as it is transported.

In the air environment, local weather patterns provide the engineer with opportunities to control the atmospheric transport of hazardous waste.

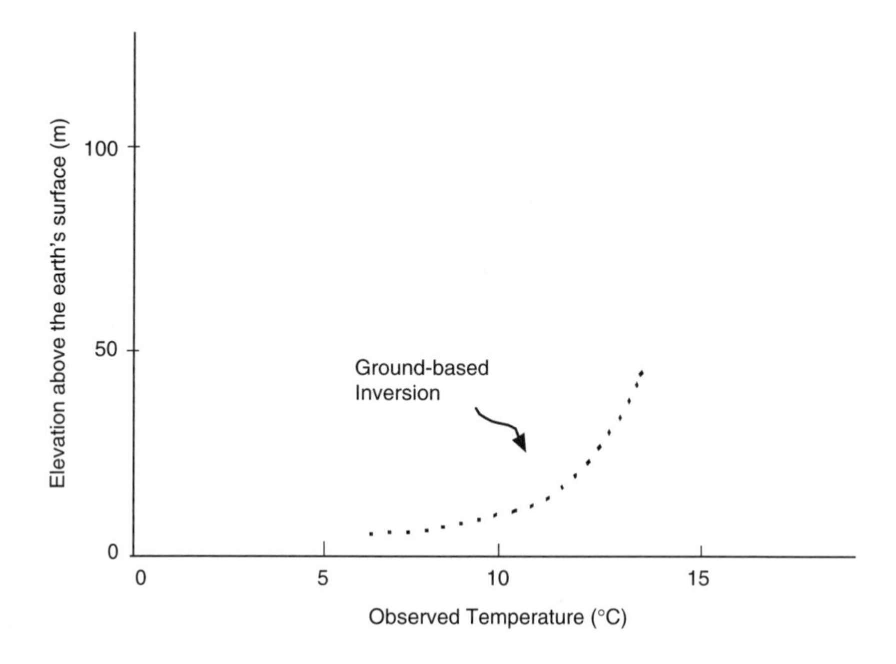

FIGURE 4-1. Ground-based inversion.

The engineer can intervene as hazardous waste generating, processing, and storage facilities are sited to avoid areas where specific local weather patterns are frequent and persistent. These avoidance areas include ground-based inversions (Figure 4-1), elevated inversions (Figure 4-2), valley winds (Figure 4-3), shore breezes (Figure 4-4), and city heat islands (Figure 4-5). In all five illustrations, the pollutants become locked into air masses with little or no chance of moving out of the respective areas. Thus the concentrations of the pollutants can quickly and significantly pose risks to public health and the environment. In the soil environment, the engineer has the opportunity to site hazardous waste generation and management facilities in areas of great depth to groundwater. The issues surrounding facility siting are discussed in more detail later in this chapter.

Intervention at the Receptor of Hazardous Waste

The receptor of hazardous waste could be a human, another animal in the general scheme of living organisms, flora, or materials or constructed facilities. In the case of humans, hazardous waste can be ingested, inhaled, or dermally contacted. Such exposure can be direct with human contact to, for example, particles of lead that are present in inhaled indoor air. Such exposure also can be indirect as in the case of human ingestion of the cadmium

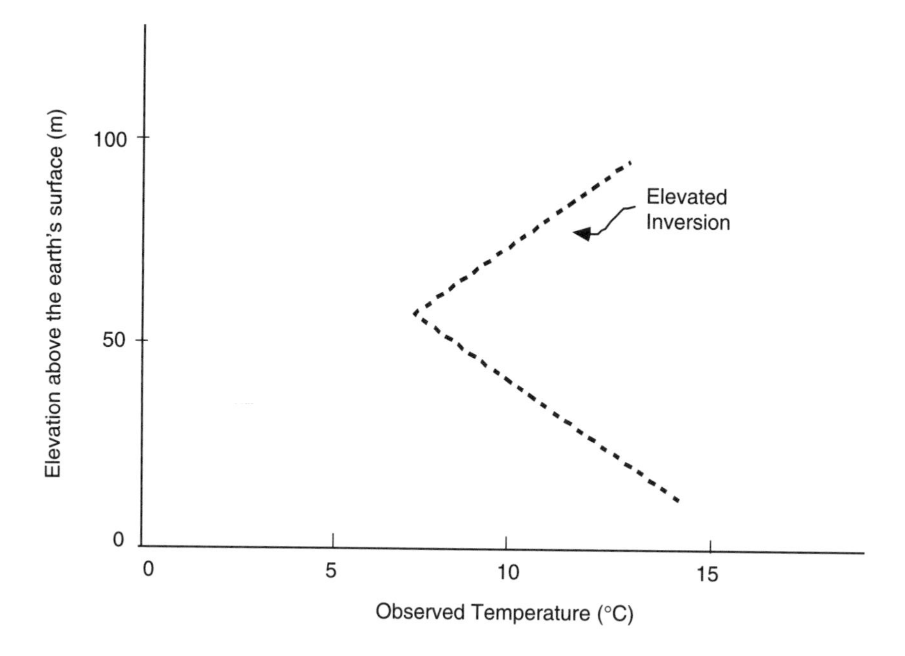

FIGURE 4-2. Elevated inversion.

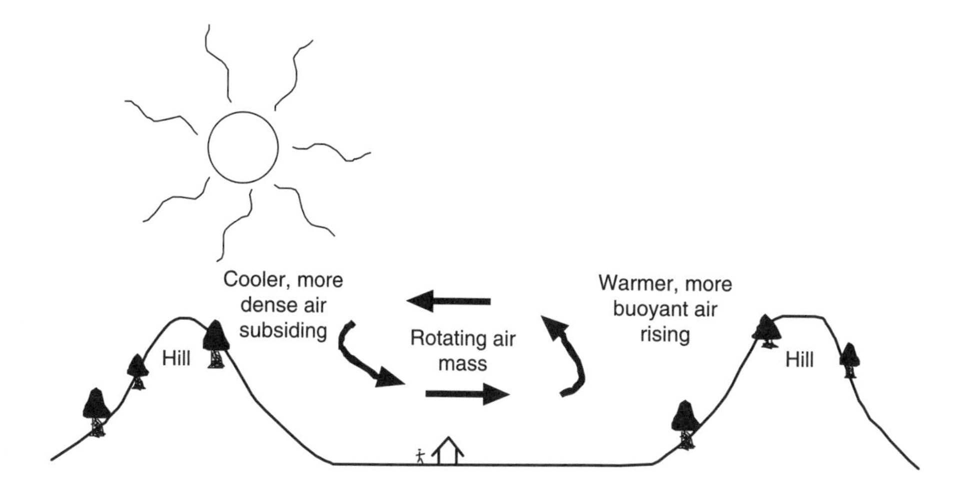

FIGURE 4-3. Valley winds.

found in the livers of beef cattle that were raised on grasses receiving nutrition from cadmium-laced municipal wastewater treatment biosolids.

Other heavy metals or chlorinated hydrocarbons similarly can be delivered to domestic animals and wild animals. Construction materials also

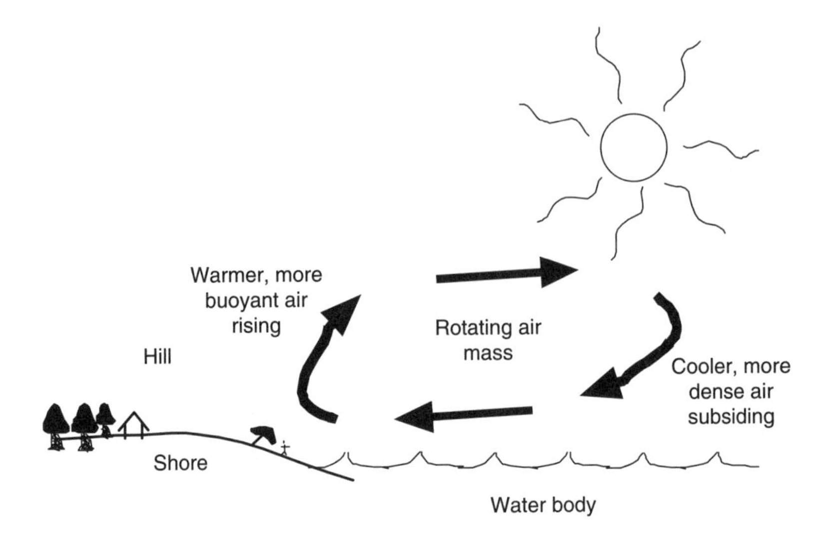

FIGURE 4-4. Shore breezes.

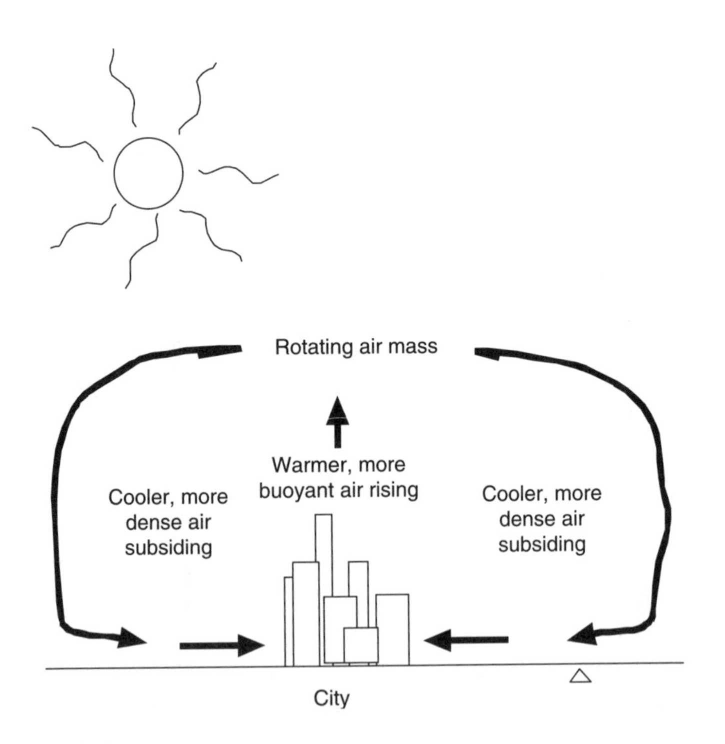

FIGURE 4-5. Urban heat island effect.

are sensitive to exposure to hazardous wastes, from the greening of statues through the dezinc process associated with low-pH rain events to the crumbling of stone bridges found in nature. To the extent the engineer can isolate potential receptors from exposure to hazardous wastes, the engineer has opportunities to control the risks to those receptors.

Intervention to Control the Dose of Hazardous Waste

Engineers have opportunities to control the dose of hazardous waste delivered to a receptor; however, these opportunities are directly associated with the engineers' abilities to control the amount of hazardous pollutants delivered to the receptor through source control and siting of hazardous waste management facilities. One solution to hazardous waste pollution could be dilution of the wastes in either the water, air, or soil environments. The science, engineering, and technologies associated with this step are discussed later in this chapter.

Intervention at the Point of Response to Hazardous Waste

The engineer cannot control the response of humans, animals, plants, or materials and facilities to the exposure to hazardous wastes with protective coatings. Scientists working on finding the prevention and cure for cancer may someday enter this sixth step and save the day.

Opportunities in Science, Engineering, and Technology to Control the Risks Associated with Hazardous Wastes

The engineers' opportunities for intervention are grounded in basic scientific principles, engineering designs and processes, and applications of proven and developing technologies to control the risks associated with hazardous wastes. Once again, any intervention scheme must be justified by the engineer as well as by the public or private client in terms of scientific evidence, sound engineering designs and processes, technologic practicality, economic realities, ethical considerations, and local, state, and national regulations. Four examples of such opportunities are (1) thermal processing of hazardous waste, (2) microbiologic processing of hazardous waste, (3) landfills as long-term repositories for hazardous waste, and (4) chemoluminescence and fluorescent in situ hybridization (FISH). These state-of-the-art measurement techniques help determine the magnitude of the risks associated with hazardous waste.

A Prerequisite Consideration: The Peirce Progression

Any and all opportunities for risk control must be evaluated in terms of the science, engineering, technology, economics, ethics, and public policy consideration; however, any and all opportunities for risk control also must consider what we will refer to as the *Peirce Progression*. Consider a house in the fictional world of a children's educational television show. The problem of the day in this setting is to rid the house of an oversupply of rats. The first solution in this particular progression is to bring in a load of cats that rid the house of rats, but the cats multiply and become a slightly bigger problem unto themselves! Depending on your point of view, cats may be a slightly greater problem and slightly more difficult to eliminate than rats. The second progression is to bring in a load of dogs that rid the house of cats, but the dogs then multiply in numbers and become a slightly bigger problem unto themselves. Another progression into a solution where that solution becomes the problem!

Now bring in a load of elephants that rid the house of dogs, but the elephants multiply and the original problem has become bigger again! Because no one knows how to rid this cartoon house of elephants, the situation has turned from a small problem into an insurmountable one. This analogy holds throughout the following introduction to the science, engineering, and technology of hazardous waste control. Engineers must avoid providing a solution to a particular hazardous waste problem where the solution progresses to a larger problem than the original.

Thermal Processing: Examples of the Science, Engineering, and Technology of Hazardous Waste Incineration

Hazardous wastes, if completely organic in structure, are, in theory, completely destructible using principles based in thermodynamics with the engineering inputs and outputs summarized as:

$$\text{Hydrocarbons} + O_2\ (+\text{energy?}) \rightarrow CO_2 + H_2O\ (+\text{energy?})$$

Hazardous wastes are mixed with oxygen, sometimes in the presence of an external energy source, and in fractions of seconds or several seconds, the by-products of gaseous carbon dioxide and water are produced to exit the top of the reaction vessel while a solid ash is produced to exit the bottom of the reaction vessel.[2] Energy may also be produced during the reaction and the heat may be recovered. One potential Peirce Progression to a second-level problem in this simple reaction could be global warming associated with the carbon dioxide.

On the other hand, if the hazardous waste of concern to the engineer contains other chemical constituents, in particular chlorine and/or heavy metals, the original simple input and output relationship is modified to a complex situation:

Hydrocarbons + O_2 (+energy?) + Cl or heavy metal(s) + H_2O + inorganic salts + nitrogen compounds + sulfur compounds + phosphorus compounds → CO_2 + H_2O(+energy?) + chlorinated hydrocarbons or heavy metal(s)inorganic salts + nitrogen compounds + sulfur compounds + phosphorus compounds

With these hazardous wastes, the potential exists for the Peirce Progression to turn into a second- or third-level problem as the hazardous waste is destructed but potentially more risky off-gases containing chlorinated hydrocarbons and/or ashes containing heavy metals are produced.

All of the incinerator systems discussed later have common attributes. All require the balancing of the three T's, which becomes the general focus of the science, engineering, and technology of incineration of any substance: *time* of incineration, *temperature* of incineration, and *turbulence* in the combustion chamber. The space required for the incinerator ranges from several square yards to possibly on the back of a flatbed truck to several acres to sustain a regional incinerator system.

The advantages of incinerators include the following:

- The potential for energy recovery.
- Volume reduction of the hazardous waste.
- Detoxification as selected molecules are reformulated.
- The basic scientific principles, engineering designs, and technologies are well understood from a wide range of other applications, including electric generation and municipal solid waste incineration.
- Application to most organic hazardous wastes, which compose a large percentage of the total hazardous waste generated worldwide.
- The possibility to scale the technologies to handle a single gallon/pound (liter/kilogram) of waste or millions of gallons/pounds (liters/kilograms) of waste.
- Land areas that are small relative to such other hazardous waste management facilities as landfarms and landfills.

The disadvantages of hazardous waste incinerators include the following:

- The equipment is capital intensive, particularly the refractory material lining the inside walls of each combustion chamber, which must be replaced as cracks form whenever a combustion system is cooled and/or heated.

- Operation of the equipment requires very skilled operators and is more costly when fuel must be added to the system.
- Ultimate disposal of the ash is necessary and particularly troublesome and costly if heavy metals and/or chlorinated compounds are found during the expensive monitoring activities.
- Air emissions may be hazardous and thus must be monitored for chemical constituents and controlled.

Given these underlying principles of incineration, seven general guidelines are suggested to engineers who are considering incineration as a method to control the risks associated with any hazardous waste problem:

1. Only liquid, purely organic hazardous wastes are true candidates for combustion.
2. Chlorine-containing organic materials deserve special consideration if they are to be incinerated at all: special materials used in the construction of the incinerator, long (many seconds) combustion time, high temperatures (>1,600°C), with lots of mixing if the hazardous waste is in the solid or sludge form.
3. Hazardous wastes containing heavy metals generally should not be incinerated.
4. Sulfur-containing organic material will emit sulfur oxides, which must be controlled.
5. The formation of nitrogen oxides can be minimized if the combustion chamber is maintained above 1,100°C.
6. Destruction depends on the interaction of a combustion chamber's temperature, dwell time, and turbulence.
7. Off-gases and ash must be monitored for chemical constituents; each residual must be treated as appropriate so the entire combustion system operates within the requirements of the local, state, and federal environmental regulators; and hazardous components of the off-gases, off-gas treatment processes, and the ash must reach ultimate disposal in a permitted facility.

The engineer must be aware that each design of an incinerator must be tailored to the specific hazardous waste under consideration, including the quantity of waste to be processed as well as the physical, chemical, and microbiologic characteristics of the waste over the planning period of the project. Laboratory testing matching a given waste to a given incinerator(s) must be conducted before the design, citing, and construction of each incinerator.

Five different types of incinerators that generally are available to the engineer are introduced as follows with accompanying text and summary diagrams: (1) rotary kiln, (2) multiple hearth, (3) liquid injection, (4) fluidized bed, and (5) multiple chamber.

Rotary Kiln

The combustion chamber in a rotary kiln incinerator (Figure 4-6) is a heated, rotating cylinder mounted at an angle, with possible baffles added to the inner face to provide the turbulence necessary for the target three T's for the hazardous waste destruction process to take place. Engineering design decisions, based on the results of laboratory testing of a specific hazardous waste, include (1) angle of the drum, (2) diameter and length of the drum, (3) presence and location of the baffles, (4) rotational speed of the drum, and (5) use of added fuel to increase the temperature of the combustion chamber as the specific hazardous waste requires. The liquid, sludge, or solid hazardous waste is input into the upper end of the rotating cylinder, rotates with the cylinder-baffle system, and falls with gravity to the lower end of the cylinder. The heated, upward-moving off-gases are collected, monitored for chemical constituents, and subsequently treated as appropriate before release, while the ash falls with gravity to be collected, monitored for chemical constituents, and treated as needed before ultimate disposal.

The rotary kiln is applicable to the incineration of most organic hazardous wastes; it is well suited for solids and sludges; and in special cases liquids and gases can be injected through auxiliary nozzles in the side of the combustion chamber. Operating temperatures generally vary from 800°C to 1,650°C (1,500°F to 3,000°F). Engineers use laboratory experiments to design residence times of seconds for gases and minutes (or possibly hours) for solid material.

Multiple Hearth

In the multiple hearth (Figure 4-7), hazardous waste in solid or sludge form is generally fed slowly through the top vertically stacked hearth; in special configurations, hazardous gases and liquids can be injected through side nozzles. Multiple hearth incinerators, which were historically developed to burn municipal wastewater treatment biosolids, rely on gravity and scrapers working the upper edges of each hearth to transport the waste through holes from upper, hotter hearths to lower, cooler hearths. Heated upward-moving off-gases are collected, monitored for chemical constituents, and treated as appropriate before release; the falling ash is collected, monitored for chemical constituents, and subsequently treated before ultimate disposal.

Most organic wastes generally can be incinerated using a multiple hearth configuration. Operating temperatures vary from 300°C to 980°C (600°F to 1,800°F). These systems are designed with residence times of seconds if gases are fed into the chambers to several hours if solid materials are placed on the top hearth and allowed to eventually drop to the bottom hearth, exiting as ash.

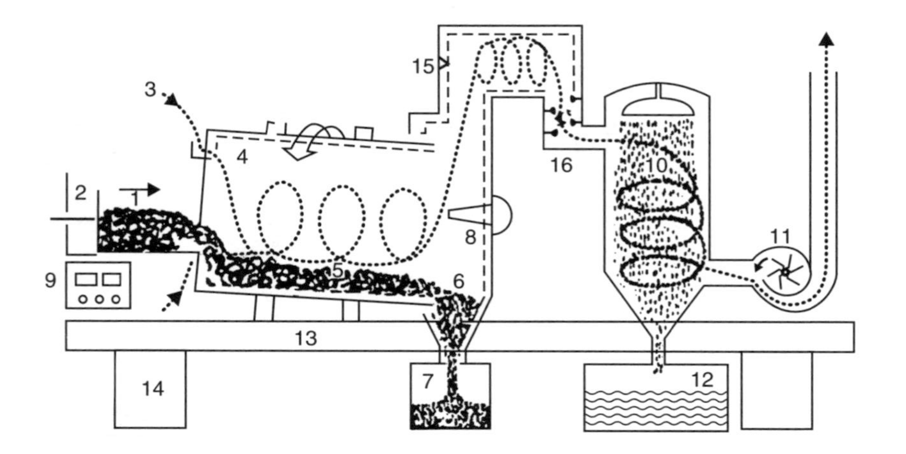

1 WASTE TO INCINERATOR
2 AUTO-CYCLE FEEDING SYSTEM:
FEED HOPPER, PNEUMATIC FEEDER, SLIDE GATES
3 COMBUSTION AIR IN
4 REFRACTORY-LINED, ROTATING CYLINDER
5 TUMBLE-BURNING ACTION
6 INCOMBUSTIBLE ASH
7 ASH BIN
8 AUTO-CONTROL PACKAGE:
PROGRAMMED PILOT BURNER
9 SELF-COMPENSATING INSTRUMENTATION-CONTROLS
10 WET-SCRUBBER PACKAGE:
STAINLESS STEEL, CORROSION-FREE WET SCRUBBER; GAS QUENCH
11 EXHAUST FAN AND STACK
12 RECYCLE WATER, FLY-ASH SLUDGE COLLECTOR
13 SUPPORT FRAME
14 SUPPORT PIERS
15 AFTER BURNER CHAMBER
16 PRECOOLER

FIGURE 4-6. Rotary kiln incinerator.

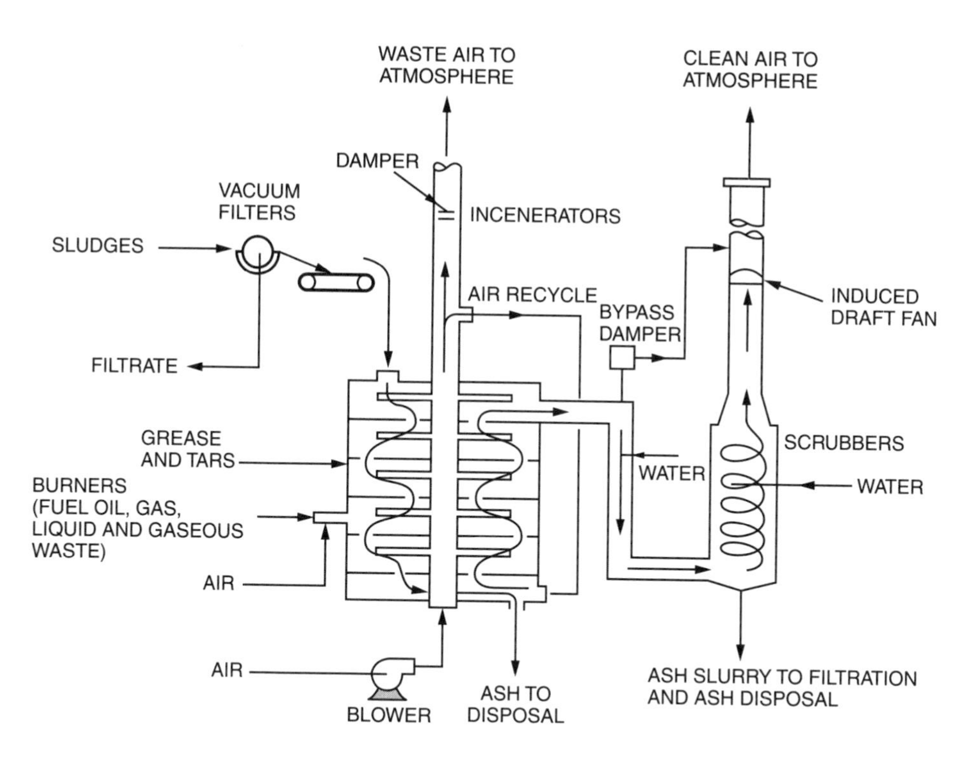

FIGURE 4-7. Multiple hearth incinerator.

Liquid Injection

Vertical or horizontal nozzles spray liquid hazardous wastes into liquid injection incinerators specially designed for the task or as a retrofit to one of the other incinerators discussed here. The wastes are atomized through the nozzles that match the waste being handled with the combustion chamber as determined in laboratory testing. The application obviously is limited to liquids that do not clog these nozzles, although some success has been experienced with hazardous waste slurries. Operating temperatures vary from 650°C to 1,650°C (1,200°F to 3,000°F). Liquid injection systems are designed with residence times of fractions of seconds as upward-moving off-gases are collected, monitored for chemical constituents, and treated as appropriate before release to the lower troposphere.

Fluidized Bed

Hazardous waste is injected under pressure into a heated bed of agitated, inert granular particles, usually sand, as the heat is transferred from the particles to the waste, and the combustion process proceeds (Figure 4-8). External heat is applied to the particle bed before injection of the waste

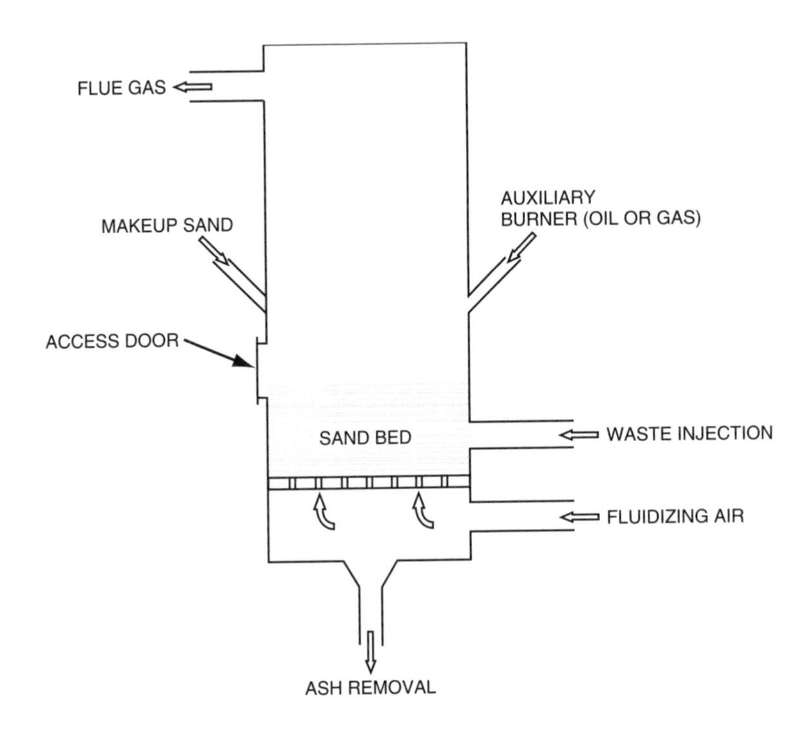

FIGURE 4-8. Fluidized bed incinerator.

and is continually applied throughout the combustion operation as the situation dictates. Heated air is forced into the bottom of the particle bed, and the particles become suspended among themselves during this continuous fluidizing process. The openings created within the bed permit the introduction and transport of the waste into and through the bed. The process enables the hazardous waste to come into contact with particles that maintain their heat better than, for example, the gases inside a rotary kiln. The heat maintained in the particles increases the time that the hazardous waste is in contact with a heated element and thus the combustion process could become more complete with fewer harmful by-products. Off-gases are collected, monitored for chemical constituents, and treated as appropriate before release, and the falling ash is collected, monitored for chemical constituents, and subsequently treated before ultimate disposal.

Most organic wastes can be incinerated in a fluidized bed, but the system is best suited for liquids. Operating temperatures vary from 750°C to 900°C (1,400°F to 1,600°F). Liquid injection systems are designed with residence times of fractions of seconds as upward-moving off-gases are collected, monitored for chemical constituents, and treated as appropriate before release to the lower troposphere.

Multiple Chamber

Hazardous wastes are turned into a gaseous form on a grate in the ignition chamber of a multiple-chamber system (Figure 4-9). The gases created in this ignition chamber travel through baffles to a secondary chamber, where the actual combustion process takes place. The secondary chamber is often located above the ignition chamber to promote natural advection of the hot gases through the system. Heat may be added to the system in either the ignition chamber or the secondary chamber as required for specific burns.

The application of multiple-chamber incinerators is generally limited to solid hazardous wastes, with the waste entering the ignition chamber through a opened charging door in batch, not continuous, loading. Combustion temperatures typically hover near 540°C (1,400°F) for most applications. These systems are designed with residence times of minutes to hours for solid hazardous wastes as off-gases are collected, monitored for chemical constituents, and treated as appropriate before release to the lower troposphere. At the end of each burn period, the system must be cooled so the ash can be removed before monitoring for chemical constituents and subsequent treatment before ultimate disposal.

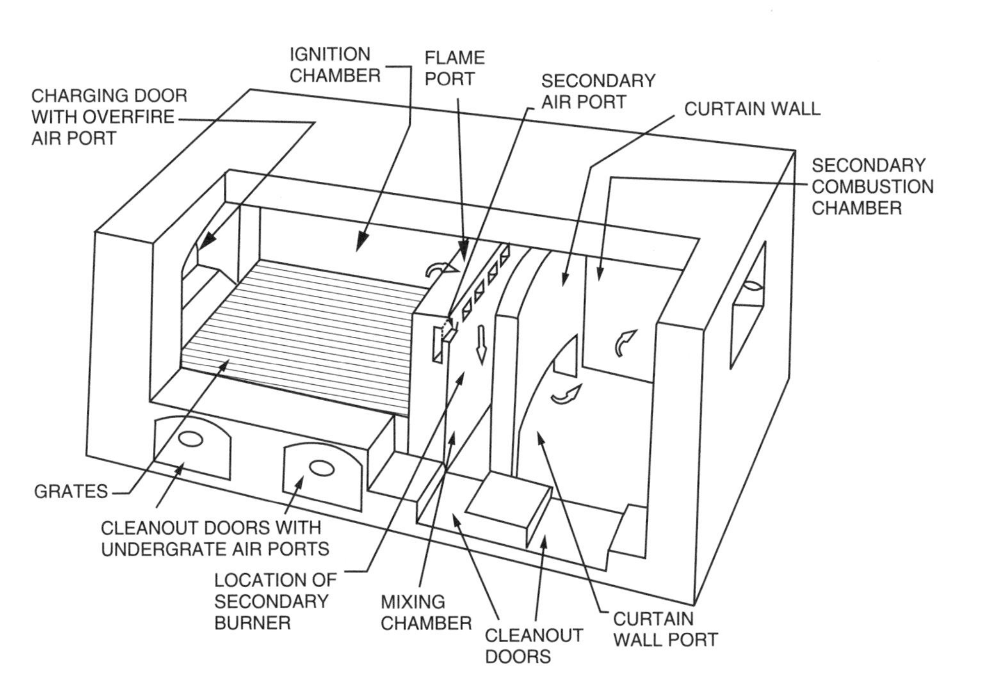

FIGURE 4-9. Multiple-chamber incinerator.

Microbiologic Processing: Examples of the Science, Engineering, and Technology of Hazardous Waste Biotreatment

Hazardous wastes, if completely organic in structure, are, in theory, completely destructible using principles based in microbiology with the engineering inputs and outputs summarized as:

$$\text{Hydrocarbons} + O_2 + \text{microorganisms (+energy)}$$
$$\rightarrow CO_2 + H_2O + \text{microorganisms (+energy?)}$$

Hazardous wastes are mixed with oxygen and aerobic microorganisms, sometimes in the presence of an external energy source in the form of added nutrition for the microorganisms, and in seconds, hours, or possibly days, the by-products of gaseous carbon dioxide and water are produced, which exit the top of the reaction vessel while a solid mass of microorganisms is produced to exit the bottom of the reaction vessel.[3] The only potential Peirce Progression to a second-level problem in this simple reaction could be global warming associated with the carbon dioxide.

On the other hand, if the hazardous waste of concern to the engineer contains other chemical constituents, in particular chlorine and/or heavy metals, and if the microorganisms are able to withstand and flourish in such an environment and not shrivel and die, the simple input and output relationship is modified to:

$$\text{Hydrocarbons} + O_2 + \text{microorganisms (+energy?)} + \text{Cl or heavy metal(s)}$$
$$+ H_2O + \text{inorganic salts} + \text{nitrogen compounds} + \text{sulfur compounds}$$
$$+ \text{phosphorus compounds} \rightarrow CO_2 + H_2O \text{ (+energy?)} + \text{chlorinated}$$
$$\text{hydrocarbons or heavy metal(s) inorganic salts} + \text{nitrogen compounds}$$
$$+ \text{sulfur compounds} + \text{phosphorus compounds}$$

If the microorganisms do survive in this complicated environment, the potential exists for the Peirce Progression to turn into a second- or third-level problem(s) that the engineer must plan for and confront as the hazardous waste initially is destructed but potentially more risky microorganism masses are produced that contain chlorinated hydrocarbons or heavy metals.

Discussion: Metal-Eating Algae[4]

Microbiologic treatment of hazardous wastes is not only the province of bacteria and fungi; recently, algae have also shown promise.

Genetically altered strains of algae are being used to clean up heavy metal–contaminated sediments in Lake Erie. Mercury, cadmium, zinc, and other metals released by industries in the Erie basin have accumulated in Lake Erie sediment.

Researchers have successfully enhanced the *Chlamydomonas reinhardtii's* natural abilities to sequester metals. The species is a commonly found one-celled plant that is readily manipulated by genetic engineering. *C. reinhardtii* is prolific, so relatively large volumes of activated algae can be readily produced, making this bioremediation even more effective, less expensive, and likely safer than the conventional chemical extraction metal approaches. The genetic enhancements allow the algal cell to bind with the metals.

The best method thus far has involved attaching the protein, metallothionen—a protein that binds heavy metals—to the algal cell membrane. The enhanced algae can take up to five times the mass of metal compared to the unenhanced cell. The enhanced cells also reproduced at a rate three times faster when grown in heavy metal–laden sediments. Engineers will soon begin testing the effectiveness of metal sequestration and removal approaches in various pilot studies.

Discussion: PCB Cleanup Efforts

The most common means of remediating waste sites contaminated with polychlorinated biphenyls (PCBs) are dredging contaminated sediments and completely removing soil layers containing PCBs, followed by thermal destruction, chemical oxidation, or storage in landfills. These methods can be very expensive and are often not completely able to destroy all PCB congeners (or even allow for the production of very toxic dioxins). These problems have led researchers to consider microbiologic approaches for destroying PCBs.

The microbes capable of detoxifying PCBs have done so by removing the Cl atoms from the PCB molecules. Until recently, only aerobic microbes have been able to detoxify PCBs by dechlorination. The aerobes have been difficult to use consistently at waste sites because they require large amounts of free oxygen, whereas soil and sediment layers of waste sites can have highly reduced conditions. PCBs are quite lipiphilic, so they tend to sorb to soil and sediment particles rather than being dissolved in water. Thus the aerobes have a difficult time coming in contact with them. The aerobes are also usually limited to PCB molecules with one or two Cl atoms, whereas most PCBs have more than two Cl atoms.

It was recently reported in *Civil Engineering*[5] that researchers at the University of Maryland's Biotechnology Institute and the Medical University of South Carolina have identified the first anaerobic microbes that can detoxify PCBs. The treatment process appears to require two steps: dechlorination by anaerobic microbes, followed by biochemical degradation by aerobic bacteria. This process may work in the field, but more efficiently in a bioreactor, wherein sediments and soil containing large PCB concentrations are first treated anaerobically so that most of the Cl is removed. Then the digested matter is moved to an aerobic reactor, where the wastes are completely broken down.

Finding these microbes may be followed by enhancements. Anaerobic microbes with an affinity for PCB destruction may be isolated and selected to improve bioremediation efficiencies in the field and in reactors.

All of the bioreactor systems discussed as follows have some similar attributes. All rely on a population(s) of microorganisms to metabolize organic hazardous waste ideally into the harmless by-products of $CO_2 + H_2O$ (+energy?). In all of the systems, the microorganisms must be either initially cultured in the laboratory to be able to metabolize the specific organic waste of concern, or target populations of microorganisms in the system must be given sufficient time—days, weeks, possibly even years—to evolve to the point where the cumbersome food (i.e., the hazardous waste) is digestible by the microorganisms.

During all treatment processes, the input waste must be monitored and possibly controlled to maintain environmental conditions that do not upset or destroy the microorganisms in the system. These monitoring and control requirements for each of the systems include but are not limited to the following:

- Temperature, possibly in the form of a heated building.
- pH, possibly in the form of lime addition.
- Oxygen availability, possibly in the form of atmospheric diffusers that pump ambient atmosphere into the mixture of microorganisms and hazardous waste.
- Presence of additional food sources and/or nutrients, possibly in the form of a secondary carbon source for the microorganisms.
- Changes in the characteristics of the input hazardous waste, including hydrocarbon availability and chemicals that may be toxic to the microorganisms, possibly including holding tanks to homogenize the waste before exposure to the microorganisms.

The engineer must tailor particular populations of microorganism(s) to the particular hazardous waste of concern and then must plan for and undertake extensive and continual monitoring and fine-tuning of each microbiologic processing system during its complete operation.

The advantages of the biotreatment systems include the following:

- The potential for energy recovery.
- Volume reduction of the hazardous waste.
- Detoxification as selected molecules are reformulated.
- The basic scientific principles, engineering designs, and technologies are well understood from a wide range of other applications, including municipal wastewater treatment at facilities across the United States.
- Application to most organic hazardous wastes, which as a group compose a large percentage of the total hazardous waste generated nationwide.
- The possibility to scale the technologies to handle a single gallon/pound (liter/kilogram) of waste per day or millions of gallons/pounds (liters/kilograms) of waste per day.
- Land areas that could be small relative to such other hazardous waste management facilities as landfills.

The disadvantages of the biotreatment systems include the following:

- Operation of the equipment requires very skilled operators and is more costly because input hazardous waste characteristics change over time and correctional controls become necessary.
- Ultimate disposal of the waste microorganisms is necessary and particularly troublesome and costly if heavy metals and/or chlorinated compounds are found during the expensive monitoring activities.

Given these underlying principles of biotreatment systems, four general guidelines are suggested whenever such systems are considered as a potential solution to any hazardous waste problem:

1. Only liquid organic hazardous wastes are true candidates.
2. Chlorine-containing organic materials deserve special consideration if they are to be biotreated at all, and special testing is required to match microbial communities to the chlorinated wastes, realizing that useful microbes may not be identifiable, and even if they are the reactions may take years to complete.
3. Hazardous wastes containing heavy metals generally should not be bioprocessed, although progress is ongoing (See this chapter's two discussion boxes).
4. Residual masses of microorganisms must be monitored for chemical constituents, and each residual must be addressed as appropriate so the

entire bioprocessing system operates within the requirements of the local, state, and federal environmental regulators.

Each application of biotechnology must be tailored to the specific characteristics of the hazardous waste under consideration, including the quantity of waste to be processed, as well as the physical, chemical, and microbiologic characteristics of the waste over the entire planning period of the project. Laboratory tests matching a given waste to a given bioprocessor(s) must be conducted before the design and citing of the system.

Three different types of bioprocessors that are generally available to the engineer are introduced as follows with accompanying text and summary diagrams: (1) trickling filter, (2) activated sludge, and (3) aeration lagoons. As a group these three types of treatment systems represent a broad range of opportunities available to engineers who are searching for methods to control the risks associated with hazardous wastes.

Trickling Filter

The classical design of a trickling filter system (Figure 4-10) includes a bed of fist-sized rocks, enclosed in a rectangular or cylindrical structure, through which is passed the waste of concern. Biofilms are selected from laboratory studies and encouraged to grow on the rocks; as the liquid waste moves downward with gravity through the bed, the microorganisms comprising the biofilm are able to come into contact with the organic hazardous waste/food source and ideally metabolize the waste into relatively harmless CO_2 + H_2O + microorganisms (+energy?). Oxygen is supplied by blowers from the bottom of the reactor and passes upward through the bed. The treated waste that moves downward through the bed subsequently enters a quiescent

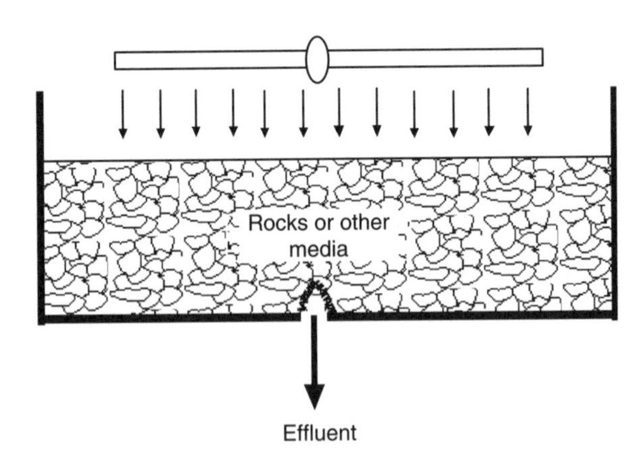

FIGURE 4-10. Trickling filter treatment system.

tank, where the microorganisms that are sloughed off the rocks are settled, collected, and ultimately disposed.

Activated Sludge

The key to the activated sludge system (Figure 4-11) is that the microorganisms that are available to metabolize the hazardous waste/food source are recycled within the system. This reuse enables this bioprocessor to actually evolve over time as the microorganisms adapt to the changing characteristics of the influent hazardous waste; with this evolution comes the potential for the microorganisms to be more efficient at metabolizing the waste stream of concern. A ready supply of tailored and hungry microorganisms is always available to the engineer operating the facility.

A tank full of liquid hazardous waste is injected with a mass of microorganisms. Oxygen is supplied through the aeration basin as the microorganisms come in contact, sorb, and metabolize the waste ideally into $CO_2 + H_2O$ + microorganisms (+energy?). The heavy, satisfied microorganisms then flow into a quiescent tank, where the microorganisms are settled with gravity, collected, and ultimately disposed. Depending on the current operating conditions of the facility, some or many of the settled and now hungry and active microorganisms are returned to the aeration basin, where they are given another opportunity to chow down. Liquid effluent from the activated sludge system may require additional microbiologic and/or chemical processing before release into a receiving stream or city sewer system.

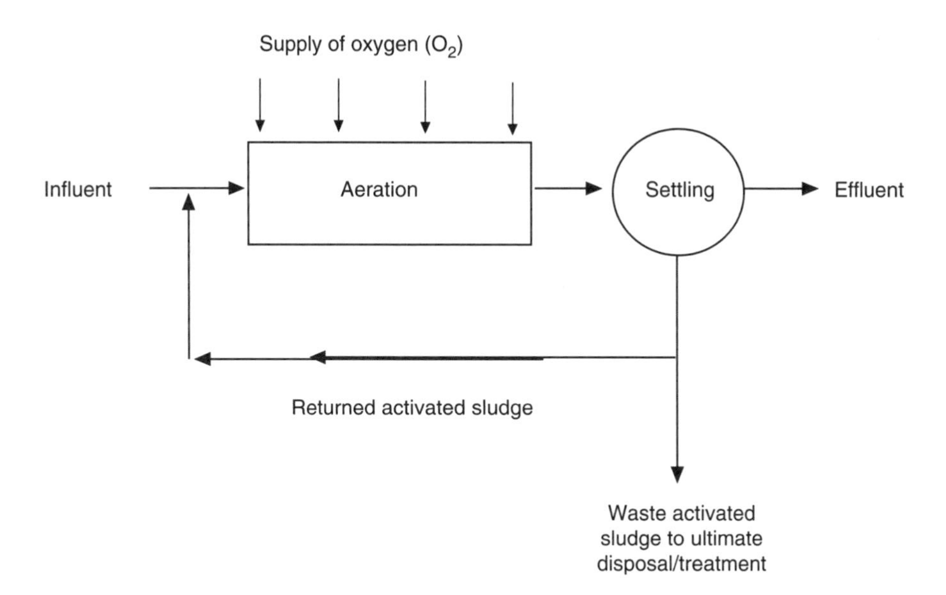

FIGURE 4-11. Activated sludge treatment system.

The activated sludge process in theory and in practice is a sequence of three distinct steps:

1. *Sorption*. The microorganisms come in contact with the food source—the organic material in the hazardous waste—and the food is adsorbed either to or through the cell walls of the microorganisms. In either case the food is now directly available to the individual microorganisms. In a correctly operating facility, this sorption phase generally takes about 30 minutes.
2. *Growth*. The microorganisms metabolize the food and biochemically break down, or destroy, the hazardous organic molecules. This growth phase, during which individual organisms grow and multiply, may take up to hours or possibly days for complete metabolism of the hazardous waste constituents. Thus the design of the activated sludge system must include a basin with a detention time adequate for the correct amount of growth to take place.
3. *Settling*. Solid (the microorganisms) to liquid (the liquid remaining from the process) separation is achieved in a settling basin, where the heavy and satisfied microorganisms sink to the bottom with gravity.

A critical design consideration of the activated sludge system is the loading to the aeration basin. *Loading* is defined as the food-to-microorganism (F:M) ratio at the start of the aeration basin. The planning is similar to the planning that precedes a Thanksgiving Day feast, with the trick being to make sure enough food is on hand for all of those in attendance. In the activated sludge system, the food shows up in the form of the organic constituents of the hazardous waste. The invited guests show up in the form of microorganisms that are returned from the settling basin to the aeration tank. With little or no control over the amount of food that may arrive during any given time period, the operating engineer must adjust the F:M ratio by adjusting the number of returned microorganisms. This balancing act between the amount of food and the numbers of microorganisms is summarized in two extreme examples suggesting ranges of F:M ratios, aeration times, and treatment efficiencies:

F:M Ratio + Aeration Time → Degree of Treatment

1. lower longer higher

(little food, lots of hungry mouths to feed, lots of time at the dinner table)

2. higher shorter lower

(smaller tanks, shortened time at the dinner table)

Sample loadings that are observed in practice range from 0.05 to greater than 2.0. The process of *extended aeration*, ranging to greater than 30 hours, might have a loading of between 0.05 and 0.20, with an efficiency of hazardous waste removal in excess of 95%. The process of *conventional aeration*, closer to six hours for aeration, might have a loading between 0.20 and 0.50, with a treatment efficiency of possibly 90%. The process of *rapid aeration*, in the range of one to three hours for aeration, might have a loading between 1.0 and 2.0, with a removal efficiency closer to 85%. For each given problem, the engineer must design an individual activated sludge facility based on laboratory testing of a specific hazardous waste; the engineer must operate that facility and select different loadings through time based on ongoing laboratory tests of the facility's input, process variables, and outputs.

Variations of the classic activated sludge system summarized previously exist to help process specific and difficult-to-treat hazardous wastes. These variations in the design and operation of such facilities include the following:

- *Tapered aeration*. The oxygen that is supplied to the aeration basin is in greater amounts at the input end of the basin and in lesser amounts at the output end of the basin (Figure 4-12), with the goal of

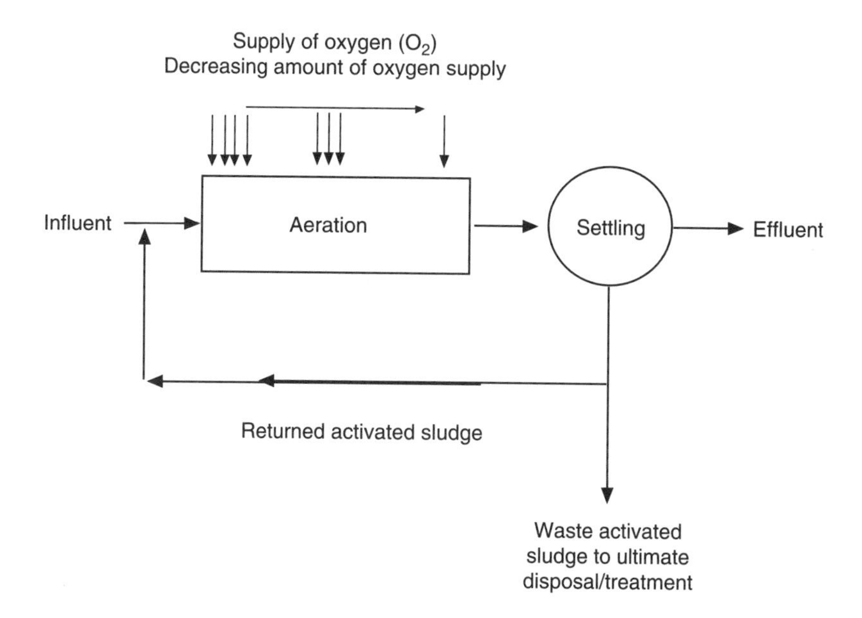

FIGURE 4-12. Tapered aeration activated sludge treatment system. Greater amount of oxygen added closer to influent because of the large oxygen demand from microbes as waste is introduced to the aeration tank.

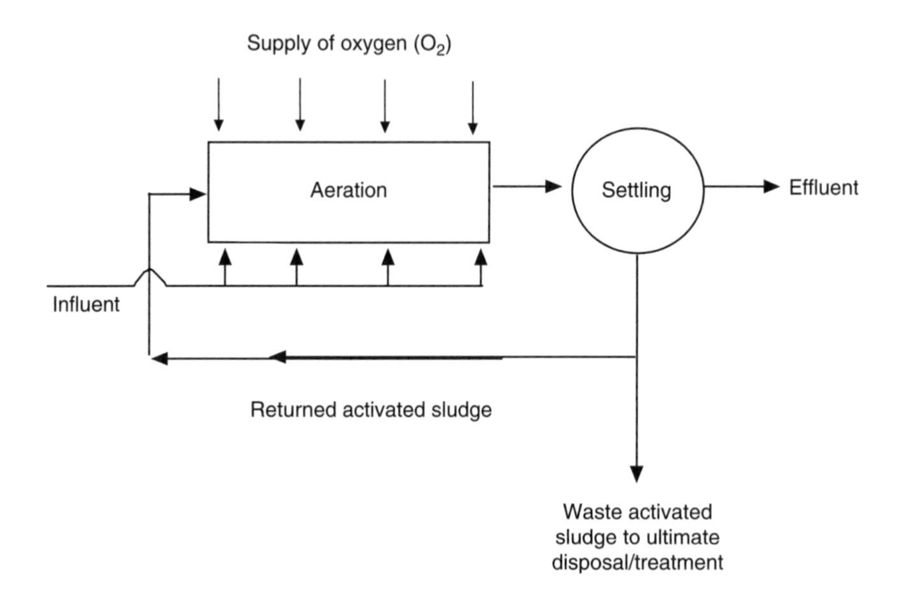

FIGURE 4-13. Step activated sludge treatment system.

supplying more oxygen where it may be needed the most to address a specific hazardous waste problem.

- *Step aeration.* The influent oxygen *and* hazardous waste is supplied to the aeration basin in equal amounts throughout the basin (Figure 4-13), with the goal of matching the oxygen demand to the location where it may be needed the most for a specific hazardous waste problem.
- *Contact stabilization or biosorption.* The sorption and growth phases of the microbiologic processing system are separated into different tanks (Figure 4-14), with the goal of achieving growth at higher solid concentrations, saving tank space, and thus saving money.

Aeration Ponds

Ponds (Figure 4-15) are available to engineers for the long-term (e.g., months to years) treatment of liquid hazardous waste. Persistent organic molecules—those not readily degraded in the trickling filter of activated sludge systems—are potentially broken by certain microbes into $CO_2 + H_2O$ + microorganisms (+energy?) if given enough time. The ponds are open to the weather, and ideally oxygen is supplied directly to the microorganisms from the atmosphere. Design decisions based on laboratory experiments include

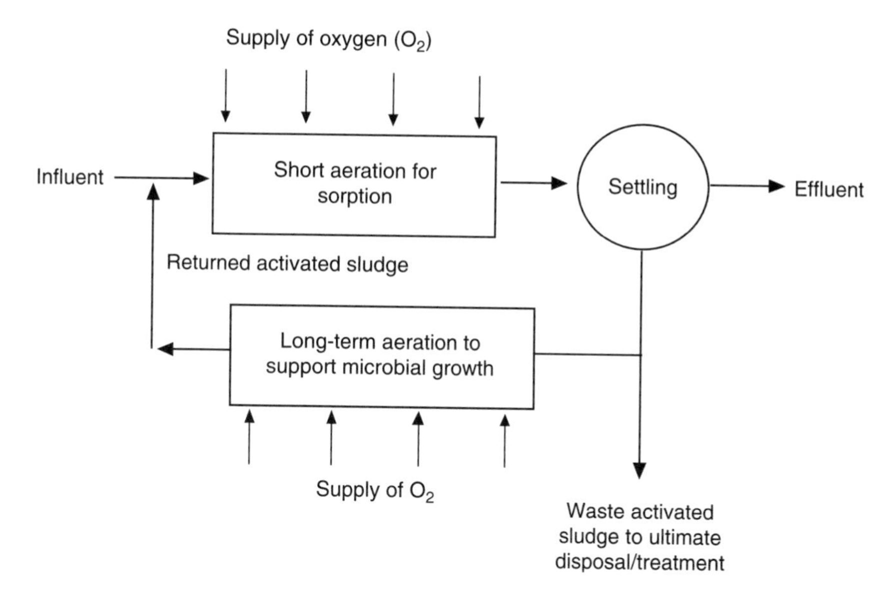

FIGURE 4-14. Contact stabilization activated sludge treatment system.

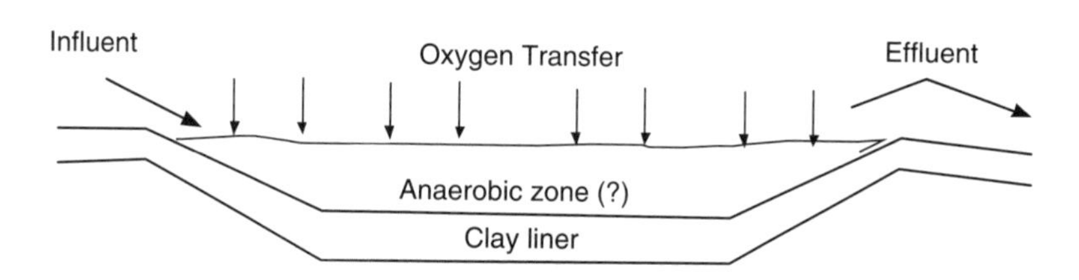

FIGURE 4-15. Hazardous waste treatment pond.

the following:

- Design: pond size of 0.5 to 20 acres.
- Design: pond depth of 1 foot to 30 feet.
- Design: detention time of days to months to possibly years.
- Operation: in series with other treatment systems, other ponds, or not.
- Operation: the flow to the pond is either continuous or intermittent.
- Operation: the supply of additional oxygen to the system through blowers and diffusers may be required.

Again the critical engineering concerns in the design and operation of this and other biotreatment facilities are the identification and maintenance of microbial populations that metabolize the specific hazardous waste of concern.

Hazardous Waste Storage Landfills: Examples of the Science, Engineering, and Technology of Long-Term Storage of Hazardous Waste

The four stages in the life of long-term storage facilities are (1) siting, (2) design, (3) operation, and (4) post-closure management. These stages offer the engineer myriad opportunities to intervene and control the risks associated with hazardous wastes. At each stage of a landfill's life, any intervention scheme must be justified by the engineer in terms of the science, engineering, and technologic aspects of the project; however, as recent U.S. history indicates, economic realities and public perceptions, as well as the laws of local, state, and national governments, particularly drive the decision-making process throughout all stages of landfill considerations. Few, if any, individuals are willing to accept a landfill in their backyards, and thus the overused expression NIMBY ("not in my backyard") has become popular.[6]

Siting

Engineering determinations are extremely important at the beginning of any process to site any facility, whether that facility is a local shopping center or a hazardous waste landfill. The game is to identify plots of land that are at least acceptable from a scientific and engineering standpoint. In no particular order, the engineer can overlay a regional landmass with at least the seven site selection criteria discussed as follows. These criteria are region-specific and must be identified on a case-by-case basis; however, the goal is always that after the selection criteria for a specific region have been established and overlaid on the region, at least one area within the region will remain standing as a contender for the landfill site.

Historically the release and transport of hazardous chemicals from a landfill has included waste moving along the surface of the earth or into groundwater supplies beneath the earth. The location of a hazardous waste landfill must consider these release and transport possibilities; thus the landfill must be sited well above historically high groundwater tables and well away from surface streams and lakes. Once the horse is out of the barn, so to speak, the damage is done because surface and subsurface drinking water supplies are jeopardized.

Climatology must also be considered when options are screened to identify sites for a hazardous waste landfill. Intensive rain events can damage the integrity of any waste barrier system found in any landfill anywhere; thus the landfill must be located outside the paths of recurring storms.

Hurricane paths in North Carolina and tornado alleys in the Midwest offer vivid examples, but microclimates exist throughout the United States that result in deluges that could and do assist in the migration of hazardous wastes from landfill impoundments to a receiving surface or subsurface body of water.

The geology of the region is similarly important as the engineer searches for a site to construct a hazardous waste landfill. The potential construction site must be stable in geologic time; thus areas of active and dormant faults must be avoided. The vertical soil profile must be composed of soil materials that are generally impervious to liquid migration; thus sandy soils and cracked bedrock must be avoided. With permeabilities ranging up to 10^{-6} cm s^{-1} and with thickness exceeding hundreds of meters, natural clay deposits could provide the most promising materials on which to site a hazardous waste landfill.

The ecology of the region poses particularly troublesome difficulties during the site selection process. Areas of low fauna and flora densities are preferred, whereas natural wilderness areas, wildlife refuges, and migration routes should be avoided. Areas supporting endangered species must also be avoided.

Transportation routes to and from a potential site raise the possibilities of local human receptors exposed to hazardous wastes if a roadside spill occurs while the waste is in transit. The need also exists for an all-weather highway that helps support adequate emergency responses should such accidental spills or catastrophic events occur at the landfill. Thus existing or possible transportation routes must enter the site selection process.

Alternatives for land resource utilization also must be considered as locations are screened to identify potential sites for a landfill within a region. The long-term storage facility should only occupy land that has low alternative land use value. There is no sense in putting the landfill where a golf course and housing development could go. Recreational areas must be avoided to help limit the accessibility of the site to the general public.

Environmental health often is the primary concern in siting these types of facilities. The landfill must not be located near drinking water wells, surface drinking water supplies, or populated areas. The goal is to avoid placing drinking water supplies and receptors in proximity to the landfill.

The real challenge to the engineer involved in the site selection process is that with numerous and often conflicting site selection criteria, few if any acceptable sites may be identified within a given region. For example, wildlife refuges have few if any human inhabitants, and thus siting a landfill in a refuge could maximize the distance from the landfill to potential human receptors of the hazardous waste. But one site selection criterion is the avoidance of wildlife areas. Engineers are often faced with the "darned if we do, darned if we don't" situation.

Design

The engineer can control the risks associated with the long-term storage of hazardous waste by incorporating sound engineering design considerations into any and all five levels of safeguard to be found in modern proposed and existing landfill designs. Starting from the top down through a landfill, these five potential levels of safeguard include (1) a cover to prevent water from entering the landfill, (2) solidification of the hazardous waste, (3) a primary barrier to liquid release with leachate collection and treatment as appropriate, (4) a secondary barrier to liquid release with leachate collection and treatment as appropriate, and (5) discharge wells downgradient from the site to pump and treat any escaped contaminants.

The landfill must be covered to prevent the movement of rainwater into and through the impoundment. The cap (Figure 4-16) must be constructed with layers of materials. The first layer should be topsoil that is graded to promote the controlled runoff of all storm events. The soil is seeded with grasses having short root systems to promote the evapotranspiration of rain that does fall on the landfill. The second layer of the cap should be composed of an impermeable material that also is graded to promote controlled runoff, avoid erosion of the cap, and prevent movement of the rainwater into the

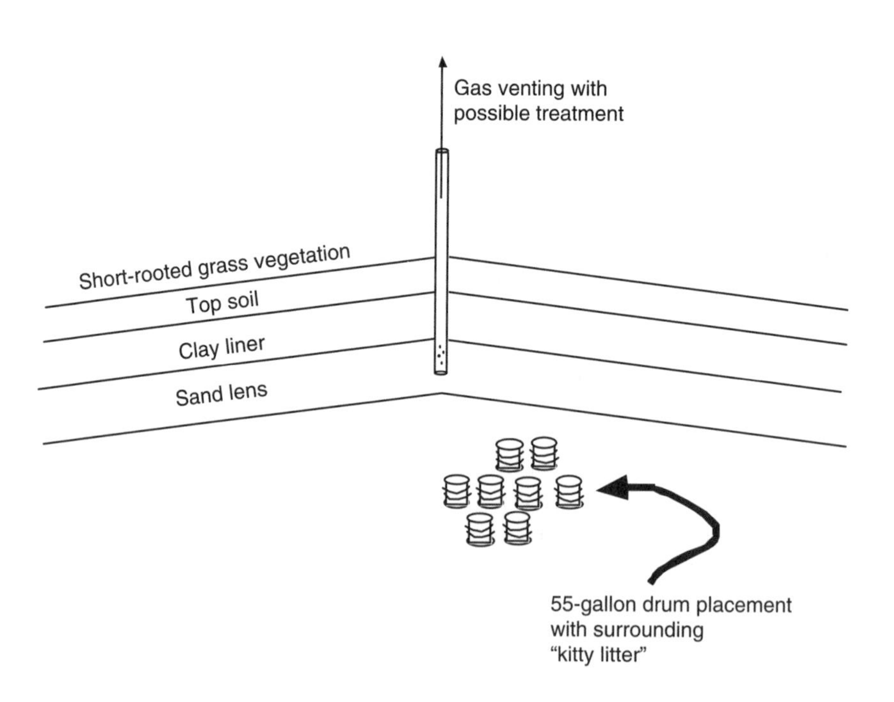

FIGURE 4-16. Engineered cover for a hazardous waste landfill.

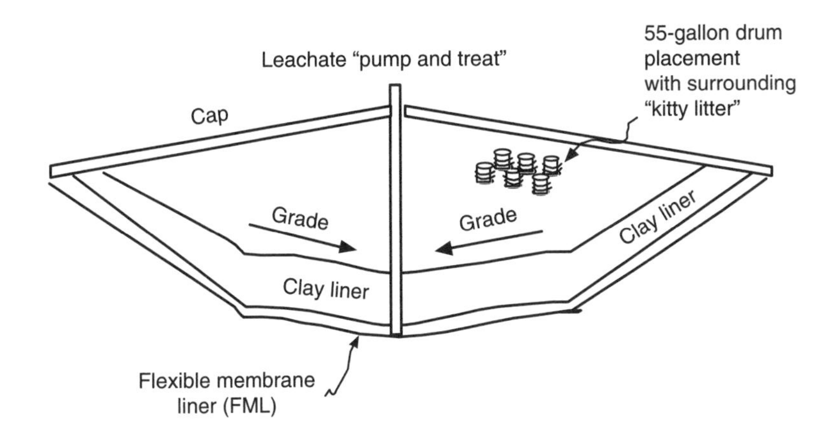

FIGURE 4-17. Leachate collection system for a hazardous waste landfill.

depths of the landfill. The third layer of the cap should be a sand lens that is graded to promote the collection, exhaust, and subsequent treatment of gases that may be produced within the landfill.

The waste within the landfill must be solidified to help preclude movement of any waste within the landfill. The first level of the solidification process is mixing all liquid and sludge wastes with sorbent material before burial. In practice the sorbent material is often an oven-dried clay taken from a nearby natural clay formation. The material is identical to kitty litter products used in households and to oil-dry products used in auto repair shops. The second level of solidification is generally filling painted 55-gallon drums with the clay/waste mixture. The third level of solidification is surrounding all of the 55-gallon drums with more clay at the time of burial.

Leachate collection systems (Figure 4-17) are another possibility; however, many regulatory agencies require two or three pairs of these systems to protect the integrity of a landfill. A primary leachate collection and treatment system must be designed like the bottom of the landfill bathtub. This leachate collection system must be graded to promote the flow of liquid within the landfill from all points in the landfill to a central collection point(s), where the liquid can be pumped to the surface for subsequent monitoring and treatment. Crushed stone and perforated pipes are used to channel the liquid along the top layer of this compacted clay liner to the pumping location(s).

Immediately below the primary leachate collection is a secondary leachate collection available in case the primary system fails. This leachate collection system also must be graded to promote the flow of liquid within the landfill from all points in the landfill to a central collection point(s), where the liquid can be pumped to the surface for subsequent treatment.

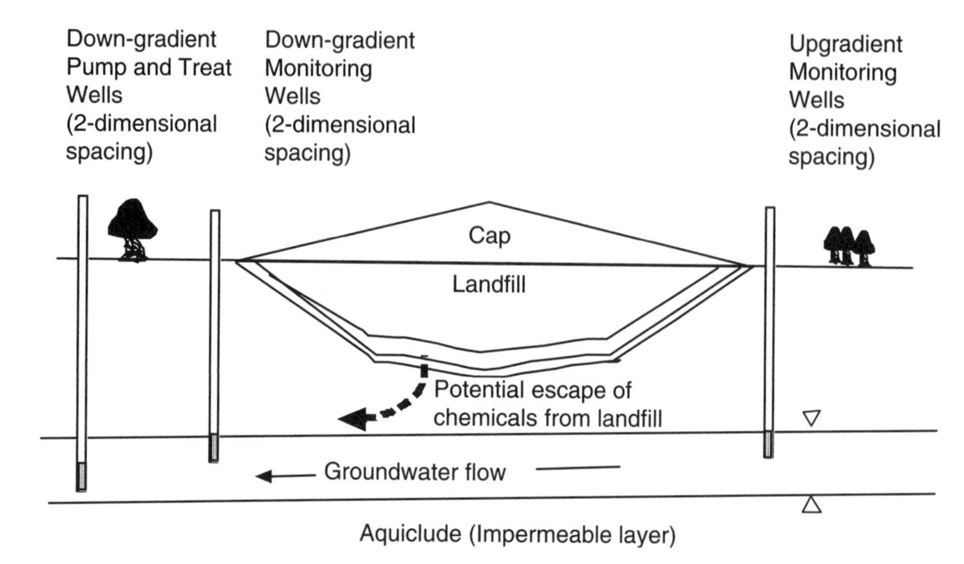

FIGURE 4-18. Monitoring and pump-and-treat wells surrounding a hazardous waste landfill.

The secondary system typically is constructed of a flexible membrane liner (FML) material, at least 2 mm thick, and an unbending plastic garbage bag.

The final barrier to liquid waste migration from the hazardous waste landfill must be a field of monitoring and extraction wells. The monitoring wells are located upgradient and downgradient from the site (Figure 4-18). The upgradient monitoring wells provide a method to identify background concentrations of the constituents in the groundwater against which to compare the information collected from the downgradient monitoring wells. If a chemical substance has been detected downgradient from the landfill that has not been detected in the upgradient monitoring wells, or if a chemical substance is detected at higher levels downgradient from the landfill, then the landfill has sprung a leak. The downgradient pump-and-treat wells can then be used to extract groundwater at rates that prohibit any additional transport of hazardous waste through the soil underlying the landfill. The entire process often becomes similar to finding a needle in a haystack; thus the location of the wells in the field becomes of paramount importance to the success of this monitoring and pump-and-treat system at the landfill.

Operation

As the landfill enters its operational phase—the phase when waste is actually buried in the facility—the engineer has additional opportunities to help control the risks associated with hazardous waste. Any leachate that is collected from the liner system(s) must be continually monitored and treated as appropriate. The groundwater monitoring wells must be continually operated with liquid samples collected and analyzed for chemical constituents, with subsequent operation of the downgradient pump-and-treat wells as appropriate. The location of the solidified waste canisters must be three-dimensionally mapped to promote excavation at a future time if and when advancing science, engineering, and recovery technologies provide economical reprocessing and recycling of the buried waste materials.

Post-Closure Management

Once the landfill is full of solidified hazardous waste, the engineer can further control the risks associated with the waste by conducting important post-closure management procedures. The filled landfill must be covered with the cap discussed previously. The liquids from the leachate and monitoring wells must be continually analyzed, and the pump-and-treat wells must be continually maintained and used when necessary. Most important, access to the site must be limited to those people responsible for post-closure management of the facility; all other people and animals must be denied entry.

Chemoluminescence and Fluorescent In Situ Hybridization (FISH): Examples of the Science, Engineering, and Technology Available to Monitor the Magnitude of the Risks Associated with a Hazardous Waste Problem

Many sophisticated instruments and procedures are available to monitor for the presence and concentration of hazardous waste in water, air, and soil. These instruments, with their attendant procedures, offer engineers many opportunities to better understand the magnitude of the associated risks. For example, these instruments and techniques are used to determine the magnitude of a hazardous waste problem, answering such questions as: How much toluene is in the soil surrounding a hazardous waste landfill? The same instruments and techniques are used to track the successes and failures of the engineers' attempts to control the risks associated with hazardous waste, answering such questions as: How much has this pump-and-treat technology, which was used for the past six months at a cost of $1,200,000, reduced the levels of toluene in the soil surrounding this hazardous waste landfill?

This text introduces the reader to two examples of measurement and monitoring instruments and procedures that are at the cutting edge of efforts to determine levels of hazardous contaminants in water, air, and soil. Chemoluminescence and fluorescent in situ hybridization (FISH) are summarized as state-of-the-art examples of the science, engineering, and technology of measurement instrumentation and techniques available to engineers addressing hazardous waste problems.

The Example Measurement and Monitoring Problem: Contaminated Soil

Soils at sites nationwide are contaminated with dense nonaqueous phase liquids (DNAPLs) that sink in the groundwater, nonaqueous phase liquids (NAPLs) that float within the groundwater, and heavy metals. These sites have become a nationwide public health and economic concern. Active soil remediation systems such as pump-and-treat and passive remediation systems such as natural attenuation require elaborate, expensive, decades-long monitoring for process control, performance measurement, and regulatory compliance. Under current monitoring practices many liquid and/or soil samples must be collected from contaminated sites, packaged, transported, and analyzed in a certified laboratory—all at great expense and with great time delay to the site owner and to state and federal regulatory agencies. In addition, waste that is generated by sample collection and by sample analyses must be properly disposed, again at great cost. Nationwide monitoring of contaminated soils will continue into the future in order to protect public health and the environment.

Chemoluminescence for Sensing the Levels of Nitric Oxide Emissions from Soil

Consider the extremely complex and only partially understood biogeochemical nitrogen cycle (Figure 4-19). The processes in soil that traditionally are suggested to contribute to the levels of nitric oxide (NO) emissions are, in order of general importance, autotrophic nitrification, respiratory denitrification, chemodenitrification, and heterotrophic nitrification. Except for chemodenitrification, all of these mechanisms are microbially mediated transformations performed by such bacteria as *Nitrosolobus* and *Nitrobacter* genera in autotrophic nitrification and *Pseudomonas* and *Alcaligenes* genera in respiratory denitrification. Autotrophic nitrification and then respiratory denitrification are suggested to be the principal sources of NO in the cycle, whereas heterotrophic nitrification and chemodenitrification can be important NO sources under extraordinary soil pH and other conditions.[7]

As levels of contamination change in a soil, chemoluminescence monitoring of NO emissions from contaminated soil can indicate to the engineer the absence or presence, including the level, of contamination of a soil pollutant. This monitoring of NO emissions from soil may be used as a surrogate indicator of the level of contamination in soils during remediation and post-remediation activities at contaminated soil sites, and thus could assist in determining when expensive soil pollution remediation activities may cease.

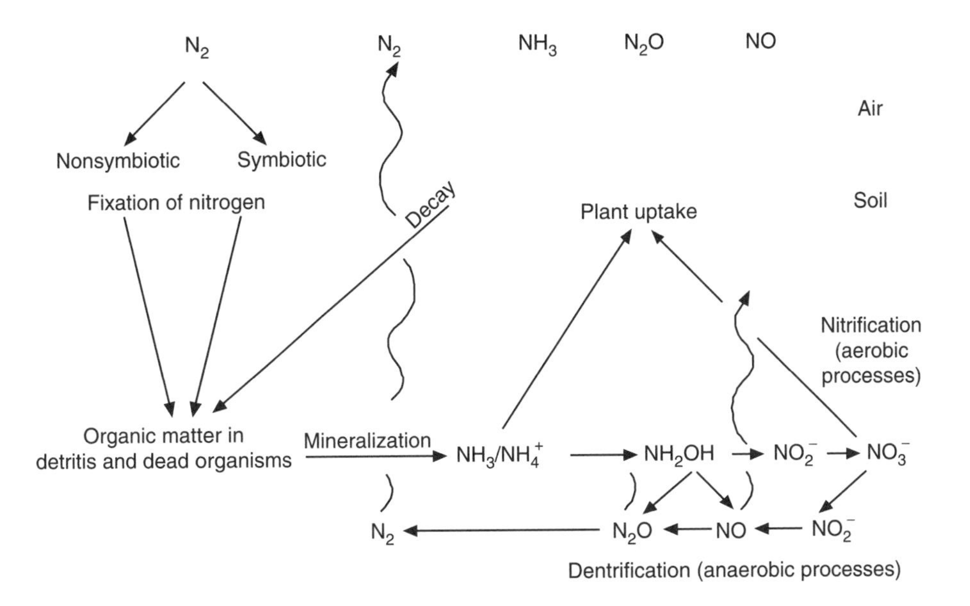

FIGURE 4-19. Biogeochemical nitrogen cycle.

Historically NO concentrations in ambient air have been determined using chemoluminescence analyzers that are inexpensive, durable, accurate, and precise. For example, these analyzers are used widely by the U.S. EPA to measure NO concentrations as precursors to ozone formation in cities and towns nationwide, contributing to decades of successful ambient air quality monitoring programs.

Chemoluminescence analyzers convert NO to electronically excited NO_2 (indicated as NO_2^*) when O_3 is supplied internally by the analyzer as summarized:

$$O_3 + NO \rightarrow NO_2^* + O_2 \qquad \text{(Reaction 1)}$$

These excited NO_2^* molecules emit light when they move to lower energy states as:

$$NO_2^* \rightarrow NO_2 + h\nu \; (590 < \lambda < 3000\,\text{nm}) \qquad \text{(Reaction 2)}$$

The intensity of the emitted light is proportional to the NO concentration and is detected and converted to a digital signal by a photomultiplier tube that is recorded.

Dynamic test chambers and systems (Figure 4-20) are available to measure the NO flux from soil. The mass balance for NO in the chamber is

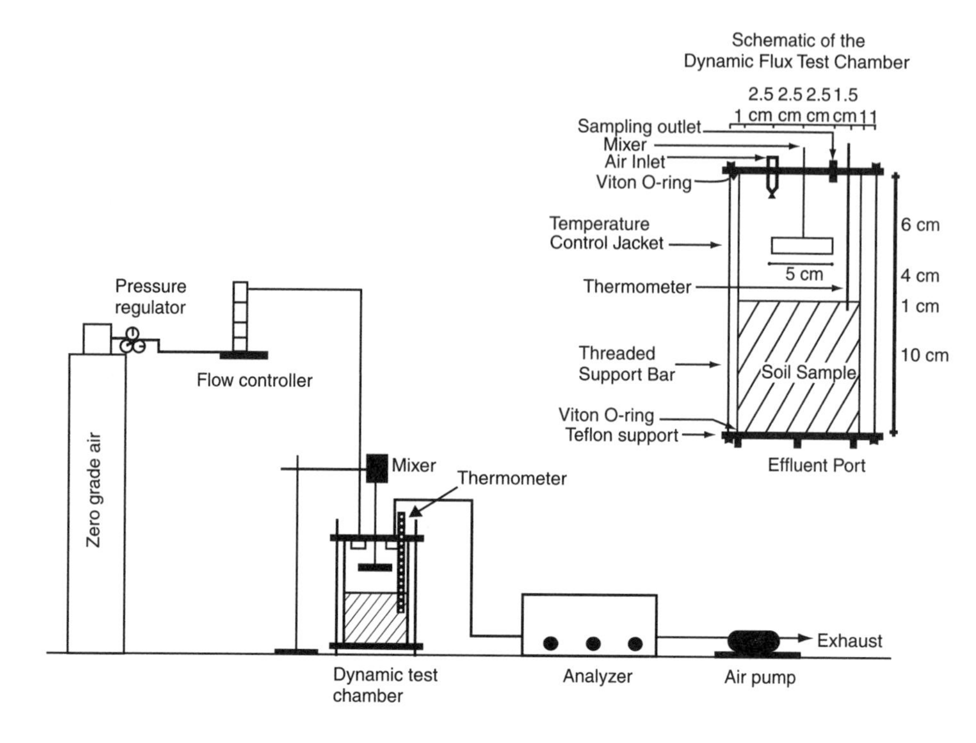

FIGURE 4-20. Test chamber and NO monitoring system.

summarized by:

$$\frac{dC}{dt} = \left(\frac{Q[C]_0}{V} + \frac{JA_1}{V}\right) - \left(\frac{LA_2}{V} + \frac{Q[C]_f}{V}\right) + R \qquad (4\text{-}1)$$

where,

A = surface area of the soil
V = volume of the chamber
Q = air flow rate through the chamber
J = emission from the soil flux per unit area
C = NO concentration in the chamber
$[C]_o$ = NO concentration at the chamber inlet
$[C]_f$ = NO concentration at the chamber outlet
L = loss of NO on the chamber wall assumed first order in [C]
R = chemical production/destruction rate for NO in the chamber

The NO emissions from soil to the headspace is calculated as J.

Using FISH to Analyze Soil Microbial Communities Exposed to Different Soil Contaminants and Different Levels of Contamination

The FISH method identifies microorganisms by using fluorescently labeled oligonucleotide probes homologous to target strains or groups of microorganisms and viewing by epifluorescent microscope in samples of soil studied in the laboratory and in the field. This technique was first applied to activated sludge cultures in 1994 but is continually undergoing modifications building on the understandings of procedures and oligonucleotide probes designed and applied to identify nitrifying bacteria in wastewater treatment systems. Methods of FISH application to soil samples are and will continue to evolve, as does every method of hazardous waste monitoring and measurement.[8]

Historically, the FISH techniques applied to the study of microbial communities in soil have not been as well developed as have the FISH techniques that are applied to the study of microbial communities in water or slurried sediment samples. The classification of active soil bacteria using FISH is a challenging research topic that appears to be developing almost exclusively outside the United States. The FISH techniques for identifying bacteria extracted from soils generally are particularly difficult to perform because of (1) the high background fluorescence signals from soil particles, (2) the exclusion of bacteria associated with soil particles, (3) the nonspecific attachment of the fluorescent probes to soil debris, (4) probing microorganisms that are entrapped in soil solids, and (5) determining the optimal stringency of hybridization.

Other obstacles to the application of the general FISH method to soils include difficulties in sequence retrieval, finding rRNA sequences of less common organisms, nonspecific staining, low signal intensity, and target organism accessibility. In addition, cells that are in the stationary phase often do not contain a sufficient cellular rRNA content to produce a detectable fluorescent image with FISH. These challenges can be overcome with the development of a variety of directed modifications to the general FISH methodologies, including altering experimental procedures for extraction and filtration of soil microbes, different selection and sequencing of oligonucleotide probes, and improving detection instrumentation, particularly the software to analyze the images obtained on a microscope with epifluorescent capability. Example FISH probes are presented in Table 4-1.

Connecting the Results of the Two Monitoring Techniques

For different soils from different contaminated sites with different levels of different contaminants, microbial activity and consequently NO production will be affected during remediation and post-remediation activities in the field. Consider, for example, a site where the soil is contaminated with NAPLs. At sampling locations at this site, observed NO emissions

TABLE 4-1
Examples of Oligonucleotide Probes

Probe	*Target Bacteria*	*Applicability in NO Studies*
EUB338	Eubacteria	All bacteria
ALF1B	α proteobacteria	Pseudomonas and Nitrobacter
BET42A	β proteobacteria	Nitrosomas
GAM42A	γ proteobacteria	Pseudomonas (for example P. putida)

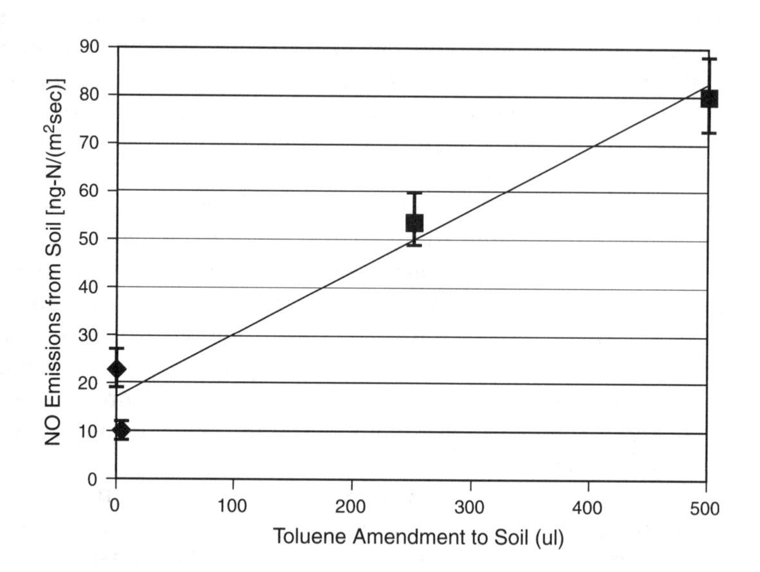

FIGURE 4-21. Nitric oxide emissions from uncontaminated soil and soil contaminated with toluene after one-day incubation. Error bars represent 1 standard deviation ($n = 3$).

measurements that are lower than representative background levels of NO emissions from the soil could indicate depressed levels of microbial activity caused by high levels of contamination in the soil that are toxic to the microorganisms in the soil. On the other hand, observed NO emission levels that are in the range of representative background levels of NO emissions from the soil could indicate normal levels of microbial activity caused by acceptably low or nonexistent levels of contamination (i.e., levels found in a successfully remediated soil or soil that was never contaminated).

Depending on the type of contaminant, the level of contamination, and the physical, chemical, and microbiologic characteristics of the soil, chemoluminescence NO emissions monitoring at different locations at the site could indicate the presence, absence, or level of contamination in the soil.

NO emissions from soil are seen as a direct indicator of microbiologic activity in the soil, which in turn can suggest the presence, absence, and/or concentration of different contaminants in soil. Consider Figure 4-21, which illustrates representative laboratory measurements of NO emissions from uncontaminated soil and soil that has been contaminated with toluene. Note that soil contaminated with toluene can produce 10 times more NO than the uncontaminated soil with a zero level of toluene. The additional production of NO is suggested to be the result of increased microbial activity in the contaminated soil.

CHAPTER 5

A Risk-Based Assessment to Support Remediating a Hazardous Waste Site

How Risk Information Is Used in Hazardous Waste Site Remediation

To illustrate the complexity and importance of risk assessment and management in hazardous waste engineering, let us consider an example. To address several risk-related considerations in site remediation, this example intentionally includes both organic and inorganic wastes. It is characteristic of both an operational and an abandoned facility, so that numerous state and federal laws apply, including the Resource Conservation and Recovery Act (RCRA), Superfund, the Clean Air Act, and the Clean Water Act. Also, we will choose a company that previously synthesized a regulated pesticide, so it would behoove the engineer to consider other legislative and regulatory provisions, especially the Federal Insecticide, Fungicide and Rodenticide Act and the Toxic Substances Control Act.

The Scenario

Your firm is in the process of bidding on the remediation of an abandoned hexachlorocyclohexane (HCH) pesticide manufacturing waste site (Figure 5-1). You are assigned the role of project engineer with the responsibility for preparing the bid, supervising the crew, managing the subcontractors, ensuring operational efficiency (including health and safety; see Appendix 4),[1] and meeting milestones and achieving remediation levels in all media (i.e., selecting and meeting measures of success for the project). This is a closed bid, so you are not certain which other engineering firms plan to compete for the job.

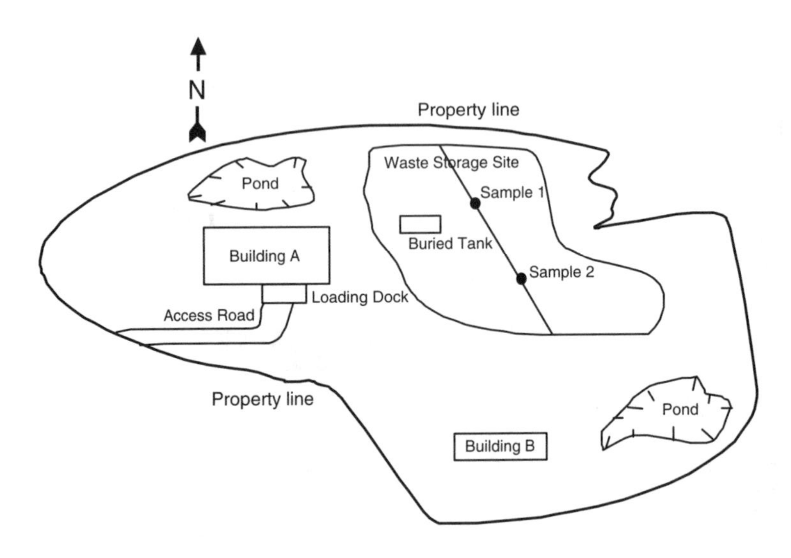

FIGURE 5-1. Site map of abandoned hexachlorocyclohexane (HCH) pesticide manufacturing waste site, showing monitoring transect and location of waste piles and storage tank.

The current owners of the site have provided you with a report prepared by a reputable environmental auditor. The report contains the following information:

- In the process of synthesizing pesticides over a 30-year period, the soil and ground water have become contaminated with organic and inorganic pollutants. The company's biggest profit-generating products during its manufacturing operations were the γ-isomer of HCH (known commercially as lindane) and the α-isomer of HCH. These pesticides were formerly used throughout the world to control insects in fruit, grains, vegetables, and forests, but recently their use has been highly restricted, limited to seed treatment[2] and medical prescriptions[3] for scabies and head lice. Interestingly, however, no lindane has been detected anywhere on the site.
- There is a 1,000-cubic-meter pile of pesticide residue (solid) that contains about 25% α-HCH, 35% γ-HCH, and the remainder unidentified "inert ingredients."[4]
- A 10-m^3 underground storage tank (Figure 5-2) has been identified on the site. The steel tank contains 3 m^3 of an oily residue floating on top of 5 m^3 of water, leaving 2 m^3 headspace. An analysis of the oily residue found total HCH to be 500 mg L^{-1}, total metals to be 5,000 mg L^{-1}, and total hydrocarbons (THC) to be 2,000 mg L^{-1}.

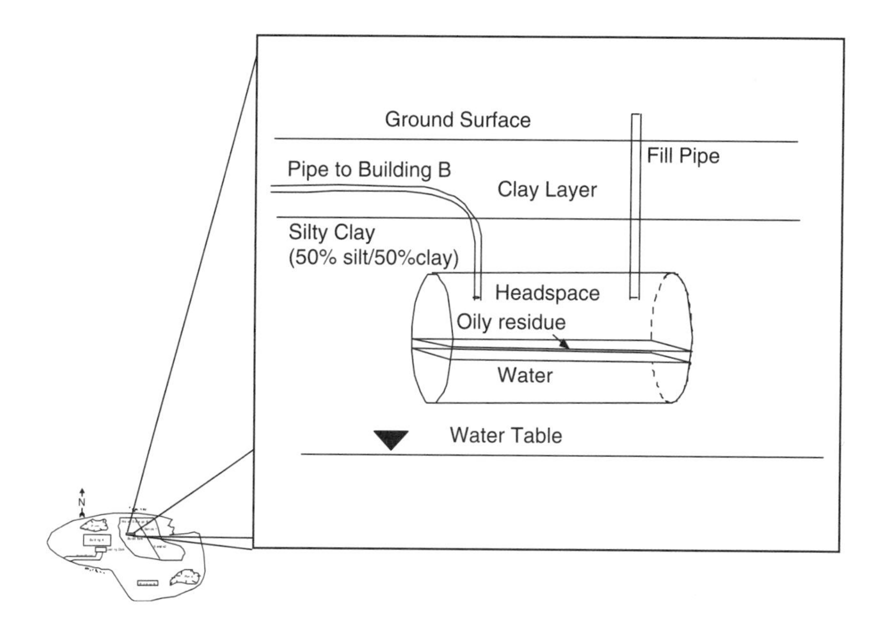

FIGURE 5-2. Buried tank profile.

The tank's water was found to contain 50 mg L^{-1} total HCH, 8,000 mg L^{-1} total metals, and 400 mg L^{-1} THC. There is evidence that other tanks and drums may be buried as well.[5]

- Trichloromethane was used throughout the operating life of the plant as a cleaning solvent. Drums were stored on site in the shed shown on the site map.
- In addition, for the first 15 years that the company operated, a small smelter/metal refinery was on site. The manufacturers processed nickel ore and sold it for use as a catalyst by other chemical companies. The smelter is gone, but soil and dust containing nickel subsulfide (Ni_2S) remain on the site. As evidence, the present owners have identified a 40 m^3 pile of slag. The environmental auditors analyzed the pile ("grab" sample) and found concentrations of 50 mg Ni_2S kg^{-1} soil.
- The auditor's report includes a preliminary investigation from a reputable laboratory requested by the owners (see Table 5-1). As shown on the site map, sampling locations 1 and 2 lie on a transect bisecting the waste site. You must evaluate the sufficiency of this data set. You may either assume it is representative of site conditions and contaminant characteristics or you may conduct additional sampling studies at your expense. If you decide that additional studies are needed, you must define their scope and data elements.

TABLE 5-1
Soil and Ground Water Concentrations of Listed Contaminants

Sampling Location	*Date*	*Substance**	*Soil [x] (mg kg^{-1})*	*Ground water [x] (mg L^{-1})*
1	1/15/98	alpha HCH**	0.025	0.040
	Winter	beta HCH**	ND	ND
		tetrachloromethane	52	80
		nickel subsulfide (Ni_2S)	100	ND
		mercury (total)	0.012	0.04
2	1/15/98	alpha HCH	0.005	0.005
	Winter	beta HCH	ND	ND
		tetrachloromethane	10	15
		Ni_2S	12	ND
		Hg (total)	0.005	0.015
1	7/15/98	alpha HCH	0.020	0.040
	Summer	beta HCH	ND	ND
		tetrachloromethane	29	12
		Ni_2S	10	ND
		Hg (total)	0.005	0.05
2	7/15/98	alpha HCH	0.005	0.005
	Summer	beta HCH	ND	ND
		tetrachloromethane	2	2
		Ni_2S	10	ND
		Hg (total)	0.003	0.004

*Other "listed" hazardous substances were not detected.
**Only isomers of HCH detected in piles, tanks, and ambient environment.

Assumptions

In planning for an actual remediation event, assumptions must be replaced with operational criteria and performance standards. Some of these measures of success will be defined by science and engineering, but they will also be influenced by legal and financial considerations (which should have been detailed in the remediation action plan). For example, a judge may have specified in a consent degree that more protective and conservative cleanup standards are necessary than the ones we will discuss in this example (e.g., more protective than a 10^{-6} cancer risk); however, because cleanup standards always highly depend on the conditions and site characteristics, we must make some assumptions to continue. Thus we will assume that your professional judgment and research have led to the following assumptions:

1. There are several ways of calculating exposure and risk. In this exercise, we will use the risk methods and exposure equations recommended in the *CRC Handbook of Toxicology*,[6] as expressed in the equations for each human exposure pathway.

2. Human exposure from ingesting contaminated water can be calculated as:

$$\text{LADD} = \frac{(\text{C}) \cdot (\text{CR}) \cdot (\text{ED}) \cdot (\text{AF})}{(\text{BW}) \cdot (\text{TL})} \tag{5-1}$$

where, LADD = lifetime average daily dose (mg kg^{-1} d^{-1}), C = concentration of the contaminant in the drinking water (mg L^{-1}), CR = rate of water consumption (L d^{-1}); ED = duration of exposure (d); AF = portion (fraction) of the ingested contaminant that is physiologically absorbed[7] (dimensionless); BW = body weight (kg); and TL = typical lifetime (d).

Drinking water is a potential source of human exposure to contaminants. Public water supplies and private wells can be polluted in many ways, including the movement of chemical compounds through the soil and reaching water supply aquifers and by runoff or discharge to reservoirs and other surface water supplies. The compounds that may be released from hazardous waste sites may also be transformed via chemical reactions with compounds that have been intentionally added for disinfection (e.g., chloramine compounds used in chlorination); and leaching of materials from distribution systems, including indoor plumbing (e.g., chelating lead from galvanized pipes).

The amount of water consumption varies by age (see Table 5-2). Currently, the U.S. EPA uses the quantity of 2 L d^{-1} for adults and 1 L d^{-1} for infants (babies weighing 10 kg body or less) as default rates of water consumption.[8] These rates include drinking water consumed in the form of juices and other beverages that are prepared using tap water (e.g., tea, cocoa, and coffee). The daily consumption of water may also vary with levels of physical activity and fluctuations in temperature and humidity.[9] It is also likely that some individuals who work in

TABLE 5-2
Summary of Tapwater Intake by Age

Age Group	*Intake (mL/day)*		*Intake (mL/kg-day)*	
	Mean	*10th–90th Percentiles*	*Mean*	*10th–90th Percentiles*
Infants (< 1 year)	302	0–649	43.5	0–100
Children (1–10 years)	736	286–1,294	35.5	12.5–64.4
Teens (11–19 years)	965	353–1,701	18.2	6.5–32.3
Adults (20–64 years)	1,366	559–2,268	19.9	8.0–33.7
Adults (65+ years)	1,459	751–2,287	21.8	10.9–34.7
All ages	1,193	423–2,092	22.6	8.2–39.8

Source: U.S. Environmental Protection Agency.

physically demanding jobs or who live and work in warm climates would have high levels of water intake.

3. Human exposure from skin contact with soil can be calculated as:

$$\text{LADD} = \frac{(\text{C}) \cdot (\text{SA}) \cdot (\text{BF}) \cdot (\text{FC}) \cdot (\text{SDF}) \cdot (\text{ED}) \cdot (10^{-6})}{(\text{BW}) \cdot (\text{TL})} \qquad (5\text{-}2)$$

where, LADD = lifetime average daily dose (mg $\text{kg}^{-1}/\text{d}^{-1}$), C = concentration of the contaminant in the soil (mg kg^{-1}), SA = skin surface area exposed (cm^{-2}); BF = bioavailability (percent of contaminant absorbed per day), FC = fraction of total soil from contaminated source (dimensionless), SDF = soil deposition, the mass of soil deposited per unit area of skin surface ($\text{mg/cm}^{-1}/\text{d}^{-1}$), ED = duration of exposure (d); AF = portion (fraction) of the ingested contaminant that is physiologically absorbed[10] (dimensionless); BW = body weight (kg); and TL = typical lifetime (d).

Although, in our example, we are limiting dermal exposure to that between soil and skin, dermal exposure can occur during a variety of activities in different environmental media and microenvironments.[11] These include:

- Water (e.g., personal hygiene, swimming).
- Sediment (e.g., waste cleanup, wading, fishing).
- Liquids (e.g., waste cleanup, use of commercial products).
- Vapors/fumes (e.g., waste cleanup, use of commercial products).
- Indoors (e.g., carpets, floors, other surfaces).
- Soil (e.g., remediation, farming, outdoor recreation, gardening, construction).

Dermal exposure is a function of the chemical concentration in contact with the skin, the potential dose, the extent of skin surface area that has been exposed, the duration of the exposure, the absorption of the chemical through the skin surface, the internal dose, and the amount of chemical that can be delivered to a target organ (i.e., biologically effective dose) (see Figure 5-3).[12] You and your remediation team will be working to remediate the hazardous wastes at the site 6 days a week and 8 hours per day, with 1 hour for lunch and 30 minutes for workers to get on and off the facility (EF = 10 h/24 h). Assume general population exposure frequency to be unity (EF = 24 h/24 h). These EFs need to be multiplied by the exposure duration factors in your LADD equations. (*Something to think about*: What do the two different EFs mean in terms of concentrations in the soil, water, and air? Are [x] values interdependent or independent of ED and EF?)

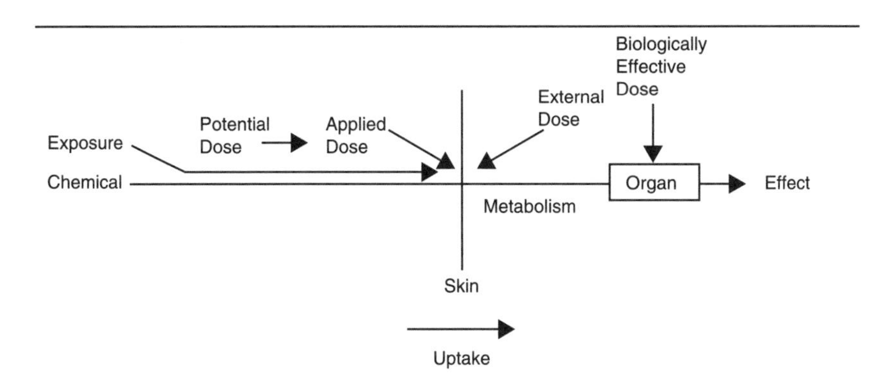

FIGURE 5-3. Exposure and dose schematic for dermal route. (*Source:* U.S. Environmental Protection Agency.)

4. Assume that residents and workers will drink well water (1 L d^{-1}) drawn from the same ground water supply (aquifer) as sampling locations 1 and 2. (How comfortable are you are with this assumption?)
5. Assume an average body weight (BW) to be 70 kg and average lifetime (TL) to be 70 years.
6. Assume that all risk comes from inhalation of contaminated air (particle and gaseous), ingestion of drinking water, and dermal contact with soil.
7. Your projected remediation period (3 years) is the exposure duration (ED).
8. Given the climate and winds in this part of the country, assume that all the air particles come from your soil. (No ambient or point sources of the listed chemicals are located near the site.)
9. Human exposure from inhaling contaminated aerosols is calculated as:

$$\text{LADD} = \frac{(\text{C}) \cdot (\text{PC}) \cdot (\text{IR}) \cdot (\text{RF}) \cdot (\text{EL}) \cdot (\text{AF}) \cdot (\text{ED}) \cdot (10^{-6})}{(\text{BW}) \cdot (\text{TL})} \qquad (5\text{-}3)$$

where, LADD = lifetime average daily dose (mg $kg^{-1}d^{-1}$), C = concentration of the contaminant on the aerosol/particle (mg kg^{-1}), PC = particle concentration in air (gm m^{-3}); IR = inhalation rate ($m^{-3}\,h^{-1}$), RF = respirable fraction of total particulates (dimensionless), EL = exposure length (h d^{-1}), ED = duration of exposure (d); AF = portion (fraction) of the inhaled contaminant that is physiologically absorbed[13] (dimensionless); 10^{-6} is a conversion factor (kg to mg); BW = body weight (kg); and TL = typical lifetime (d).

The particle concentration in the air is 100 mg m^{-3}. Assume 90% of this is fine particulate matter (respirable fraction = 90%).

10. Human exposure from inhaling contaminated gases is calculated as:

$$\text{LADD} = \frac{(\text{C}) \cdot (\text{IR}) \cdot (\text{EL}) \cdot (\text{AF}) \cdot (\text{ED})}{(\text{BW}) \cdot (\text{TL})} \tag{5-4}$$

using the same factors as those in Equation 5-3.

For the gas phase, aliphatic compounds with few carbons in the chain are lighter, and hence, have high vapor pressure (v.p.). For example, v.p. of tetrachloromethane (carbon tetrachloride) is 15 kP at 25°C. Many chlorinated organic pesticides have low vapor pressure. For example, the γ-isomer of HCH v.p. $= 10^{-3}$kP). We will make a gross assumption that volatile organic compounds are 50% particle/ 50% vapor phase. Assume that semivolatile pesticides and metals are 99% particle/1% vapor phase (disregarding temperature, absolute pressure, and partial pressure effects). In actual remediations, we will need to apply phase partitioning to each compound. This is known as *fugacity*, or the likelihood that the compound will be transported to the atmosphere. This is determined either by calculating Henry's law constants from the compound's solubility and vapor pressure or by modeling (e.g., by computation chemistry).

11. Although your local health department generally uses a moderate value of 2.5 $m^3 h^{-1}$ inhalation rates (IR) to protect residential areas, you decide to use heavy inhalation rates (IR $= 4.8\ m^3 h^{-1}$) to be more protective of your workers. This adheres to the so-called precautionary principle to provide an added measure of worker protection,[14] but it would be highly unlikely that a worker would maintain these ventilation rates over an entire workday.

12. A lipophilic compound is one that is fat soluble, and usually not water soluble, and can be persistent and can bioaccumulate in the human body. Lipophilicity (as represented by K_{ow}) can be used to estimate the bioaccumulation factor (BF). Because organic compounds are usually lipophilic, and some like PCBs, dioxins, and chlorinated pesticides can be very lipophilic, assume they have BF $= 100\%$ and the absorption factor (AF) is also 100%. For a more exact estimate of bioaccumulation, follow the guidance in Chapter 3 on physicochemical characteristics of each chemical and apply these characteristics to toxicologic models and frameworks.[15] (For metallic and many other inorganic compounds, which can be more water soluble (hydrophilic), you would use a smaller BR and AF because much of the compound is more readily eliminated from the human body.)

13. We will strive for redundancy to ensure a high level of worker protection, but as a worst-case scenario, we will also assume that some breakage and misuse will occur, exposing 150 cm^2 of the skin surface area (SA $= 150\ cm^2$) to possible contamination.

14. Deposition depends on the contaminant and atmospheric conditions. A substance with a high deposition velocity is more likely to settle onto surfaces than one with a low deposition velocity. In this example, we will also assume that the soil deposition factor (SDF) is 5 mg $cm^{-2}d^{-1}$.
15. We will assume that all soil comes from the contaminated source (FC = 100%). SDF in exposure is actually a "flux" term (mass per surface area per time), but note that you will often see "deposition velocities" (mass per length per time) in environmental engineering literature. Be certain which factor is reported so that your units are correct.
16. We will use "potential dose" as the metric for "exposure." The units for potential dose, such as those for lifetime average daily dose, are mass of contaminant per mass of receptor per time. Some other models use mass contaminant per time (such as mg d^{-1}) for exposure for a normalized population. Be prepared to see these and other representations of exposure in various studies, journal articles, and texts.

The Charge to the Engineer

The following questions must be answered before remediation of the site can begin. Information and data gaps regarding science, engineering, and technology must be identified upfront and must be incorporated into the remediation plan:

1. What toxic endpoints (cancer or noncancer?) should be considered for the HCHs, nickel subsulfide, tetrachloromethane, and mercury? Sketch the shape and label the dose-response curves for these substances. Given our assumptions, calculate the cancer risk from HCHs from four exposure pathways: air particles, air vapors, drinking water, and dermal (soil). Calculate the total HCH risk to you and your remediation crew at the site for the three-year cleanup period. List and discuss the sensitivity of your total risk calculation in terms of what you think are the three most important inputs of these four risk calculations.
2. What is the best means for remediating the pesticide residue? Is thermal destruction appropriate for any of the wastes at the site? Are either of the wastes good candidates for clearinghouses or exchanges? Discuss any trends in the data between summer and winter and any differences between the monitoring results in Sites 1 and 2. Suggest why trends and differences may occur from a scientific and engineering perspective. Explain, where possible, the nondetects in the data.
3. Based on the site specifications, waste characteristics, and risk assessment, should the consulting engineers accept a contract to clean this site? What would be the advantages and disadvantages of taking

this job? Note any information that would be preferable to have before making this decision, as compared to the information that is essential to your decision.

4. Give two scientifically sound reasons for lindane not being detected at the site.

Answers and Explanations

To answer these questions, we will need to consider several hazardous waste risk topics covered in this text, including the following:

- Movement and change among various environmental compartments.
- Source characterization and processes that lead to the formation, degradation, and transformation of hazardous substances.
- Hazard identification and exposure assessment (population and occupational).
- Pathways and routes of exposure.
- Acute, subchronic, chronic, and intermittent toxicity (various outcomes).

The example provides a situation where an engineer must calculate the baseline and target contaminant release concentrations in the process of remediating an abandoned waste site.

Question 1

What toxic endpoints (cancer or noncancer?) exist for HCHs, nickel subsulfide, tetrachloromethane, and mercury? *HCHs, nickel subsulfide, and tetrachloromethane are carcinogens, but mercury is neurotoxic (but carcinogenicity has not been proven).*

Sketch the shape and label the dose-response curves for these substances.

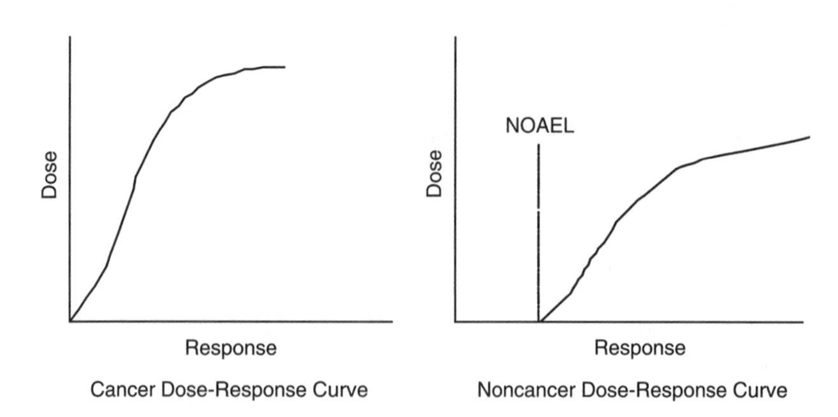

The principal differences between the dose-response curves of the carcinogens and the noncarcinogens are (1) there is a threshold for noncancer where the substance elicits no response (NOAEL), but none for cancer; and (2) most of the cancer curve is linear, although the slopes below the observable range (range of inference) will vary by the model selected (multistage, one-hit, Weibull, or log-probit). Similarities include (1) the need for confidence intervals, (2) increased dose results in increased response (except the deficiency for essential elements), (3) the need for extrapolation from animal and epidemiologic studies, (4) an observable and inferred range, and (5) the need to extrapolate using models in the range of inference.

The potency of a carcinogen is reflected by the slope of its dose-response curve. The larger the slope factor, the more potent the carcinogen. We shall apply the U.S. EPA slope factors given in Table 5-3.

Neither of the metals are known to be essential to humans, so both Ni and Hg dose-response curves would look like the noncancer curve, but for essential metals like Se, Fe, CrIII, and Zn, the curve would be U-shaped with an optimal concentration range between deficiency at the low end and toxicity at the high end:

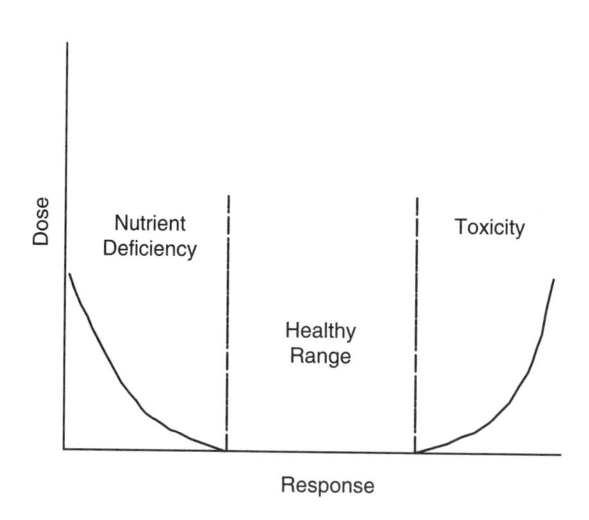

Essential Element (e.g., Cr, Se, Zn) Dose-Response Curve

> Although we are only talking about human health endpoints, this curve would also apply to Ni, which is a micronutrient for some bacteria (such as some anaerobes in sludge digestion).

TABLE 5-3
Cancer Potency Slope Factors and Categories for Various Chemicals (2 significant figures)

Chemical	*Classification*	*Carcinogen Slope Factor* $(mg\ kg^{-1}d^{-1})^{-1}$
Acrylonitrile	B1	0.24
Aldrin	B2	17
Benzidine	A	230
Bis(2-choroethyl)ether	B2	11
1,3-Butadiene	B2	1.8
Cadmium (inhalation)	B1	6.1
Carbon Tetrachloride	B2	0.13
Chromium – Hexavalent (Inhalation)	A	41
1,2-Dichoroethane	B2	0.091
1,1-Dichoroethylene	C	0.60
Dichloromethane	B2	0.0075
Deithyl nitrosamine	B2	150
Dimethly nitrosamine	B2	51
1,2-Diphenylhydrazine	B2	0.80
Epichlorohydrin	B2	0.0099
Heptachlor	B2	4.5
Heptachlor expoxide	B2	9.1
Hexachlorobutadiene	C	0.078
α-Hexachlorocyclohexane	B2	6.3
β-Hexachlorocyclohexane	C	1.8
Hexachlorodibenzo-para-dioxin (HxCDD)	B2	62000
Hexachloroethane	C	0.014
Nickel refinery dust (inhalation)	A	0.84
Nickel subsulfide (inhalation)	A	1.7
N-Nitrosodi-N-propylamine	B2	7.0
N-Nitrosoodiethanolamine	B2	2.8
N-Nitrosoodiphenylamine	B2	0.0049
N-Nitrosoomethyethylamine	B2	21.8
N-Nitrosopyrrolidine	B2	2.1
1,1,2,2-Tetrachorethane	C	0.20
Trichloroethylene	B2	0.11
2,4,6-Trichlorophenol	B2	0.020

Source: U.S. Environmental Protection Agency, Risk Assessment Guidelines, 1988.

Given the listed assumptions, we can begin to calculate the cancer risk from the somewhat lipophilic HCH from four exposure pathways: air particles, air vapors, drinking water, and dermal (soil). To begin, let us calculate the *total HCH risk* to the remediation crew at the site for the three-year cleanup period. *You need to use both the alpha and beta isomers to calculate total HCHs, although the alpha isomer was the only one detected.*

Warning: Remember that this may not be true! The lab may not have looked for all isomers (so you should ask). Although gamma and alpha isomers were produced intentionally, the other isomers, especially beta, were not. So, it is likely other isomers would be there if you knew how to look! For our purposes, let us calculate the sum of alpha and beta to represent the HCHs (known as a "surrogate" or structure activity relationship approach).

HCH Cancer Risk Calculations

We shall assume that all risk comes from the three major pathways (inhalation, ingestion, and dermal). We find the cancer potency in the slope factor table (Table 5-3). The HCHs are somewhat lipophilic, so we shall assume AF = 1 and BF = 1 (however, the absorption and bioaccumulation factors may be better defined by modeling using the solubility, the K_{OW} and K_{OC}). The ED = 3 yr. and 6 days per week = 936 d.

We shall use equation 10 of the *CRC Handbook of Toxicology* for HCH sorbed to airborne particles. Because most persistent, chlorinated pesticides, including HCH, are semivolatile, we will assume that 99% stays in solid phase, sorbed to particles. In fact, HCH can be rather volatile, but because the pile has been in place for a while, the high estimate of HCH in the solid phase may not be without justification. The more volatile HCHs may have already escaped, so that the balance of HCHs remaining are sorbed tightly to particles.

We will use the highest concentration in soil. One could break it into seasonal exposures because the winter concentrations in soil are much lower than those in summer (possibly because of increased vapor pressure at higher temperatures). Also keep in mind that when you handle the pile of residue, these concentrations are 60% HCH (250,000 mgkg^{-1} alpha and 350,000 mgkg^{-1} beta), so workers near these piles can be exposed at rates seven orders of magnitude higher than the contaminated soil!

$$\alpha + \beta \text{ HCH (Air)} = 99\% \text{of HCH found on particles (versus vapor form).}$$

$$\begin{aligned}\text{LADD alpha (air-particulate)} &= (0.99) \times \text{Equation 5-3} \\ &= (0.99) \times [(0.025 \text{ mg kg}^{-1}) \times (100 \text{ mg m}^{-3}) \\ &\quad \times (0.9) \times (4.8 \text{ m}^3\text{h}^{-1}) \times (10 \text{ h d}^{-1}) \times (1) \\ &\quad \times (936 \text{ d}) \times (10^{-6})]/[(70 \text{ kg}) \times (25,550 \text{ d})] \\ &= 6 \times 10^{-8} \text{ mg kg}^{-1} \text{ d}^{-1}\end{aligned}$$

The β-isomer was not detected in the soil. For this calculation, we will assume it is not detected because of low concentrations, but one must be sure that it is not an analytic error (the chromatographer looked for the peak at the right retention time coming off the column), or that it may have been lost in the extraction procedure (stayed sorbed to the soil particle). This is possible because the beta isomer is a much smaller molecule than the other HCH isomers and is tightly held in lipids. Remember, "not detected" does not mean "zero."

$$
\begin{aligned}
&\text{Thus, risk (air particles) from HCH} \\
&\quad = \text{LADD} \times \text{Slope Factor} \\
&\quad = (6 \times 10^{-8}\ \text{mg kg}^{-1}\ \text{d}^{-1}) \times (6.3\ \text{mg kg}^{-1}\ \text{d}^{-1})^{-1} \\
&\quad = \underline{4 \times 10^{-7}}
\end{aligned}
$$

Use Equation 5-4 for vapor phase airborne HCH.

HCH (air-vapor) = 1% from vapor phase. However, we also need to know from where the vapor is being released. Our decision to assume that it is released from surface soil, from the suspended particles, or both will affect the exposure calculation. Because we have a value of 100 mg m^{-3} concentration of the contaminant on the particles, it is realistic to assume that most of the volatilization will come from these particles, but remember there is also some volatilization from the soil surface (we will ignore this in these calculations).

We must have the units of mg m^{-3} for C in Equation 5-4, so let us calculate a phase distribution from the particles. Soil particle concentration is in weight per weight (mg kg^{-1}); PC is 100 mg m^{-3}; so if we multiply by the inverse units of C (kg mg^{-1}), we will have the units of mg m^{-3} that we need in Equation 5-4.

$$
\begin{aligned}
\text{LADD (gas)} &= (0.01) \times [(0.025\ \text{mg kg}^{-1}) \times \text{Phase Distribution} \\
&\quad \times (4.8\ \text{m}^3\text{h}^{-1}) \times (10\ \text{h d}^{-1}) \times (1) \times (936\ \text{d})]/ \\
&\quad [(70\ \text{kg}) \times (25{,}550\ \text{d})] \\
&= (0.01) \times [(0.025\ \text{mg kg}^{-1}) \times (100\ \text{mg m}^{-3}) \\
&\quad \times (10^{-6}\ \text{kg mg}^{-1}) \times (4.8\ \text{m}^3\ \text{h}^{-1}) \times (10\ \text{h d}^{-1}) \\
&\quad \times (1) \times (936\ \text{d})]/[(70\ \text{kg}) \times (25{,}550\ \text{d})] \\
&= 6 \times 10^{-10}\ \text{mg kg}^{-1}\ \text{d}^{-1}
\end{aligned}
$$

$$
\begin{aligned}
\text{Risk (air vapor) from HCH} &= \text{LADD} \times \text{Slope Factor} \\
&= (6 \times 10^{-10}\ \text{mg kg}^{-1}) \times (6.3\ \text{mg kg}^{-1}\ \text{d}^{-1})^{-1} \\
&= 4 \times 10^{-9}
\end{aligned}
$$

$$
\text{Total HCH air risk (particles + vapors)} = 4 \times 10^{-7} + 4 \times 10^{-9} = \underline{4 \times 10^{-7}}
$$

HCH (drinking water) uses equation 5-6. Again use the higher winter concentrations (the lower concentrations in summer may again be caused by increase volatilization, considering Henry's law, in the warmer seasons; however, this may also indicate that the sampling is suspect, because groundwater concentrations would not be expected to show such intra-annual variation. These show an order of magnitude difference between summer and winter. Your problem is that you do not have another year for comparisons and to determine temporal trends. You really need at least three years of data for a trend. Again, beta was not detected, but the same warnings for soil apply here.

We do not need an exposure frequency (EF) for this calculation because it is based on daily water intake, so the 10 h d^{-1} do not matter:

$$\begin{aligned}\text{LADD (drink)} &= [(0.04\ \text{mg L}^{-1}) \times (1\ \text{L d}^{-1}) \times (936\ \text{d}) \times (1)]/\\ &\quad [(70\ \text{kg}) \times (25,550\ \text{d})]\\ &= 2 \times 10^{-5}\ \text{mg kg}^{-1}\text{d}^{-1}\\ \text{Risk (drink) from HCH} &= (2 \times 10^{-5}\ \text{mg kg}^{-1}\ \text{d}^{-1}) \times (6.3\ \text{mg kg}^{-1}\ \text{d}^{-1})^{-1}\\ &= 1 \times 10^{-4}\end{aligned}$$

This is the highest risk so far! We need to go two orders of magnitude just to be at 1 per million population risk.

HCH (dermal-soil) uses Equation 5-2. We will use the same assumptions regarding which data point to use as we did in the airborne particles risk calculations.

$$\begin{aligned}\text{LADD (soil)} &= [(0.025\ \text{mg kg}^{-1}) \times (150\ \text{cm}^2) \times (1) \times (1)\\ &\quad \times (5\ \text{mg cm}^{-2}\ \text{d}^{-1}) \times (10\ \text{h}/24\ \text{h}) \times (936\ \text{d})\\ &\quad \times (10^{-6})]/[(70\ \text{kg}) \times (25,550\ \text{d})]\\ &= 4 \times 10^{-9}\ \text{mg kg}^{-1}\ \text{d}^{-1}\\ \text{Risk (dermal-soil) from HCH} &= (4 \times 10^{-9}\ \text{mg kg}^{-1}\ \text{d}^{-1})\\ &\quad \times (6.3\ \text{mg kg}^{-1}\text{d}^{-1})^{-1}\\ &= 3 \times 10^{-8}\end{aligned}$$

Total Risk from HCH exposure during project

$$\begin{aligned}&= (\textit{air} + \textit{drinking water} + \textit{soil}) = 4 \times 10^{-7} + 1 \times 10^{-4} + 3 \times 10^{-8}\\ &= 1 \times 10^{-4}\end{aligned}$$

Because the cumulative risks are two orders of magnitude above $\mathbf{10^{-6}}$, protection is needed for the workers; however, because these risks are mainly caused by drinking water, if you were to prohibit and enforce the use of bottled water, the risks would be below the threshold, and mainly caused by airborne particles. To reduce this risk, you would require at least masks, and probably as self-contained air systems and "moon suits."

Risks from Piles and Tank Remediation
In addition, remember that your workers will be removing the higher concentrations of contaminants in the piles and the tank, as well as newly discovered tanks and barrels on the site. So let's do a high-exposure scenario. This also introduces the possibility of exposing your crew to two carcinogenic isomers of HCH.

To be realistic, this will be a short-term exposure because the 1,000 m^3 pile will be removed more rapidly than the whole three-year project period. Let us assume that we are able to remove five truckloads (5 ton/8 yd) per day. For safety and to ensure that we minimize fugitive dust during transport, the trucks are not filled completely, are watered down, and are sealed with tops. This allows us to ship only 6 m^3 per load. Thus we can move 30 m^3 per day, so our new ED is 1,000/30 = 33 days for this high potential exposure.

LADD alpha (air-particulate) in pile

$$= (0.99) \times [(2.5 \times 10^5 \text{ mg kg}^{-1}) \times (100 \text{ mg m}^{-3}) \times (0.9) \times (4.8 \text{ m}^3 \text{ h}^{-1}) \times (10 \text{ h d}^{-1}) \times (1) \times (33 \text{ d}) \times (10^{-6})]/[(70 \text{ kg}) \times (25{,}550 \text{ d})]$$

$$= 2 \times 10^{-2} \text{ mg kg}^{-1} \text{ d}^{-1}; \text{ and}$$

LADD beta (air-particulate) in pile

$$= (0.99) \times [(3.5 \times 10^5 \text{ mg kg}^{-1}) \times (100 \text{ mg m}^{-3}) \times (0.9) \times (4.8 \text{ m}^3 \text{ h}^{-1}) \times (10 \text{ h d}^{-1}) \times (1) \times (33 \text{ d}) \times (10^{-6})]/[(70 \text{ kg}) \times (25{,}550 \text{ d})]$$

$$= 3 \times 10^{-2} \text{ mg kg}^{-1} \text{ d}^{-1}$$

Thus the airborne particle **risk** of total HCH is the sum of the two isomers:

$[(2 \times 10^{-2}) \times (6.3)] + [(3 \times 10^{-2}) \times (1.8)] =$ **0.2 or 200,000 per million risk!**

HCH (air-vapor) in pile = 1% from vapor phase.

LADD (gas) alpha

$$= (0.01) \times [(2.5 \times 10^5 \text{mg kg}^{-1}) \times (100 \text{ mg m}^{-3}) \times (10^{-6} \text{ kg mg}^{-1}) \times (4.8 \text{ m}^3 \text{ h}^{-1}) \times (10 \text{ h d}^{-1}) \times (1) \times (33 \text{ d})]/[(70 \text{ kg}) \times (25{,}550 \text{ d})]$$

$$= 2 \times 10^{-4} \text{ mg kg}^{-1}\text{d}^{-1}; \text{ and}$$

LADD (gas) beta

$$= (0.01) \times [(3.5 \times 10^5 \text{ mg kg}^{-1}) \times (100 \text{ mg m}^{-3}) \times (10^{-6} \text{ kg mg}^{-1}) \times (4.8 \text{ m}^3 \text{ h}^{-1}) \times (10 \text{ h d}^{-1}) \times (1) \times (33 \text{ d})]/[(70 \text{ kg}) \times (25{,}550 \text{ d})]$$

$$= 3 \times 10^{-4} \text{ mg kg}^{-1} \text{ d}^{-1}$$

$$\textit{Total HCH vapor risk} = [(2 \times 10^{-4}) \times (6.3)] + [(3 \times 10^{-4}) \times (1.8)] = 2 \times 10^{-3}$$

Another very high risk! So special protective measures to prevent inhalation risk are needed. *These values tell us that the risks are very high from both phase releases, and that protection for workers must be high. We also can reduce this number by requiring a maximum ED for all workers (no one can do this job more than a total of 5 days) and reducing the volatilization by wetting, and so on; however, this process is obviously going to need self-contained air systems. Finally, this is one area where estimated calculations are insufficient. Even though the pile's concentration is high, it does not mean the air will contain like concentrations, so actual measurements are needed in and around the pile, under working conditions (e.g., digging, loading).*

The drinking water risk is not affected (since it is the same source). The dermal risk will also be increased:

LADD (soil) alpha

$$= [(2.5 \times 10^5 \text{ mg kg}^{-1}) \times (150 \text{ cm}^2) \times (1) \times (1) \times (5 \text{ mg cm}^{-2} \text{ d}^{-1}) \times (10 \text{ h}/24 \text{ h}) \times (33 \text{ d}) \times (10^{-6})]/[(70 \text{ kg}) \times (25{,}550 \text{ d})]$$

$$= 3 \times 10^{-2} \text{ mg kg}^{-1} \text{ d}^{-1}$$

LADD (soil) beta

$$= [(3.5 \times 10^5 \text{ mg kg}^{-1}) \times (150 \text{ cm}^2) \times (1) \times (1) \times (5 \text{ mg cm}^{-2} \text{ d}^{-1}) \times (10 \text{ h}/24 \text{ h}) \times (33 \text{ d}) \times (10^{-6})]/[(70\text{kg}) \times (25{,}550 \text{ d})]$$

$$= 5 \times 10^{-2} \text{ mg kg}^{-1} \text{ d}^{-1}$$

$$\textit{Total HCH dermal risk} = [(3 \times 10^{-2}) \times (6.3)] + [(5 \times 10^{-2}) \times (1.8)] = \mathbf{0.3}$$

Our highest risk! *Here, one must reconsider that the SDF at 150 is far too high, so more durable suits and a closer inspection program is needed to bring this closer to zero.* ***Remember, if any term in the numerator is 0, the risk is 0. This 33-day period can make or break you in terms of liability, profit, and successful remediation.***

The calculations for the tank are not possible from the data provided because the actual species of compounds is not given; however, the information is useful to you because you know that HCHs have similar (1.8 versus 6.3) cancer slope factors, that the tank provides evidence that metals and organics coexisted, and that liquid wastes were present (and may still be present) on site. The protective measures used in the piles should be used in tank removal, plus you must ensure that when you extract the tank you check for leaks to see if and where that leachate migrated from the tank. You probably should analyze the contents in the two layers for the four substances you are remediating.

WE CAN STOP OUR CALCULATIONS HERE BECAUSE YOU WERE ONLY ASKED TO CALCULATE HCH EXPOSURE TO WORKERS DURING THE PROJECT.

Let us now list and discuss the sensitivity of your total risk calculation in terms of what you think are the three most important inputs of these four risk calculations. *Note that these are all linear equations, so mathematically, all factors have equal weight; however, scientifically, they can vary considerably. For example, you can protect your workers with self-contained air devices (lowers the IR), full-body protection (lowers SA and SDF), and bottled drinking water (exponentially reduces ingestion pathway exposure). For the population as a whole, however, you cannot assume this level of protection, but you could rationalize using lower IR values and reduce the ED and EF, if access were restricted.*

You can also work at this mathematically and toxicologically. What if new research decreased the SF? This would greatly reduce the calculated risk values. You could compare the slopes from different chemical compounds to see if your remediation goes to something less than carbon dioxide and water. What if you could knock off a few chlorines and open the ring structures to get a new compound like hexachloroethane? You have reduced the risk by eight orders of magnitude.

For the HCH that is bound to soil particles, you would have to remove it first to make such changes. The groundwater, however, might be a good candidate for pump-and-treat technologies that do change the chemical structure of the HCH to less toxic compounds, but you would still have to do something with all carcinogens, even the less toxic ones. The higher concentrations in the piles can be an advantage because at least the alpha is still used as a pesticide. One may be able to give this to a processor through an exchange because the pile has been "assayed," although the single grab sample is not sufficient to conclude that the pile is a commercially viable source of the pesticide. Plus, any buyer would have to contend with all the impurities.

Question 2

Is thermal destruction an appropriate waste technology for the oily residue? *We can consider combusting these wastes in the HCH pile. The inerts are usually surfactants and clay-like substances. The surfactants allow the lipophilic chlorinated compounds to be suspended in water, and the clay makes handling easier and reduces potential exposures to the applicators. If these are the only inerts, and if the assay is correct, the pile may be a good candidate for thermal breakdown.*

The tank wastes are not likely to be good candidates for thermal destruction. The metals present a problem. Combustion can give us strange compounds, and thermal processes will not remove the metals (they will end up in the ash or as metal oxides and other gases and particles leaving the stack). Some type of mechanical separation would be needed. Because this is a dilute liquid, some of the metals will be cations in the water and others will be sorbed to negatively charged particles. So alum will remove some, but not all metals. An additional step would be needed before dealing with the sludge. Then, the supernatant and sludge should be analyzed to see if all of the metals have gone to the sludge, which will be much like the oily residue discussed previously in terms of its hazardous properties. So, at least a two-step process is needed: some kind of separation (physical/chemical), followed by a process to break down the organics. You do have some BTUs in the organic (especially the THC) content, but you should consider thermal processes only after the metals are removed. Generally, thermal destruction is probably not a good technology for the tank waste.

The nickel subsulfide pile is not a candidate because of the high metal content.

Overall, the mercury is an additional complication because it will change forms and volatilize, so that regulators will want to know what the emissions will be.

Are any of the wastes good candidates for clearinghouses or exchanges? Give the pros and cons. *The organic wastes in the tank and the contaminated soil are not good candidates for at least two reasons. Each contains metals, and the organics are halogenated. The high HCH concentrations are problematic enough, but because tetrachloromethane has been found in high concentrations at the two monitors, one should suspect that the THC content in the water and residue are also heavily chlorinated.*

> The most likely candidate for exchange is the nickel subsulfide pile. It has been somewhat assayed and shown to have a certain percentage of Ni. The reduced form may not be in demand, but are there processors who can change it to usable forms? For example, are the Ni catalysts still being used? Are there new uses for Ni? The exchanges and clearinghouses can help answer these questions.

Discuss any trends in the data between summer and winter and any differences between the monitoring results in Sites 1 and 2. Suggest why trends and differences may occur from a scientific and engineering perspective. *The first seasonal trend appears to be higher tetrachloromethane concentrations in winter. This finding is plausible because volatility is temperature dependent, and the concentrations in soil and water may be lower in the well and in the soil probes as the volatiles are driven off. Because HCHs are less volatile and more persistent, their seasonal variability is slight. An important data gap, however, is whether these changes are, in fact, seasonal or whether they are permanent decreases. The only way to know this is to have at least one more year's data, preferably two years. If the volatiles increase again, you have an active and continuous source of tetrachloromethane. If they stay low, you have a "spike" or "plug" flow, where your source is no longer contributing mass.*

> The change in mercury (Hg) is relatively small, possibly because of sampling and analytical variability, and one may suspect that the Hg has speciated (changed chemical forms) to become almost nonvolatile.

The Ni values are interesting. The higher Ni concentrations were found in sampling site 1. This is farther from the Ni pile than site 2. Is it possible that there was a road where the Ni was moved to the plant, from which the Ni was dropped? An engineer should find out where the original smelter was in relation to the pile and the plant. There is a large seasonal variability in [Ni_2S]. This may mean that the speciation is still active (it may be changing valence states or may be more mobile in the environment, so that the snapshot taken at the two times are different). One would expect that, at least near the surface, the reduced form of Ni (subsulfide) would be oxidized to oxides and sulfates of Ni (i.e., the valence of Ni increases over time). This may be because forms of the metal are more water soluble and reactive than the organics. We can analyze for total Ni (like the Hg) to see if it is changing as well. If the total Ni is not changing significantly, it would indicate that the Ni is simply changing forms and not moving out from the site. The other observation one could make from an engineering perspective is that the HCHs and metals are here to stay, because of their persistence, so remediation is needed. Another thing that an engineer would likely want to do would be to monitor the water quality of the pond. This may give some clues about whether the wastes are moving from the tank or whether the whole aquifer is contaminated.

The nondetects in the data set are troublesome. We must ask for more documentation. Are these values below detection limits? Were they even analyzed for? If they are below detection limits, what are these limits?

Include a calibration curve and list of standards. This information, combined with the lack of any findings of lindane (gamma-HCH), even though it was produced there for years, may indicate that there were some detection or separation problems.

Another explanation of the lack of lindane is an ongoing debate within the scientific community on how HCH isomerizes and speciates, discussed extensively in Chapter 3. Some say that the ratios of the isomers change as a result of photochemistry and microbial processes. Others totally dismiss the role of photochemistry and consider physical processes, like rain washout, to be most important. There may have been sufficient time, depending on when the manufacturer stopped making lindane, that it has all changed to the other isomers or has been removed and microbially degraded.

Question 3

Based on the site specifications, waste characteristics, and risk assessment, would you recommend that your company accept a contract to clean this site? What would be the advantages and disadvantages of taking this job, without a significant investment of time and resources? *Given the assumptions and demands of the health department and worker risk, you probably would have a hard time being successful in this job. In addition to your calculations for worker risk, which should probably scare you, you need a good idea of the ultimate cleanup requirements before you accept this job.*

Cleanup Standards

> Note: This exercise is hypothetical and assumes that cleanup success is based entirely on risk reduction. In actual site cleanup, the standards should be defined in the remedial action plan or other hazardous waste cleanup plans. They should have been based on risk reduction, but other factors—including legal requirements, government regulations, and feasibility—will also determine the target cleanup levels.

You weren't specifically asked, but you would ordinarily have to calculate target values for cleanup before you could start work. The measure of success must be at most a 10^{-6} total cancer risk to the population. This means the ED must equal lifetime of exposure (with an EF of 24 hd^{-1}); therefore, TL and ED cancel. What is the target concentration to meet this risk?

So set each of the three pathways = 10^{-6} to begin, and solve for a target concentration for each (remember, because this is an additive risk, protection to one in a million for each would give us three in a million, *which is well within the "noise"—we haven't talked much about "false precision"—but risk assessors are often accused of this). In fact, it is probably best to report the target concentration for each pathway and aim to reduce the concentration to be the most protective:*

$$\underline{\text{Target conc.@ population } 10^{-6}} \text{ risk} = \text{SF} \times \text{LADD};$$

$$C = 1/(\text{SF})(\text{LADD}/C)(10^{-6})$$

$$[\alpha\text{-HCH}] \text{ sorbed to particles} = [(70 \text{ kg}) \times (10^{-6})]/[(6.3) \times (0.99) \times (100 \text{ mg m}^{-3}) \times (4.8 \text{ m}^3 \text{ h}^{-1}) \times (0.9) \times (1) \times (1) \times (10^{-6})] = \underline{2 \times 10^{-2}} \ mg\ kg^{-1} [\alpha\text{-}HCH] \text{ in soil.}$$

This is the α-HCH cleanup concentration for soil at the site as dictated by airborne particle exposures.

$$[\beta\text{-HCH}] \text{ sorbed to particles} = [(70 \text{ kg}) \times (10^{-6})]/[(1.8) \times (0.99) \times (100 \text{ mg m}^{-3}) \times (4.8 \text{ m}^3 \text{ h}^{-1}) \times (0.9) \times (1) \times (1) \times (10^{-6})] = \underline{9 \times 10^{-2}} \ mg\ kg^{-1} [\beta\text{-}HCH] \text{ in soil.}$$

This is the β-HCH cleanup concentration for soil at the site as dictated by airborne particle exposures.

$$[\alpha\text{-HCH}] \text{ in vapor phase} = [(70 \text{ kg}) \times (10^{-6})]/[(6.3) \times (0.01) \times (100 \text{ mg m}^{-3}) \times (10^{-6} \text{ kg mg}^{-1})(4.8 \text{ m}^3 \text{ h}^{-1}) \times (1)] = \underline{2.3} \ mg\ kg^{-1}.$$

This is the α-HCH cleanup concentration for soil at the site as dictated by airborne exposures from the vapor phase.

$$[\beta\text{-HCH}] \text{ in vapor phase} = [(70 \text{ kg}) \times (10^{-6})]/[(1.8) \times (0.01) \times (100 \text{ mg m}^{-3}) \times (10^{-6} \text{ kg mg}^{-1})(4.8 \text{ m}^3 \text{ h}^{-1}) \times (1)] = \underline{8} \ mg\ kg^{-1}.$$

This is the β-HCH cleanup concentration for soil at the site as dictated by airborne exposures from the vapor phase.

$$[\alpha\text{-HCH}] \text{ soil (dermal exposure)} = (70 \text{ kg}) \times (10^{-6})/(6.3) \times (150 \text{ cm}^2) \times (1) \times (1) \times (5 \text{ mg cm}^{-2} \text{ d}^{-1}) \times (10^{-6}) = \underline{0.01} \ mg\ kg^{-1}.$$

Because particle emissions of α-HCH require a lower concentration, we will use particle flux from soil to set the cleanup standard for α-HCH for all soils at the site.

$$[\beta\text{-HCH}] \text{ soil (dermal exposure)} = [(70 \text{ kg}) \times (10^{-6})\text{ d}^{-1}]/[(1.8) \times (150 \text{ cm}^2)] \times (1) \times (1) \times (5 \text{ mg cm}^{-2} \text{ d}^{-1}) \times (10^{-6})] = \underline{0.05} \; mg \; kg^{-1}.$$

Because particle emissions of β-HCH require a lower concentration, we will use soil particle flux potential to set the cleanup standard for β-HCH for all soils at the site.

Thus, target soil $[\alpha\text{-}\boldsymbol{HCH}] = \underline{2 \times 10^{-2}}\; mg\; kg^{-1}[\alpha - HCH]$ *in soil; and*

Target soil $[\beta\text{-}HCH] = \underline{9 \times 10^{-2}}\; mg\; kg^{-1}[\beta - HCH]$ *in soil.*

This means you must clean the soil to 20 parts per billion (ppb) of soil for alpha and 90 ppb for beta. You will need elaborate monitoring and analytical equipment to detect HCH isomers at these concentrations. The good news is that the soil samples given are at these levels, so not a lot of remediation is needed to get a margin of safety. The uncertainty is in how representative these samples are, what the ND for beta means, and how the dynamic will change when you start cleaning the piles and tanks. Also, what happens when you find some buried drums or other tanks?

You can argue that the assumptions for inhalation are overly conservative. People do not ordinarily breath at the high rate. Another argument is that of a reasonably exposed person. We are assuming that people will be living at the site's boundary. You can apply some type of dispersion model (e.g., the concentration in the air decreases exponentially with distance from the source). So we can put a distance term in the numerator, which helps identify a target receptor site to be protected downwind. We know that the town's water well is 1,500 m downwind. If we assume a deposition and dispersion rate that finds compound concentrations reduced by the square of the distance every 100 meters from the source to the receptor (for illustrative purposes only) and apply it to our calculation for airborne particles:

$$[(70 \text{ kg}) \times (10^{-6}) \times (15)^2]/[(6.3) \times (0.99) \times (100 \text{ mg m}^{-3}) \times (4.8 \text{ m}^3 \text{ h}^{-1}) \times (0.9) \times (1) \times (1) \times (10^{-6})] = \underline{6}\; mg\; kg^{-1}$$

This puts us in the parts per million cleanup range for alpha, which is practical.

Also, if you assume a moderate IR, fewer hours near the site (i.e., no one would live there), use better absorption and bioaccumulation data, and fewer respirable particles, this concentration could increase further. But don't go overboard with manipulation of factors or you will be accused (correctly) of trying to "define the problem away!"

Now for the risk-based water cleanup:

$$[\alpha\text{-HCH}] \textbf{ in groundwater} = [(70 \text{ kg}) \times (10^{-6})]/[(6.3) \times (1 \text{ L d}^{-1}) \times (1)]$$
$$= \underline{\mathbf{1 \times 10^{-5}}} \mathbf{\ mg\ L^{-1}}$$

This is the target concentration for groundwater alpha concentrations. This is in the parts per trillion range!

$$[\beta\text{-HCH}] \textbf{ in groundwater} = [(70 \text{ kg}) \times (10^{-6})]/[(1.8) \times (1 \text{ L d}^{-1}) \times (1)]$$
$$= \underline{\mathbf{4 \times 10^{-5}}} \mathbf{\ mg\ L}$$

This is the target concentration for groundwater beta concentrations. Again, this is in the range of ppt.

You probably could not measure these concentrations. Here, the best you can do is restrict access (even prohibit) to nearby wells and drinking water supplies. Prohibition would mean zero exposure and no risk (note that the numerator goes to 0); however, you must remove the source before you know that the contamination will not spread. You need to find out if any HCH is showing up in the town water supply. If so, this is very bad news for the waste site because it is probably the source, and the drinking water aquifer is probably already contaminated. Even if there is no HCH identified at the town supply, you need to find out if they are even analyzing for it and what their levels of detection are. You should also do some research to see if there are other potential sources of HCH in the area.

Realistically, if you assume that the only way this water makes it to drinking water supplies is by migration off-site—say everyone gets 0.001 liter per day—you would have a target concentration of 0.01 mg L^{-1}, which is below the concentrations measured in the groundwater. You would want to model the migration and estimate the gradient from the site to other parts of the aquifer to calculate this risk (nearby homes would have a higher risk).

Therefore, you must now design a sampling plan (including project period and long term) to collect water and soil samples to monitor the success of your remediation efforts. You must demonstrate that the processes you have selected will achieve these target concentrations. *There is a large difference between the concentrations you have measured and what you need to get with cleanup. If you cannot sell the modeling numbers showing dilution down aquifer, you will have to clean the entire site to the ppt level. So, your likelihood of success, given all the assumptions, is low, unless you can convince the powers that be that your cleanup levels and models are sufficiently protective.*

FYI: You weren't asked specifically, but *if you still want to take the job*, you would need to calculate cancer risks for tetrachloromethane, mercury, and nickel subsulfate because the health endpoint of concern is total cancer for all contaminants, except mercury, which is also neurotoxic.

Cancer Risk from Tetrachloromethane (CCl_4):

Lipophilic, so assume BF and AF = 1.

Volatile Organic Compound (VOC).

CCl_4(Air) = 50% from particles, so use equation 5-3.

ED = 3 yr.

$$\begin{aligned}\text{LADD (part)} &= (0.5) \times [(52 \text{ mg kg}^{-1}) \times (100 \text{ mg m}^{-3}) \times (0.9) \times (4.8 \text{ m}^3 \text{ h}^{-1}) \\ &\quad \times (10 \text{ h d}^{-1}) \times (1) \times (936 \text{ d}) \times (10^{-6})]/[(70 \text{ kg}) \times (25{,}550 \text{ d})] \\ &= 6 \times 10^{-5} \text{ mg kg}^{-1} \text{ d}^{-1}\end{aligned}$$

$$\begin{aligned}\text{Risk (air particles) from } CCl_4 = \text{LADD} \times \text{SF} &= (6 \times 10^{-5} \text{ mg kg}^{-1} \text{ d}^{-1}) \\ &\quad \times (0.13 \text{ mg kg}^{-1} \text{ d}^{-1})^{-1} \\ &= 8 \times 10^{-6}\end{aligned}$$

CCl_4 (Air) = 50% from vapor phase, so use vapor inhalation equation:

$$\begin{aligned}\text{LADD (gas)} &= \frac{(C) \times (IR) \times (EL) \times (AF) \times (ED)}{(BLO) \times (TL)} \\ &= (.5) \times [(52 \text{ mg kg}^{-1}) \times (100 \text{ mg m}^{-3}) \times (10^{-6} \text{ kg mg}^{-1}) \\ &\quad \times (4.8 \text{ m}^3 \text{ h}^{-1}) \times (10 \text{ h d}^{-1}) \times (1) \times (936 \text{ d})]/ \\ &\quad [(70 \text{ kg}) \times (25{,}550 \text{ d})] \\ &= 7 \times 10^{-5} \text{ mg kg}^{-1} \text{ d}^{-1}\end{aligned}$$

$$\begin{aligned}\text{Risk (air vapor) from } CCl_4 = \text{LADD} \times \text{SF} &= (7 \times 10^{-5} \text{ mg kg}^{-1}) \\ &\quad \times (0.13 \text{ mg kg}^{-1} \text{ d}^{-1})^{-1} \\ &= 8 \times 10^{-6}\end{aligned}$$

$$\text{Total } CCl_4 \text{ air risk} = 8 \times 10^{-6} + 7 \times 10^{-5} = 8 \times 10^{-5}$$

CCl_4 (drinking water) uses Equation 5-1.

$$\begin{aligned}\text{LADD (drink)} &= [(80 \text{ mg L}) \times (1 \text{ L d}^{-1}) \times (936 \text{ d}) \times (1)]/[(70 \text{ kg}) \times (25{,}550 \text{ d})] \\ &= 0.04 \text{ mg kg}^{-1} \text{ d}^{-1}\end{aligned}$$

$$\begin{aligned}\text{Risk (drink) from } CCl_4 &= (0.04 \text{ mg kg}^{-1} \text{ d}^{-1}) \times (0.13 \text{ mg kg}^{-1} \text{ d}^{-1})^{-1} \\ &= 5 \times 10^{-3}\end{aligned}$$

CCl_4(dermal-soil) uses Equation 5-2.

$$\text{LADD (soil)} = [(80\ \text{mg kg}^{-1}) \times (150\ \text{cm}^2) \times (1) \times (1) \times (5\ \text{mg cm}^{-2}\text{d}^{-1}) \times (10\text{h}/24\text{h}) \times (936\text{d}) \times (10^{-6})]/[(70\text{kg}) \times (25{,}550\text{d})]$$

$$= 1 \times 10^{-5}\text{mg kg}^{-1}\text{d}^{-1}$$

$$\text{Risk (soil) from } CCl_4 = (1 \times 10^{-5}\ \text{mg kg}^{-1}\text{d}^{-1}) \times (0.13\ \text{mg kg}^{-1}\text{d}^{-1})^{-1}$$

$$= 2 \times 10^{-6}$$

Total Risk from CCl_4 during project $= 8 \times 10^{-5} + 5 \times 10^{-3} + 2 \times 10^{-6}$

$= \mathbf{5 \times 10^{-3}}$

These are significant risks, but the additional protection for the workers needed for HCH will be protective for the VOCs as well. Also, because drinking water is the biggest risk, there is no doubt that bottled water will be needed.

Again, the measure of success must be at most a 10^{-6} total risk to the population. This means the ED must equal lifetime of exposure (ED = 25,550 d); TL and ED cancel.

$$\text{Target conc.} = 10^{-6} = [CCl_4] \text{ sorbed to particles}$$

$$= [(70\ \text{kg}) \times (10^{-6})]/[(0.13) \times (0.5) \times (100\ \text{mg m}^{-3}) \times (4.8\ \text{m}^3\text{h}^{-1}) \times (0.9) \times (1) \times (1) \times (10^{-6})]$$

$$= 2.5\ \textit{mg kg}^{-1} [CCl_4] \textit{ in soil.}$$

This is the cleanup concentration for soil at the site as dictated by airborne particle exposures. This is only a factor of 20 cleanup (52 to 2.5), so this is practical. Not surprisingly, HCH is the big obstacle for soils, not the volatiles.

$$[CCl_4] \text{ in vapor phase} = (70\ \text{kg}) \times (10^{-6})/(0.13 \times (0.5) \times (100\ \text{mg m}^{-3}) \times (10^{-6}\ \text{kg mg}^{-1}) \times (4.8\ \text{m}^3\text{h}^{-1}) \times (1)$$

$$= \underline{2}\text{mg kg}^{-1}.$$

This is the cleanup concentration for soil at the site as dictated by airborne exposures for vapor phase CCl_4.

$[CCl_4]$ soil (dermal exposure) :

$$(70\ \text{kg}) \times (10^{-6})/(0.13) \times (150\ \text{cm}^2) \times (1) \times (1) \times (5\ \text{mg cm}^{-2}\text{d}^{-1}) \times (10^{-6})$$

$$= \underline{0.7}\ \text{mg kg}^{-1}.$$

So the target cleanup level for HCH compounds in the soil is slightly more stringent that that of the airborne particles and vapor phase HCH.[16]

Target Soil [CCl_4] = $\underline{0.7\ \text{mg kg}^{-1}}$. So, for this project to be a success, you must clean the soil to about 700 ppb [CCl_4]. Compare this to the soil cleanup level for HCH. *The volatiles in the soil should be an order of magnitude easier to deal with than the semivolatiles (often the case).*

$$[CCl_4] \text{ in groundwater} = (70\ \text{kg}) \times (10^{-6})/[(0.13) \times (1\ \text{L d}^{-1}) \times (1)]$$
$$= \underline{5 \times 10^{-4}\ \text{mg L}}$$

is the target concentration for groundwater tetrachloromethane. *Again, this is near HCH water cleanup levels. Air stripping might be a good choice here.*

<u>Nickel Subsulfide (Ni_2S) :</u>

Salts and oxides can be more hydrophilic than the organics (although many inorganic compounds are extremely insoluble in water and precipitate from the water column), so assume BF and AF = 0.5. Also, assume that the Ni salts are semivolatile.

<u>Ni_2S(Air) = 99% from particles, so use Equation 5-3.</u>

$$\text{LADD (part)} = (0.99) \times [(100\ \text{mg kg}^{-1}) \times (100\ \text{mg m}^{-3}) \times (4.8\ \text{m}^3\text{h}^{-1})$$
$$\times (0.9) \times (10\ \text{h d}^{-1}) \times (0.5) \times (936\text{d}) \times (10^{-6})]/$$
$$[(70\text{kg}) \times (25,550\text{d})]$$
$$= 1 \times 10^{-4}\ \text{mg kg}^{-1}\text{d}^{-1}$$

$$\text{Risk (air particles) from } Ni_2S = \text{LADD} \times \text{Slope Factor}$$
$$= (1 \times 10^{-4}\ \text{mg kg}^{-1}\ \text{d}^{-1})$$
$$\times (1.7\ \text{mg kg}^{-1}\ \text{d}^{-1})^1 = 2 \times 10^{-4}$$

<u>Ni_2S (Air) = 1% from vapor phase, so use Equation 5-4.</u>

$$\text{LADD (gas)} = (0.01) \times [(100\ \text{mg kg}^{-1}) \times (100\ \text{mg m}^{-3}) \times (10^{-6}\ \text{kg mg}^{-1})$$
$$\times (4.8\ \text{m}^3\ \text{h}^{-1}) \times (10\ \text{h d}^{-1}) \times (0.5) \times (936\ \text{d})]/$$
$$[(70\ \text{kg}) \times (25,550\ \text{d})]$$
$$= 1 \times 10^{-6}\ \text{mg kg}^{-1}\ \text{d}^{-1}$$

$$\text{Risk (air vapor) from } Ni_2S = \text{LADD} \times \text{SF}$$
$$= (1 \times 10^{-6} \text{ mg kg}^{-1}) \times (1.7 \text{ mg kg}^{-1}\text{d}^{-1})^{-1}$$
$$= 2 \times 10^{-6}$$
$$\text{Total } Ni_2S \text{ air risk} = 1 \times 10^{-4} + 2 \times 10^{-6} = 1 \times 10^{-4}$$

Ni_2S (drinking water) uses Equation 5-1.

$$\text{LADD (drink)} = [(1.5 \text{ mg L}) \times (1 \text{ L/d}^{-1}) \times (936 \text{ d}) \times (0.5)]/$$
$$[(70 \text{ kg}) \times (25,550 \text{ d})]$$
$$= 4 \times 10^{-4} \text{ mg kg}^{-1}\text{d}^{-1}$$
$$\text{Risk (drink) from } Ni_2S = (4 \times 10^{-4} \text{ mg kg}^{-1}\text{d}^{-1}) \times (1.7 \text{ mg kg}^{-1}\text{d}^{-1})^{-1}$$
$$= 7 \times 10^{-4}$$

Ni_2S (dermal-soil) uses Equation 5-2.

$$\text{LADD (soil)} = [(100 \text{ mg kg}^{-1}) \times (150 \text{ cm}^2) \times (0.5) \times (0.5)$$
$$\times (5 \text{ mg cm}^{-2} \text{ d}^{-1}) \times (936 \text{ d}) \times (10^{-6})]/$$
$$[(70 \text{ kg}) \times (25,550 \text{ d})]$$
$$= 1 \times 10^{-5} \text{ mg kg}^{-1}\text{d}^{-1}$$
$$\text{Risk (soil) from } Ni_2S = (1 \times 10^{-5} \text{ mg kg}^{-1} \text{ d}^{-1}) \times (1.7 \text{ mg kg}^{-1} \text{ d}^{-1})^{-1}$$
$$= 2 \times 10^{-5}$$

Total Risk *from* Ni_2S ***during project*** $= 2 \times 10^{-4} + 7 \times 10^{-4} + 2 \times 10^{-5}$
$$= \mathbf{1 \times 10^{-3}}$$

Again these are significant risks, reemphasizing the need to protect yourself and workers during remediation, with all four pathways having similar risks. So all of the inhalation, dermal, and ingestion protective measures mentioned earlier should be used.

With a measure of success equal to 10^{-6} total Ni_2S risk to the population, let us find our heavy metal remediation targets.

$$\text{Target conc.} = 10^{-6} = [Ni_2S] \text{ sorbed to particles}$$
$$= [(70 \text{ kg}) \times (10^{-6})]/[(1.7) \times (0.99) \times (100 \text{ mg m}^{-3}) \times (4.8 \text{ m}^3\text{h}^{-1})$$
$$\times (0.9) \times (1) \times (1) \times (10^{-6})] = \underline{0.1} \text{ mg kg}^{-1} \text{ [}Ni_2S\text{] in soil.}$$

This is the soil cleanup level dictated by airborne particles.

$$\text{Target conc. for } [Ni_2S] \text{ in vapor phase} = [(70\ \text{kg}) \times (10^{-6})]/[(1.7) \times (0.01) \times (100\ \text{mg m}^{-3}) \times (10^{-6}\ \text{kg mg}^{-1}) \times (4.8\ \text{m}^3\ \text{h}^{-1}) \times (0.5)] = \underline{17}\ \text{mg kg}^{-1}.$$

This is the cleanup concentration for soil at the site as dictated by airborne exposures for vapor phase Ni_2S.

$$[Ni_2S]\ \text{soil (dermal exposure)} = [(70\ \text{kg}) \times (10^{-6})]/[(1.7) \times (150\ \text{cm}^2) \times (0.5) \times (0.5) \times (5\ \text{mg cm}^{-2}\ \text{d}^{-1}) \times (10^{-6})] = \underline{0.2}\ \text{mg kg}^{-1}.$$

Skin absorption requires a less protective concentration than breathing the vapors.

Target Soil $[Ni_2S]$ = $\underline{0.1}$ mg kg^{-1}.

So, for this project to be a success, you must clean the soil to about 100 ppb $[Ni_2S]$. The dermal and inhalation (particles) are your problem. This is logical because virtually none of this should volatilize because it has been stabilized in the environment (we are assuming 1% volatilizes) and is more likely to stay in a mineral form in the soil.

Again, the measured soil concentrations should be low, at least near the surface, because the Ni_2S should have been oxidized to sulfates and oxides, so you may want to look at total Ni before making any decisions.

$$[Ni_2S]\ \text{in groundwater} = [(70\ \text{kg}) \times (10^{-6})]/[(1.7) \times (1\ \text{L d}^{-1}) \times (1)] = \underline{4 \times 10^{-5}\ \text{mg L}}$$

is the target concentration for groundwater nickel subsulfide.

Since the Ni_2S concentrations in groundwater is not known (ND in the table), you do not know how far you have to treat to reach this very low concentration. Similar to the others, though, you may want to argue with the liter per day level, since no one should be tapping in this part of the aquifer. *I wouldn't hazard a guess as to whether you would be successful with this argument.*

If the ND really means that nickel subsulfide levels are very low, it is likely that most of the Ni is oxidized. Therefore, you need to find out if any of these Ni compounds are carcinogenic (or otherwise toxic). *Pump and treat with physical settling and chemical separation might be a good choice here.*

Bottom line: *The HCH, CCl_4, and Ni_2S are all cancer risk problems. This project is fraught with many problems—technical, legal, and financial—but the cleanup levels appear to be feasible.*

One last thing: Because you found Hg, you need to be concerned about chronic neurologic toxicity, especially in children. Treat Hg as a non-carcinogen, so eliminate the lifetime factor and exposure duration in the calculations to derive a maximum daily dose (rather than LADD).

We will not address the treatment of mercury, except to say that it is very difficult to measure and that it speciates in air, water, soil, and biota, so you must have extremely intricate and careful monitoring programs to find it. You should question the company about how they obtained these total Hg measurements. You should also ask your regulatory agencies what cleanup levels they are expecting for each media.

Noncancer Rick Calculations

Hg (Air) = All from particles, so we should use a maximum daily dose (MDD) equation, but in this case we will use Equation 5-3 *as* our MDD.

$$\mathrm{MDD(air)} = [(0.012\ \mathrm{mg\ kg^{-1}}) \times (100\ \mathrm{mg\ m^{-3}}) \times (0.9) \times (4.8\ \mathrm{m^3\ h^{-1}}) \times (10\ \mathrm{h\ d^{-1}}) \times (0.5) \times (10^{-6})]/[(70\ \mathrm{kg})] = 3 \times 10^{-7}\ \mathrm{mg\ kg^{-1}}$$

Assuming daily intake to average 1 μg kg^{-1}:

$$\mathrm{Risk\ (air)\ from\ Hg} = \mathrm{MDD/ADI} = (3 \times 10^{-7}\ \mathrm{mg\ kg^{-1}})/(0.001\ \mathrm{mg\ kg^{-1}}) = 3 \times 10^{-4}$$

In addition, some forms of Hg (including elemental Hg) are very volatile, so they will end up in the vapor phase in addition to the particles-laden Hg (but we are assuming most is on particles because it has had some time to reach chemical equilibrium).

Hg (drinking water) uses Equation 5-1 as our MDD equation.

$$\mathrm{MDD(drink)} = [(0.05\ \mathrm{mg\ L^{-1}}) \times (1\ \mathrm{L\ d^{-1}}) \times (0.5)]/[(70\ \mathrm{kg})] = 3.6 \times 10^{-4}\ \mathrm{mg\ kg^{-1}}$$

$$\text{Risk (drink) from Hg} = \text{MDD/ADI}$$
$$= (3.6 \times 10^{-4}\ \text{mg kg}^{-1})/(0.001\ \text{mg kg}^{-1}) = 0.36$$

Hg (soil) uses Equation 5-2.

$$\text{MDD(dermal-soil)} = [(0.012\ \text{mg kg}^{-1}) \times (150\ \text{cm}^2) \times (0.5) \times (1) \times (5\ \text{mg cm}^{-2}\ \text{d}^{-1}) \times (10^{-6})]/[(70\ \text{kg})]$$
$$= 6.4 \times 10^{-8}\ \text{mg kg}^{-1}$$
$$\text{Risk (soil) from Hg} = \text{MDD/ADI}$$
$$= (6.4 \times 10^{-8}\ \text{mg kg}^{-1})/(0.001\ \text{mg kg}^{-1})$$
$$= 6.4 \times 10^{-5}$$
$$\text{Total Risk from Hg} = 3 \times 10^{-4} + 0.36 + 6.4 \times 10^{-5} = 0.36$$

This indicates that the Hg risk is not likely to be significant and that the most risk comes from the groundwater. Therefore the [Hg] in groundwater will probably not have to be reduced to achieve the hazard index less than 1. So, the three-year project should use protective equipment and ensure that no additional risks from Hg are encountered.

(Author's note: With the great concern about the many potential impacts associated with Hg exposure, including neural, endocrine, and immune system effects, this conclusion is not likely, even with the risk less than 1. The ADI used is high, and newer calculations of these noncancer endpoints are just as conservative as those for cancer; however, the metal provides examples of how noncancer risks are calculated. Also, additional precision can be provided by using specific MDD values for noncarcinogens. Such MDD values may be published by federal and state agencies, and can be plugged into the equations we used in this example.)

CHAPTER 6

The Role of the Engineer in Emergency Response

Up to this point, we have generally addressed situations for which ample time is available to assess the conditions and situation associated with possible exposures to hazardous substances. Assessments often take months or years to complete, and remedial actions take years to plan and even longer to complete. This "luxury" is not available in emergency situations created by natural disasters, such as the floods in the Mississippi River and Red River basins, earthquakes on the West Coast of the United States, or even sinkholes in Florida. These natural disasters are often associated with the release of chemical and biologic contaminants to the environment. For instance, recall the scenes of pesticide tanks floating down the Mississippi River that had been displaced from farms or the fears of submerged electrical equipment as possible sources of PCBs.

Disasters with widespread contamination can be precipitated by human-caused events as well, such as hazardous plumes from the huge oil fires intentionally set by the Iraqis during the Gulf War of the 1990s. However, there arguably has never been a more dramatic disaster than the recent attacks on the World Trade Center (WTC) towers and the Pentagon. The loss of life from the intentional crash and during rescue operations was followed by potential threats to human health from the impending fires, resuspended dust, and subsequent exposures as people returned to homes and businesses near Ground Zero.

Lessons from the Emergency Response at the World Trade Center

The September 11, 2001 attack on the WTC resulted in an intense fire (more than 1,800°F) and the subsequent, complete collapse of the two main structures and adjacent buildings, as well as significant damage to many surrounding buildings within and around the WTC complex. This 16-acre area has become known as Ground Zero. The collapse of the buildings and the fires created a large plume comprising both particles and gases that were transported into the New York City air shed. The plume began at elevation (80- to 90-story height) with the initial combustion of the jet fuel and building materials. After the collapse of the buildings, aerosols were emitted from ground level, moving downwind and reaching many outdoor and indoor locations downwind. For the first 12 to 18 hours after the collapse, the winds transported the plume to the east and then to the southeast toward Brooklyn, New York.

The collapse of the WTC towers was unprecedented. Most building implosions are performed under controlled conditions in which many sources of contamination are not present. For example, in controlled demolitions, carpets, furniture, wallboard, and other flammable and aerosol-producing materials are removed before implosion. The primary differences between the WTC incident and that of other building fires and implosions was the simultaneous occurrence of many events: the intense fire, the extremely large mass of material (more than 10^6 tons) reduced to dust and smoke, and the previously unseen degree of pulverization of the building materials.

Using Ambient and Exposure Data to Support Cleanup and Emergency Response[1]

Characterizing the possible exposures during and after the attacks was one of the mandates of the emergency response team (see Figure 6-1). To begin assessing the exposure to dust and smoke among the residential and commuter population during the first few days, samples of dust particles that initially settled in downtown New York City were taken from three undisturbed protected locations to the east of the WTC site. Two samples were taken on day 5 (September 16, 2001), and the third sample was taken on day 6 (September 17, 2001) after the terrorist attack. The purposes for collecting the samples were to (1) determine the chemical and physical characteristics of the material that was present in the dust and smoke that settled from the initial plume, and (2) to determine the absence or presence of contaminants that could affect acute or long-term human health by inhalation or ingestion. It was anticipated that the actual compounds and materials present in the plume would be similar to those found in building fires or implosion of collapsed buildings. The primary differences would be the simultaneous occurrence of each type of event, the intense fire (greater than 1,000°C), the extremely large mass of material (greater than 10^6 tons) reduced to dust and smoke, and the previously unseen degree of pulverization of the building materials.[2]

The dust and smoke were inhaled directly by individuals or inhaled after the resuspension of the settled aerosol by turbulence. The dust particles could also be ingested after being deposited on surfaces inside homes.

FIGURE 6-1. Members of the World Trade Center emergency response team (*author on far-right*) siting air quality monitoring equipment near Ground Zero. The equipment, including personal exposure monitors, gas canisters, and saturation monitors, had to be mobile and battery-powered because alternating current (AC) electricity was not available. The equipment was used to measure particulate matter and volative organic compounds. More sensitive and complex equipment was used farther from the site, where AC power was available, providing data on background air quality levels and measuring a wider array of pollutants, including semivolative organic compounds, such as polycyclic aromatic hydrocarbons, dioxins, and furans.

The residuals of dust and smoke would remain on surfaces and in building ventilation systems not properly cleaned before people moved back into buildings. Children and adults were then at risk of exposure to the dust via nondietary ingestion and by inhalation after being resuspended from the ventilation system. Larger particles (larger than 2.5 μm diameter) could also be ingested following inhalation once they have been cleared from the lung's upper airways by mucocilia.

Several initial measurements made by various organizations focused on the general composition of the dust and smoke, with a primary concern being asbestos.[3] The approach employed here for analyzing the three dust and smoke samples includes detailed measurement of the inorganic and organic components of the mass and a general characterization of the percentage distribution by mass or volume of various materials present in each sample.

Samples of the total settled dust and smoke were collected at three different locations. The first was collected from protected external ledges around the entrance of a building on Cortlandt Street, which is one block east of the WTC building complex. The initial direction of the plume was from west to east, thus the other samples were collected at locations to the east of Cortlandt Street. These two samples were collected from 10- to

15-cm-thick deposits that were on the top of two cars about 0.7 km from the WTC site. The automobiles were in locations protected from rain that occurred on Friday, September 15, 2001. One automobile was located one city block and the other was two city blocks west of the East River between the Manhattan and Brooklyn bridges, on Cherry Street and Market Street, respectively. These cars appeared to have been in their respective locations since September 11th, but it is possible that each could have been moved from FDR Drive, an adjacent thoroughfare on the east side of New York City.

One of the reasons for collecting samples from these locations was to determine whether chemical composition and physical morphology of the particles changed with distance from the WTC site. Because each sample comprised a complex mixture of materials, different techniques for examining chemical and physical characteristics were needed. For example, particles were analyzed by microscope to identify major components and the particles' shapes (i.e., morphology). Stereomicroscopy was employed to identify larger particles, and polarized light microscopy was used to identify minerals, building products, and fibers greater than 1 μm in diameter. Scanning electron microscopy with X-ray elemental analysis identified metal fragment and particles and fibers smaller than 1 μm. Transmission electron microscopy, with electron diffraction and X-ray elemental analysis, was used to identify the smallest fraction of particles, including single asbestos fibrils and carbon soot. The samples were also extracted and analyzed for metals and organic constituents, as well as for pH, corrosion, aerodynamic particle size for fine and coarse particle fractions, percentage of mass by particle sieving, general radiation levels, and asbestos.

The composition of material collected from the WTC site was complex. The aerosol released and deposited onto surfaces downwind of Ground Zero included pulverized building debris and products of incomplete combustion (PICs) produced by the explosion that ignited the thousands of liters of jet fuel. The mass of material deposited was extremely high, and in many indoor locations the deposited particle loadings were 1 to 3 cm thick. The outdoor dust and smoke loadings in some places exceeded 10 cm thickness. So, in the initial days following the attack on the WTC, more than 70% of the mass of deposited aerosols was from construction materials, such as pulverized cement, wallboard, and office furnishings. A small percentage of the carcinogen asbestos was found in these samples, about 0.8% by volume. The PICs formed in the intense combustion of building materials, including furnishings, equipment, debris, wiring, metal, wood, and so on. The PICs that were present at levels from five to hundreds of $\mu g\ g^{-1}$ in the samples were the polycyclic aromatic hydrocarbons (PAHs). The individual compounds [e.g., benzo(a)pyrene] were above 20 $\mu g\ g^{-1}$, and the total mass of PAHs present was more than 0.1% of the mass.

Compared to the vast amounts of other material present in the air during the first day after the collapse, these levels were sufficiently high to indicate significant short-term inhalation exposure. In fact, based on the

PAH results obtained from air samples after September 25, the PAH species released into the atmosphere at that time were similar to those detected in the settled dust and smoke samples collected in the first week after the collapse. Dioxin and furan concentrations were similar to levels found in other studies,[4] but the levels of 2,2′4,4′,5,5′-hexabromobiphenyl were elevated, probably because of its use during the construction of the WTC in the 1970s.[5] The levels of lead ranged from 100 to greater than 600 parts per million (ppm), which are not very high levels compared to the levels found in typical urban soils; however, the actual levels of dust and smoke deposited in individual buildings and businesses need to be assessed for cleanup based on the actual surface loading of lead and asbestos. A systematic effort is needed to properly clean indoor locations to avoid exposures to persistent levels of lead, asbestos from indoor surfaces, and air.

The high pH of the samples is likely caused by the presence of cement and other basic materials associated with the construction debris in the deposited particles. This factor, along with the presence of long and thin glass fibers (nonasbestos) and attached agglomerated fine particles, was a consideration of possible sources of initial lung irritations reported by residents and workers in the days and initial weeks after the collapse of the WTC buildings.

After the initial rain on September 15, the heavy rains that occurred on September 24 carried away much of the material from outdoor surfaces; however, because of the extremely dry weather pattern in the Northeast during the fall of 2001, dust remained on some outdoor surfaces and rooftops through November. The WTC site itself was continually sprayed with water to reduce the resuspendable dust levels during recovery operations. The quantities of settled and resuspendable dust are of concern indoors. Re-entrained dust can lead to adverse health effects if the toxic constituents present on the indoor surfaces are not cleaned properly and if the heating, ventilating, and air-conditioning (HVAC) system of each structure is not concurrently cleaned or cleaned before the cleanup of the indoor surfaces and reentry into the residence or office. The EPA and other health and environmental agencies recommended a so-called HAZMAT-type residential cleanup before rehabitation of residences or offices to ensure that rehabitation clearance values are achieved for contaminants, such as lead ($40\ \mu g\ ft^{-2}$) on floors.[6]

Some types of material that were released are similar to materials that we are exposed to during our daily lives, but there were extraordinarily high quantities of coarse and fine particles released and dispersed after the WTC collapse. Future analysis needs to be completed on the health consequences of the exposure among the commuters, workers, and residents. The results from the different samples vary among aerosol materials released on September 11 and during the subsequent weeks. This is not surprising given the large amounts of different materials present in each of the collapsed and burning structures.

Estimates of human exposure to the materials are ongoing. The results for composition and particle size, with and without agglomerates on glass fiber and other fibrous particles, are being used to assess short- and long-term effects among various populations, including sensitive subgroups. The people potentially exposed to the initially suspended dust and smoke, or subsequently settled dust and smoke, include unprotected rescue workers, residents, and workers in downtown Manhattan immediately after and in the first few weeks after the collapse. The settled dust and smoke could be resuspended and expose unprotected residential cleanup workers and workers and residents in poorly or inefficiently cleaned buildings weeks to months afterward. Finally, the levels of exposure encountered will have to be placed within the context of the materials that have been released from the diminishing smoldering fires that continued to burn until December 14, 2001.

Measurements of the hazardous chemicals in the atmospheric plume were also an important component for assessing possible human health risk. Local people in Lower Manhattan and emergency responders were exposed to gases and particles released directly from the site, as well as from previously settled particles that have become resuspended by air turbulence. The major pathways of exposure likely were inhalation, ingestion of deposited particles, and dermal exposure. Some chemical species that have been associated with human health effects include carcinogenic compounds [e.g., benzo(a)pyrene and other PAHs from smoldering fires], endocrine disruptors (e.g., phthalates and styrene derivatives from plastics), and neurotoxins (such as dioxins from incomplete combustion).

Calculation of Endocrine Risk in a Cleanup: The 1,3-Diphenyl Propane Example

Table 6-1 provides the emergency team's measurements of concentrations of various nonpolar semivolatile compounds found in the air around Ground Zero. A high-capacity Integrated Organic Gas and Particle (HiC IOGAP) sampler with a 2.5 μm cyclone inlet for particle discrimination was used to collect semivolatile gases and particles for speciation of organic compounds (see Figure 6-2). Note that the concentrations of the compound 1,3-diphenyl propane [1′,1′-(1,3-propanediyl)bis-benzene] appear to be elevated. This chemical species had not previously been reported from ambient sampling. It has been associated with polyvinyl chloride (PVC) materials, which are believed to be in abundance at the WTC site. Although 1,3-diphenyl propane has been shown to be estrogenic, its binding affinity for the human estrogen receptor has been shown to be relatively low (about 5 orders of magnitude less than the estrogenicity of estradiol).[7]

Thus, the presence of 1,3-diphenyl propane in the WTC air provides an example of two occurrences the engineer may encounter in an emergency setting. First, little data may be available about a compound from the usual sources, such as from material safety data sheets (MSDS) or the Integrated Risk Information System (IRIS).[8] To estimate the possible risks from the WTC associated with this chemical, the average daily dose ($mg\,kg^{-1}\,day^{-1}$) for inhalation of a noncarcinogenic species may be used:

$$ADD = \frac{C(mg\,m^{-3}) \cdot IR(m^3 day^{-1})}{BW(kg)} \qquad (6\text{-}1)$$

where C is concentration of the species in inhaled air, IR is the inhalation rate, and BW is body weight. The default inhalation rate[9] for an adult (70 kg) is $20\,m^3\,day^{-1}$. Given the mean 24-hour concentration of 1,3-diphenyl propane at approximately $5 \times 10^{-4}\,mg\,m^{-3}$, the average daily dose (ADD) is $1.4 \times 10^{-4}\,mg\,kg^{-1}\,day^{-1}$. Little data exist on this compound, so there is much uncertainty of what these findings mean in terms of risk. In particular, no pharmacokinetic models are available to compare cellular scale studies to possible effects in humans.

Along with the presence of 1,3-diphenyl propane, there is further evidence that the plume contained emissions of burning and remnant materials from the WTC site. The molecular markers for these emissions include retene and 1,4a-dimethyl-7-(methylethyl)-1,2,3,4,9,10,10a,4a-octahydrophenanthrene that are typically biogenic in origin. For example,

TABLE 6-1(a)
Gas, Particle, and Total Concentrations (ng m^{-3}) of Selected Semivolative Organic Compounds Measured at a Site near Ground Zero

	09/26–09/27			*10/04–10/05*		
	Gas	*Particle*	*Total*	*Gas*	*Particle*	*Total*
Indane*	16	9	25	4	4	8
1-Octanol*	179	44	223	201	104	305
1-Nonanal*	30	38	68	114		114
Diphenyl ether*	7		7			
Dibenzofuran	98	1	99	135	3	138
Bibenzyl*	22	1	23	33	1	34
1,3-Diphenyl propane*	187	5	192	591	5	596
Pristane	43	4	47	35	4	39
Phytane	35	4	39	28	3	31
1,4a-dimethyl-7-(methylethyl)-1,2,3,4,9,10,10a,4a-octa-hydrophenanthrene* #1	3		3	3		3
1,4a-dimethyl-7-(methylethyl)-1,2,3,4,9,10,10a,4a-octa-hydrophenanthrene* #2	4		4	4		4

*Estimated concentrations based on calibrations of similar compounds.

TABLE 6-1(b)
Total Concentrations (ng m^{-3}) of Selected Semivolatile Organic Compounds Measured at a Site near Ground Zero.

	09/26–09/27	*10/04–10/05*	*10/06–10/07*	*10/12–10/13*	*10/20–10/21*	*LA[a]*
	Total	*Total*	*Total*	*Total*	*Total*	*Ave (Range)*
Indane*	25	8	3	1	1	
1-Octanol*	223	305	15	11	84	
1-Nonanal*	68	114	369	22	11	
Diphenyl ether*	7			9	9	
Dibenzofuran	99	138	9	106	97	20 (2-57)
Bibenzyl*	23	34		42	43	
1,3-Diphenyl propane*	192	596	5	693	541	
Pristane	47	39	16	68	48	68 (0-392)
Phytane	39	31	13	55	40	65 (13-188)
1,4a-dimethyl-7-(methylethyl)-1,2,3,4,9,10,10a,4a-octa-hydrophenanthrene* #1	3	3		8	7	
1,4a-dimethyl-7-(methylethyl)-1,2,3,4,9,10,10a,4a-octa-hydrophenanthrene* #2	4	4		8	7	

*Estimated concentrations based on calibrations of similar compounds.
[a]Fraser et al., 1997; Fraser et al., 1998.

retene is a known marker for smoke from burning wood and was seen in all samples analyzed from WTC.[10]

When Is It Safe to Move Back?

As environmental professionals in emergency response situations, we first look at what the data tell us and interpret those results to determine when and if it is safe to return to a site. At the WTC site, the air and settled dust measurement results indicated that after the initial destruction of the WTC the remaining air plumes from the disaster site consisted of many pollutants and classes and represented a complex mixture. This mixture includes compounds that are typically associated with fossil fuel emissions. The molecular markers for these emissions include the high levels of PAHs observed, the n-alkanes Carbon Prefix Index ~ 1 (odd carbon:even carbon ~ 1), as well as pristane and phytane as specific markers for fuel oil degradation. These results are not unexpected considering the large number of diesel generators and outsized vehicles used in the removal phases.

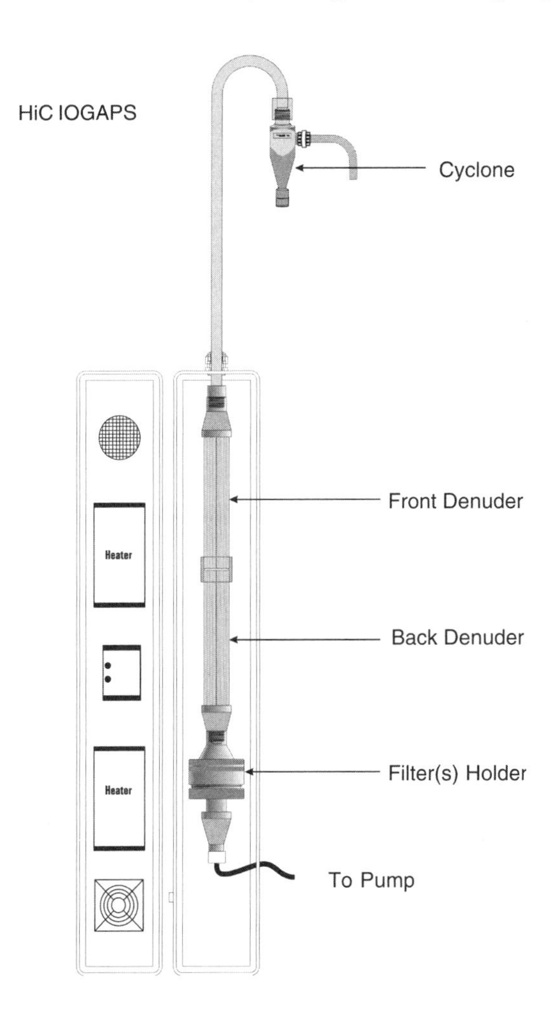

FIGURE 6-2. Schematic of the high-capacity Integrated Organic Gas and Particulate (HiC IOGAP) sampler with a 2.5 μm cyclone inlet for particle discrimination, which was used in Lower Manhattan, New York, following the September 11 attacks to collect semivolative gases and particles for speciation of organic compounds. The sampler utilizes two sorbant (XAD-4) coated eight-channel annular denuders (52 mm outer diameter, 285 mm length) to collect the gas-phase species and a prebaked quartz filter followed by three XAD-4 impregnated quartz filters to collect the particle phase. The XAD-4 impregnated quartz filters were used to collect those compounds that desorb from the particles on the quartz filters and/or those not removed by the denuders. The volumetric flow rate of the HiC IOGAP sampler was set to 85 L min^{-1} and temperature was controlled at 4°C above ambient to prevent condensation of water. (*Source:* Drawing used with permission of URG Corporation.)

The WTC dust mixture also included emissions of burning and remnant materials from the WTC site. The molecular markers for these emissions include retene and 1,4a-dimethyl-7-(methylethyl)-1,2,3,4,9,10,10a,4a-octahydrophenanthrene, which are typically biogenic in origin. Another potential marker is 1,3-diphenyl propane [1′,1′-(1,3-propanediyl)bis-benzene] mentioned previously. It was found in significant concentrations. This species has not previously been reported from ambient sampling. It has been associated with PVC materials, which are believed to be in abundance at the WTC site. These emissions lasted for at least three weeks (September 26 through October 20, 2001) after initial destruction of the WTC.

The findings underline the importance of sound science to support engineering and management actions in an emergency situation. It also points out the need for engineers to be flexible in setting up monitoring and measurement equipment, using what is available at the time, because public concern is immediate and major emergencies require as rapid a deployment as possible.

A lesson from the WTC response is that accurate and reliable data are difficult to obtain in real time. Even if much data are available, not every chemical species of concern can be monitored reliably, nor can risk assessments be prepared that are completely unassailable. There is always a great deal that one does not know, but the public will demand some information regarding the likelihood that they have been or will be exposed to hazardous chemicals and substances. The engineer must find the appropriate balance of conveying what should be said without saying things that the data cannot support. Prematurely declaring something safe may be worse than finding later that the exposures and risks were in fact much less than predicted. This is the precautionary principle at work. Of course, declaring everything potentially dangerous erodes the engineer's credibility, so eliminating some of the obvious nonproblems can help both the engineer and the public to begin to set cleanup priorities.

A principle lesson learned at the WTC is the importance of risk communication and risk perception. We have seen some of these problems in other emergency situations before, but the unsettling nature of the attacks, the size of the area of devastation, and the unprecedented nature of the WTC site led to an environment of uncertainty, and even distrust. Numerous times, the scientists and engineers who collected data and shared results fell short in conveying their message to the public. People were extremely concerned about possible health effects and needed reliable information to decide when it would be safe to move back to their homes and businesses. Very tough questions were asked. Scientists and engineers are often ill equipped to handle them. The next chapter provides some guidance on how we can communicate effectively with those who have placed their trust in us as professionals.

Discussion: Choosing the Correct Monitoring Equipment

The WTC response highlighted the importance of selecting and siting the appropriate monitoring equipment. Let us consider the need to measure the concentrations of airborne dioxins and furans. As discussed in Chapter 3, these compounds are generated under certain conditions when chlorine-containing substances are combusted. As a result of the burning jet fuel, building material and furnishings made of polyvinyl, polybutyl plastics, polymers, and other chlorinated substances subsequently burned. Shortly after the attack, environmental engineers and scientists decided that these conditions created a sufficient likelihood that chlorinated dioxins and furans may be present in the smoke from the WTC fire.

As is often the case, the first step in monitoring at the WTC was a high-level screening step. The first step was to collect 12-hour air samples and analyze these samples for the dioxin congener suite. Most of these samples were found to be below detection limits of the analytic equipment. Had the monitoring stopped at this point, there would have been a high probability of a false negative, which means that even though the testing showed no dioxins, they were present because the tests were not sufficiently sensitive.[11] Thus at the WTC another screening step consisted of collecting settled particles, extracting and analyzing the particles for a suite of dioxin and furan congeners. These tests actually showed the presence of dioxin compounds, but most at levels considered background for most U.S. urban areas.

Both of the screening steps were limited by science and technology. The airborne screening technique had too short of a time and too little air volume for measuring dioxins and furans. The dust collection technique included only settled particles that were formed at unknown times. That is a dramatic example of why selecting the appropriate monitoring method for the situation at hand is so important. This is one of the major reasons that regulatory agencies publish methods that must be used for the data to be acceptable.[12]

The methods for measuring compounds are often published according to the type of environmental media in which they are found. Thus individual methods are available for water, soil, and air. In the case of dioxins, a soil method[13] was used for the settled dust, and an air method[14] was used for airborne dioxins. The soil method selection was straightforward because settled dust is physically similar to soil, and the means for collecting samples and bringing them to the laboratory were almost identical to those for soil.

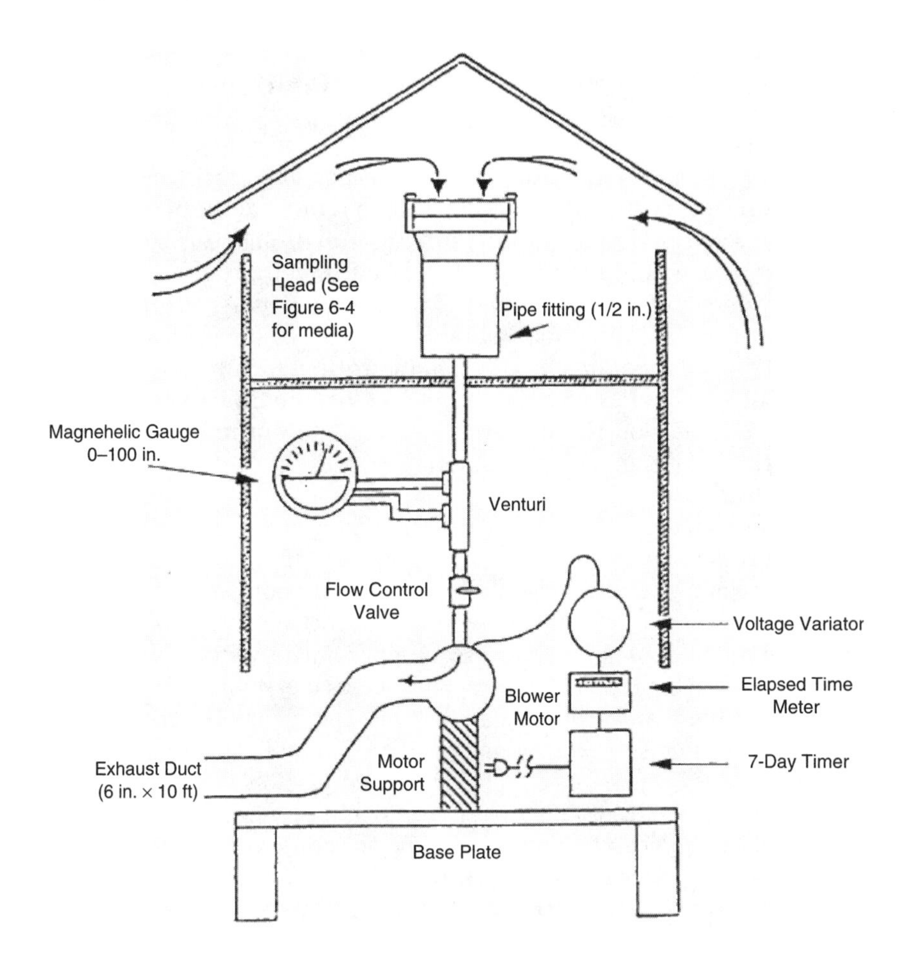

FIGURE 6-3. Typical dioxin/furan high-volume air sampler. (*Source:* U.S. EPA, Report No. EPA/625/R-96/010b, 1999.)

The airborne dioxin method was complicated. First, the EPA method requires the use of a high-volume air sampler equipped with a quartz-fiber filter and polyurethane foam (PUF) adsorbent for sampling 325–400 m^3 ambient air in a 24-hour sampling period (see Figures 6-3 and 6-4). Unfortunately, these high-volume samplers require AC power, which was not available near Ground Zero. Thus it was decided to locate dioxin samplers as close as possible to the WTC fire, where AC power was available. It was also decided that three samplers would be deployed to ensure that the plume would be followed even in the event

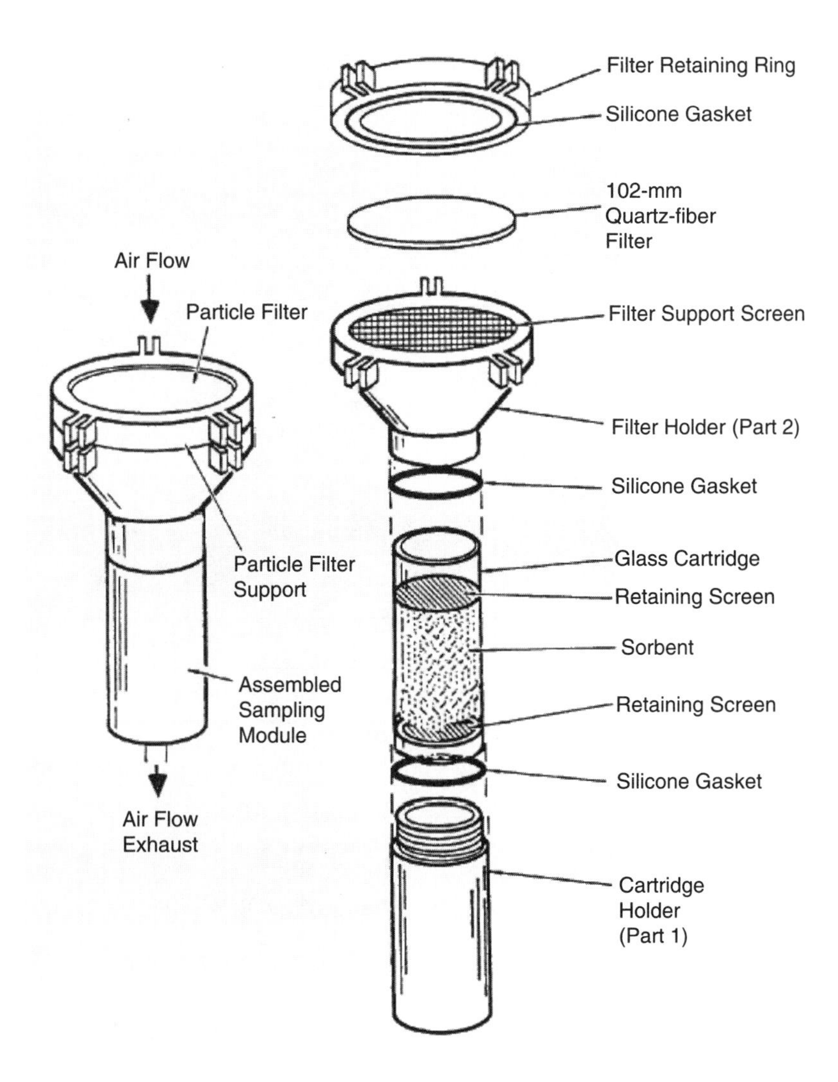

FIGURE 6-4. Typical absorbent cartridge assembly for sampling dioxin/furans. (*Source:* U.S. EPA, Report No. EPA/625/R-96/010b, 1999.)

of major wind shifts (i.e., sited according to the wind rose). Thus the samplers were sited on the rooftops of three lower Manhattan buildings.

The sampler uses the PUF to trap gas-phase dioxins and the filter to collect liquid and solid particles. The two media must be extracted and analyzed by high-resolution gas chromatography/high-resolution mass spectrometry (HRGC-HRMS). The published method provides

COMPENDIUM METHOD TO-9A
FIELD TEST DATA SHEET
GENERAL INFORMATION

Sampler I.D. No.: ______________
Lab PUF Sample No.: ______________
Sample location: ______________

Operator: ______________
Other: ______________

PUF Cartridge Certification Date: ______
Date/Time PUF Cartridge Installed: ______
Elapsed Timer: ______
Start ______
Stop ______
Diff. ______
Sampling

M1 ______ B1 ______
M2 ______ B2 ______

	Start	Stop
Barometric pressure ("Hg)	______	______
Ambient Temperature (°F)	______	______
Rain	Yes____	Yes____
	No____	No____

Sampling time
Start ______
Stop ______
Diff. ______

Audit flow check within ±10 of set point
____ Yes
____ No

TIME	TEMP	BAROMETRIC PRESSURE	MAGNEHELIC READING	CALCULATED FLOW RATE (scmm)	READ BY
Avg.					

Comments

____
____
____

FIGURE 6-5. Example field test data sheet. (*Source:* U.S. EPA, Report No. EPA/625/R-96/010b, 1999.)

chromatographic information, including the mass-to-charge ratios for each dioxin and furan congener, as well as the exact masses and elemental compositions of the congeners.

In the WTC case, measuring the total (solid, liquid, and gas phases) dioxin concentrations was sufficient. There will be other situations, however, for which the engineer needs to know the concentrations in each physical phase (i.e., the so-called phase distribution or phase partitioning).

In a response situation, the engineer must evaluate the available information, including any screening level data, and decide what additional data will be needed. This includes deciding which compounds must be measured, the intensity of sampling, the spatial coverage, the sampling frequency, and the minimum levels of detection needed.

As in every type of hazardous waste effort, reliable documentation and credible record-keeping practices are essential to project management. The published methods can help direct the engineer to the type and level of documentation needed for environmental monitoring at a site, even providing field data sheets (see Figure 6-5).

CHAPTER 7

Risk Perception: What You Say May Not Be What They Hear

What Are People's Perceptions of Risks Posed by Hazardous Waste?

Back in 1942, Elmo Roper, the famous pollster, said that "many of us make two mistakes in our judgment of the common man. We overestimate the amount of information he has; and underestimate his intelligence." Roper was surprised that the general public often has too little information to decide on important matters. Roper was even more surprised that despite this lack of sufficient information, the common person's "native intelligence generally brings him to a sound conclusion." This is important for the engineer to remember when dealing with people who will potentially be affected by environmental actions (and inactions). We, as experts, must provide people with ample information and credible science, while respecting the intelligence of these people. They may not care much for science and engineering, but when their lives, livelihoods, and peace of mind are threatened, one can expect them to be keen on the engineer's every word.

The hazardous waste challenge for the environmental professional goes beyond the credible assessment of risk and a well-engineered remedy discussed so far. The risks and proposed remedies must be properly articulated and openly shared with the concerned community. This process can be complicated and difficult. Because risk assessment, as applied to environmental science and engineering, is relatively young, people are naturally skeptical about its application to important decisions that will affect their health, their neighborhood, their property values, and their livelihood. The former Administrator of the U.S. Environmental Protection Agency, William D. Ruckelshaus, points out that the word *risk* was seldom mentioned in the early 1970s, and as a result played no major role in clean water and clear air legislation developed during that period.[1] Ruckelshaus attributes much of the increased focus on risk to the public's concern about PCBs and asbestos and their growing associations with cancer. Today, certain substances are recognized to cause cancer, but the public was just beginning to make these connections in the 1970s.

For decades, people have experienced a growing concern that the public health was being threatened in a clandestine manner. The new contaminants were odorless, tasteless, and invisible. So, the only way to get a handle on the risks posed by these substances was to begin to build a new scientific framework to characterize and predict the risks. In other words, people could not trust their own senses to assess hazards and risks. They were completely dependent on the scientific community to tell them about possible outcomes under various exposure scenarios.

This daunting challenge for scientists and engineers has often worked well, so long as the scientific community can gain and maintain trust. Unfortunately, environmental risks have a checkered track record in such trust. The public's trust depends on several variables. A good place to begin to identify how to win and maintain public confidence in environmental remedies is for the engineer to compare how environmental experts may differ from the public in perceiving risks. This is a function of differing methods of thinking and intuition. Vincent Covello developed a list of factors that may account for these differing perceptions.[2] These factors serve as an outline of why an engineer may confront problems in dealing with the public even when—at least from the engineer's vantage point—all of the technical considerations have been met.

What Is the Possibility of a Severely Negative or Catastrophic Outcome?

The public is less likely to trust the engineer, even in a well-designed remediation effort, if the possible negative outcomes are centralized in time and space, compared to those that are more scattered and random. This is problematic for hazardous waste engineers because they are usually called in after contamination has been observed in some manner at a specific site and at a certain time. Ironically, the engineer may increase the public's concern by properly investigating and characterizing the site. The engineer has grouped the negative outcomes in space (e.g., the site's location and the extent of contamination of soil, water, and air have been characterized) and time (e.g., the source has been documented and the movement of the contaminants has been modeled retrospectively and prospectively).

The engineer should be clear, careful, and sensitive when describing the site and possible remedies. Even when the potential prognosis for site remediation is good, exposures to possible contaminants can be effectively eliminated, and existing technologies have worked well in other similar situations, the public will not automatically be reassured. When the professionals describe what is to be done, the community members may perceive something different from what the engineers and scientists are trying to convey. The community members' perception may be that they are living near another "Love Canal" or "Times Beach."

How Familiar Are the Situation and the Potential Risks?

People fear what they do not understand. Covello asserts that the public is generally more concerned about unfamiliar risks. This may help explain why so large a segment of the population is comfortable with cigarette smoking but terrified of the storage of spent fuel from nuclear power plants, even when the former hazard accounts for much more disease and death. The nature of nuclear science and information is mysterious to many people. The same is true for hazardous wastes.

One major problem with hazardous wastes involves the associated nomenclature and vernacular. Although a few hazardous wastes can be well understood by a broad audience, such as the leaking of leaded gasoline from an underground storage tank, most hazardous wastes are mixtures of ominous-sounding compounds. Many are organic compounds, with various congeners and isomers. The challenge for the professional is to describe the compounds sufficiently so that all parties understand what is at risk.[3]

Another problem is the complex, or at least complex sounding, methods used to test, model, and characterize the actual and predicted movement and change of these compounds under different remediation scenarios. The public must understand the difference between no action and the other remediation alternatives. For example, the engineer must explain that without intervention, the plume of X, Y, and Z contaminants will move 10 meters per year vertically and 100 meters per year horizontally. The engineer will also need to explain how this relates to sensitive receptor sites, such as drinking well intakes and stream inputs. In addition, the chemical, physical, and biologic transformation processes must also be explained, so that what may have been released has changed, in part, to other compounds. So, in addition to X, Y, and Z, other degradation products X', X'', and Z' in the water, soil, and air at various times must be measured. Such equilibrium chemistry is complicated for engineers and scientists, let alone those members of the community who do not confront it frequently (if at all).

Likewise, all alternative approaches to remedy the situation must explain these same processes and models, including the uncertainties involved in predicting success. These descriptions are further complicated as a function of available engineering controls and remediation steps, each of which must also be explained to the community's satisfaction. So, if a pump-and-treat alternative is being proposed, then all of the chemistry, physics, and biology associated with this technique must be explained. In addition, the public must completely understand how the approach will be evaluated in terms of success. The success is not only to be explained in terms of engineering performance standards like the total volume of water treated and the target level of contaminant removal (e.g., 99.99% removal efficiency), but the quality of the environment following the removal must also be described (e.g., the aquifer's water will contain x ng L^{-1} X, Y, and Z, the soil following treatment will contain y ng kg^{-1} X, X′, Y, and Z).

Can the Engineer Succinctly Explain the Processes and Mechanisms Being Proposed or Undertaken?

The likelihood of engendering public trust decreases in relation to the complexity of the processes and mechanisms of exposure and risk. When such systems are not well understood, and the engineer is unable to clarify them, the community is more likely to be concerned about the problem and the proposed remedies. As mentioned in the discussion on familiarity, hazardous waste processes can be highly complex and involve numerous variables.

The successful engineering solution is one that can be explained properly by the engineer and comprehended by the affected public. This is no easy task, but it *must* be done. The best engineering solutions are worthless if they are left in the design phase and never implemented.

How Certain Is the Science and Engineering?

People tend to lose confidence in science and engineering when, in their view, there is too much uncertainty in outcomes and risks associated with a remedial action or any other important public health or environmental endeavor. Uncertainty in science arises from several sources. Even the most carefully conducted test of a chemical has some degree of variability in the data it provides. If the data are produced from different studies conducted by different laboratories, this will add uncertainty to the data.

Environmental measurement and other technologies continue to change and improve, so comparing historical data may add uncertainty, at least as perceived by the public.[4] In reality the old "non detects" may have been just as high or higher than more recent analyses. For example, if a table of findings shows that a certain pesticide's concentration in soil was found in the 1980s to be undetected, but was increasingly found in the soil at about 10 micrograms per kilogram ($\mu g\ kg^{-1}$) in the 1990s, the first question the engineer should ask is what were the detection limits for the pesticide in soil, and how have these limits changed over time? The concentration of the pesticide may have not changed, or may even have fallen, over the two decades, but the retrospective data could neither confirm nor reject this finding.

Multiple measurements by different laboratories will give varying results. The quality assurance/quality control (QA/QC) plan will define data quality objectives that must be met for any study. For example, preliminary screening level studies may satisfy the data quality objectives by simply seeing whether a chemical exists in an environmental medium (a so-called detect/nondetect study), whereas a hazardous waste site investigation will require more stringent data. The former is sometimes referred to as a qualitative or semiquantitative evaluation, whereas the latter, more rigorous study requires quantitation.

In dealing with the public, the hazardous waste engineer should be clear about the uncertainties of the data and information from which decisions are being made. Again, full disclosure is required.

How Much Personal Control Is Perceived?

People are generally more comfortable when they have a modicum of control. Unfortunately, when dealing with hazardous wastes, the public can be alienated by the sophistication of the physics of remedies being conducted by a cadre of outsiders. The public's input must be sought and incorporated into all remediation efforts.

Is the Exposure Voluntary or Involuntary?

Cigarette smoking has shown us that scientific research can provide important, even sound advice, but predicting how the public will incorporate this advice into their daily lives is difficult (see the discussion: "Choose Your Route of Exposure"). Surely, one important factor in the public's acceptance or rejection of even the most sound scientific advice is whether it interferes with their choices in the matter. If people consider being exposed to dangerous chemicals to be their choice, they are more likely to accept the risks associated with those exposures. Conversely, the public may reject sound scientific and engineering advice that detracts from their freedom to select the "best" option.

Again, the engineer must ensure to the best extent possible that a wide array of the public is included in the earliest stages of hazardous wastes responses. This goes beyond the required public meetings and includes notices, letters, and other forums *asking* which of the possible options the public would choose to solve the problem.

Discussion: Choose Your Route of Exposure

An interesting phenomenon seems to be taking place on today's college campuses. From some anecdotal observations, it would appear that students are more concerned about some exposure pathways and routes than others. It is not uncommon at Duke University, for example, to see a student smoking a cigarette and carrying bottled water. For some reason, the student is not as concerned about the potential carcinogens in tobacco smoke as the contaminants found in tap water. Or is it simply taste—or mass marketing?

This observation does demonstrate at least two of Covello's principles regarding increased concern about risk, whether the student maintains some control over risk decisions and whether the exposures and risks are voluntary or involuntary.

Are Children or Other Sensitive Subpopulations at Risk?

Children are particularly sensitive to many environmental pollutants. They are growing, so tissue development is in its most prolific stages. In addition, society has (and certainly should have!) special levels of protection for infants and children. For example, regulations under the Federal Food Quality Protection Act mandate special treatment of children, evidenced by the so-called 10X Rule (see Discussion). This rule recommends that, after all other considerations, the exposure calculated for children should include 10 times more protection (thus the exposure is multiplied by 10) when children are exposed to toxic substances.

Discussion: Children's Safety Factor

Engineers are familiar with factors of safety, which are usually applied to equations to address uncertainties. For example, an equation accounting for a roof design not only includes material properties and load, but also includes a factor of safety. Similar factors are needed to protect public health. The Federal Food Quality Protection Act (FQPA)[5] requires that risk assessments related to children include a safety factor regarding the potential for prenatal and postnatal effects. Prenatal and postnatal toxicities are often included when calculating a reference dose (RfD) or margin of exposure (MOE) from the prenatal or postnatal adverse effects in the offspring and traditional uncertainty factors (see Chapter 2 for a discussion of NOAEL, LOAEL, and RfDs and how risks associated with chronic effects can be extrapolated from subchronic studies and from incomplete toxicology data); however, uncertainties or an elevated concern for children are not always sufficiently addressed using uncertainty factors in the RfD and MOE.

Thus the FQPA requires an additional evaluation of the weight of all relevant evidence. This involves examining the level of concern for how children are particularly sensitive and susceptible to the effects of a chemical and determining whether traditional uncertainty factors already incorporated into the risk assessment adequately protect infants and children. This evaluation is accomplished mathematically in the exposure assessment. The U.S. EPA has prepared guidance on how data deficiency uncertainty factors should be used to address the FQPA children's safety factor. The final decision to retain the default 10X FQPA safety factor or to assign a different FQPA safety factor is made during the characterization of risk and not determined as part of the RfD process. The weight-of-evidence approach, therefore, includes both hazard and exposure considered together for the chemical being evaluated. The FQPA safety factor for a particular chemical must have the level of confidence in the hazard and exposure assessments and an explicit judgment of the possibility of other residual uncertainties in characterizing the risk to children.

By extension, other sensitive strata of the population also need protection beyond those of the general population.[6] Elderly and asthmatic members of society are more sensitive to airborne particles. Pregnant women are at greater risk from exposure to hormonally active agents, such as phthlates and several pesticides. Pubescent females undergo dramatic changes in their endocrine systems and, consequently, are sensitive to exposures during this time.

When Are the Effects Likely to Occur?

People may not like acute effects, but they are more likely to accept them than those that manifest themselves only after a protracted latency. Therefore, people will endure some short-term risks to prevent future problems. The engineer should clearly state the acute and chronic outcomes that may result from all phases of remediation.

Are Future Generations at Risk?

If there is any risk to future generations, the public will be concerned. This partly explains many people's discomfort with nuclear power generation and nuclear wastes (which have half-lives of hundreds of thousands of years) that will leave a dangerous legacy. Hazardous wastes, especially the so-called persistent, bioaccumulating toxic substances (PBTs), like dioxins, are also perceived by people as something that they do not want to pass along to future generations. In fact, the engineer should be prepared for such concerns from the news media and public forums.

Are Potential Victims Readibly Identifiable?

The public, according to Covello, is more concerned about real victims than about statistical victims; however, it is difficult to explain what a one-in-a-million risk means and even more difficult if this risk is described using engineering notation (i.e., risk = 10^{-6}).

Vivid examples of Chernobyl and Hiroshima have provided graphic images of real victims of radiation exposure. Love Canal has done the same for hazardous wastes. So, when the engineer attempts to characterize risks by the numbers, there may not be a complete appreciation of what those numbers mean.

How Much Do People Dread the Outcome?

The concept of dread is important in risk perception and communication. The more dread that is associated with a hazardous substance, the more concern the public will have about dealing with it. The health effects associated with toxic substances dictate the public's concern. Arguably, the most dreaded effect is cancer. Most hazardous waste sites are contaminated with carcinogens, so the engineer should be prepared to address people's concerns about these carcinogens (and should not expect the participants to be coldly objective about the various remediation efforts). Carcinogens are not the only chemicals associated with large dread factors. For example, witness the mothers who have expressed at public hearings their dread of the possible learning disabilities and central nervous system problems in their children when they find out that their drinking water or air has been contaminated by lead emissions from a nearby smelter. This extended dread is particularly important when it also includes risks to children and future generations.

Telling people that the success of a remedy is to reduce the cancer risk to less than 10^{-6} is not sufficient to allay their fears. The engineer must be sensitive to the possible misinterpretation of data and recommended actions and find ways to make this information more understandable.

Do People Trust the Institution Responsible for Assessing the Risk and Managing the Cleanup?

All institutions have issues. No matter how good the engineer's reputation in dealing with hazardous wastes, the association with the government agencies and firms that are involved in the project will influence the public's acceptance of remediation plans or even in people's willingness to trust the data presented at public forums. Individual medical doctors and scientists have high trust levels with the public, but the public's distrust of government agencies and corporations has been growing.[7] The engineer should expect to be "guilty by association" in planning for public involvement (i.e., hazardous waste siting and remediation is not likely to begin without a certain amount of skepticism and resistance from the public).

What Is the Media Saying?

If the reports in the newspapers and other parts of the news media have documented a history of problems at a site, and if there is much media attention, the public's concern will be heightened. All the engineer can do to address this concern is to be accessible to the press (adhering to the communication strategy developed by the government agency and other parties) and deal openly and honestly with all inquiries. This is not the time or the place for "spin."

What Is the Accident History of This Site or Facility or of Similar Sites or Facilities?

If the company responsible for cleanup has a poor history of accidents or a track record of incidents related to hazardous chemicals, the public concern can be expected to be heightened. If the types of corrective and remedial actions being proposed have a checkered past or have undocumented success, this will also carry over to the plans proposed for a specific site or facilities. People do not appreciate being used as "guinea pigs."

This does not mean that actions that have not worked elsewhere should be dismissed out of hand. It does require, however, an accounting of why the previous plan failed and why one would not expect similar failures at this particular site. After the engineer is satisfied that the conditions are sufficiently different to warrant recommending an action, the reasons for expected success should be thoroughly explained. For example, the public is owed a discussion of how their situation is matched to the engineering solution.

Is the Risk Distributed Equitably?

The history of environmental contamination has numerous examples in which certain segments of society are exposed inordinately to chemical hazards. This issue has been particularly problematic for communities of low socioeconomic status. A landmark study showed that landfill siting and the presence of hazardous waste sites in a community was disproportionately higher in African American communities.[8] Hispanic workers can be exposed to higher concentrations of hazardous chemicals where they live and work, largely because of the nature of their work (e.g., agricultural chemical exposures can be high shortly after fields are sprayed).

Even a scientifically sound remedial action will be resisted in neighborhoods that have had to deal with past injustices. Sensitivity to these experiences should be part of any risk communication plan.

Are the Benefits Clear?

The engineer and the planning team may be well aware of why the remediation is being undertaken. In fact, the benefits of risk and exposure reductions may be so obvious that the engineer is tempted to give merely a short consideration and attention to this topic in meetings with the public and move directly to the more "technical" discussions, such as target clean-up levels. This is a mistake. In order for the public to comprehend the plan of action fully, the expected benefits must be clearly articulated. This includes outlining the improvements resulting from hazard reduction, exposure reduction, and prevention of health and environmental effects. It may also call for "listening sessions" with neighbors, without any recommendations from the professionals, before moving to technical meetings.

If There Is Any Failure, Will It Be Reversible?

The potential irreversibility of damages is akin to other public concerns about future generations and controllability; however, the public is also looking to the experts to provide reassurance that the site will be monitored during and well after remediation to prevent catastrophes or at least to catch problems before they become large and irreversible. The monitoring component of the plan should stress that this is why measurements are taken before and after completion of the remedies.

What Is the Personal Stake of Each Person?

Each person's interest in and concern about the project is unique. When dealing with a person, the engineer should be sensitive to that person's particular concerns. For example, a person living adjacent to the site may have a greater personal stake in the health issues than someone living a mile away; however, the person living a mile away may own property that could become more or less valuable, depending on the remedial actions selected.

What Is the Origin of the Problem?

Members of the public are generally more tolerant of and patient with solutions needed to address natural disasters than they are about those problems caused or exacerbated by humans. All hazardous waste problems should be considered human-derived, even if they are worsened by natural causes. For example, if a tank is ruptured during an earthquake and hazardous chemicals contaminate an aquifer, this should be treated as a human-caused problem (because humans built and installed the tank in the first place). If the chemicals are the same, the exposure and toxicological considerations are the same for either anthrogenic or natural chemical hazards. However, the public's perception of how and when to take action is likely to be different.

What Is the Bottom Line about Risk Perception?

The bottom line is that assessing risks is complicated but follows a fairly linear, step-wise scientific approach. Risk management is less linear but is driven by the assessment. Risk perceptions, however, are highly variable, even unpredictable, so when communicating risks and involving the public in hazardous waste decisions, the engineer must be sensitive to a myriad of concerns about the hazardous waste site.

The engineers and planners must be open and must fully disclose the pros and cons of any action. Great care should be taken when sharing information and ideas with the public. A word or phrase may be perfectly clear to the person using it but might completely unsettle the already nervous and skeptical neighbors of a hazardous waste facility.[9]

The Enigma of Risk Perception

You may have seen the Müller-Lyer illusion (Figure 7-1). Usually, if people unfamiliar with this illusion are asked which of the two line segments, the top or the bottom, is longer, they will quickly select the top one. They will hold to this conclusion until or unless they can be convinced by sound scientific principles, such as the use of a measuring device (i.e., a ruler), or until the line segments undergo other investigation, such as comparing each segment to the length of one's finger. It is only then that they agree that the line segments are the same length.

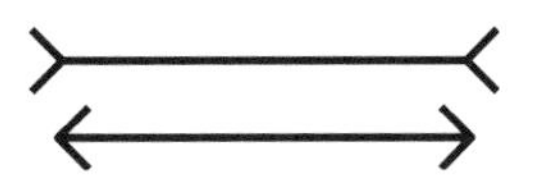

FIGURE 7-1. The Müller-Lyer illusion.

What does this tell us about data and our expectation as scientists that people's minds will change if only we can present sound scientific and engineering arguments? For one thing, when a person is confronted with the need to decide whether the person's intuition or scientific analysis is to be trusted, the person is likely to have greater confidence in the intuition.[10] The scientist and engineer should not be discouraged by this quandary. It simply means that the science is necessary, but seldom sufficient in public environmental decisions.

Using Benchmarks to Explain Exposures and Risks

When we share data and findings, it is best to present them within a context understandable to the people affected or potentially affected by our proposed actions. This means we will need to compare what we have found in their neighborhood or at their waste site to other neighborhoods or waste sites. Or, we may be able to compare the concentrations of pollutants at their site to national or other standards. Such standards have the advantage of having gone through rulemaking processes, public hearings, and scientific peer review panels. They are also usually in some way based on protecting public health and the environment. So, when data such as those for lead concentrations in airborne particles near the World Trade Center site (see Figure 7-2) are presented and compared to national standards, there is likely to be less concern, or at least more certitude, about the benchmark itself than simply comparing it to other waste sites around the country.

Unfortunately, sometimes there are no standards available or the standards have been written with regard to environmental media or situations different from the one of concern to a particular neighborhood. In this instance, the measurements may need to be compared to "ordinary" concentrations. For, example, a pneumonia patient in the hospital may be breathing an air mixture of nearly 100% oxygen. It is instructive for safety reasons to inform the health care personnel that the ordinary or mean oxygen content of the troposphere (the part of the atmosphere where we live and breath) is about 21%. The higher oxygen content is associated with greater flammability hazards. Thus, when the people living and working near Ground Zero in Lower Manhattan wanted to know about their exposures to dioxins, the benchmarks were more problematic. Thus, the airborne concentrations shown in Figure 7-4 are compared to a so-called "screening level." This level is very conservative. It is the total dioxin concentration representing the cancer risk of 10^{-6} for a population living at the monitoring site 24 hours per day for 30 years. The fire actually burned for months, but certainly not 30 years. Another dioxin benchmark is the "action level," which is a level above which health concerns are raised by health agencies. This level for dioxins, 300 parts per billion, is not much greater than one would expect to find in the soil of many back yards in the United States.

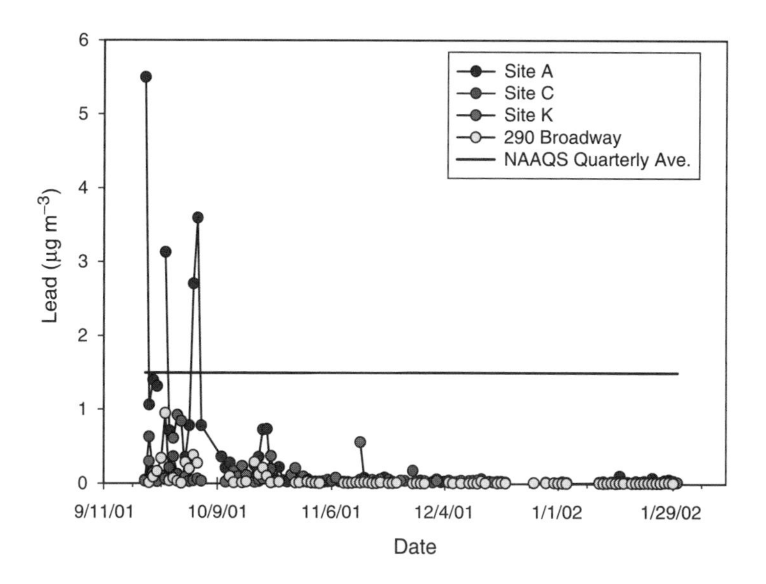

FIGURE 7-2. Lead (Pb) concentrations of fine particle matter less than 2.5 µm ($PM_{2.5}$) found at sites A, C, and K and a background site at 290 Broadway Federal Building in Lower Manhattan, NY (see Figure 7-3). The measured concentrations are compared to the National Ambient Air Quality Standard (NAAQS) for quarterly mean concentrations of airborne Pb. (*Source:* A. Vette, M. Landis, E. Swartz, R. Williams, D. LaPosta, M. Kantz, J. Filippelli, L. Webb, T. Ellestad and D. Vallero, 2002, "Concentrations and Speciation of PM at Ground Zero and Lower Manhattan Following the Collapse of the WTC," International Society of Exposure Analysis Annual Conference, August 2002.)

Risk Perception Lessons in Other Fields

The environmental scientist and engineer should not feel that their challenge is unique. Others in fields ranging from security to health care to transportation planning must deal with the dichotomy of actual versus perceived risk. For example, in 1992 three modes of transportation in the United States accounted for about the same number of accidents: 775 from air travel, 755 from rail, and 722 from bicycles. The public, however, perceived that the risk associated with air travel to be much higher than the risk of rail, and much less for bikes. The reasons are explained in this chapter, but the greatest were the dramatic, catastrophic events with the accompanying media coverage, and the lack of control in air travel compared to the perception of almost complete control in biking.[11]

Another recent example of skewed public perception is that of the Homeland Security proposal to immunize the entire U.S. population against smallpox. The Centers for Disease Control and Prevention's estimated risk of deaths from the vaccine is one in a million vaccinations. Since the studies

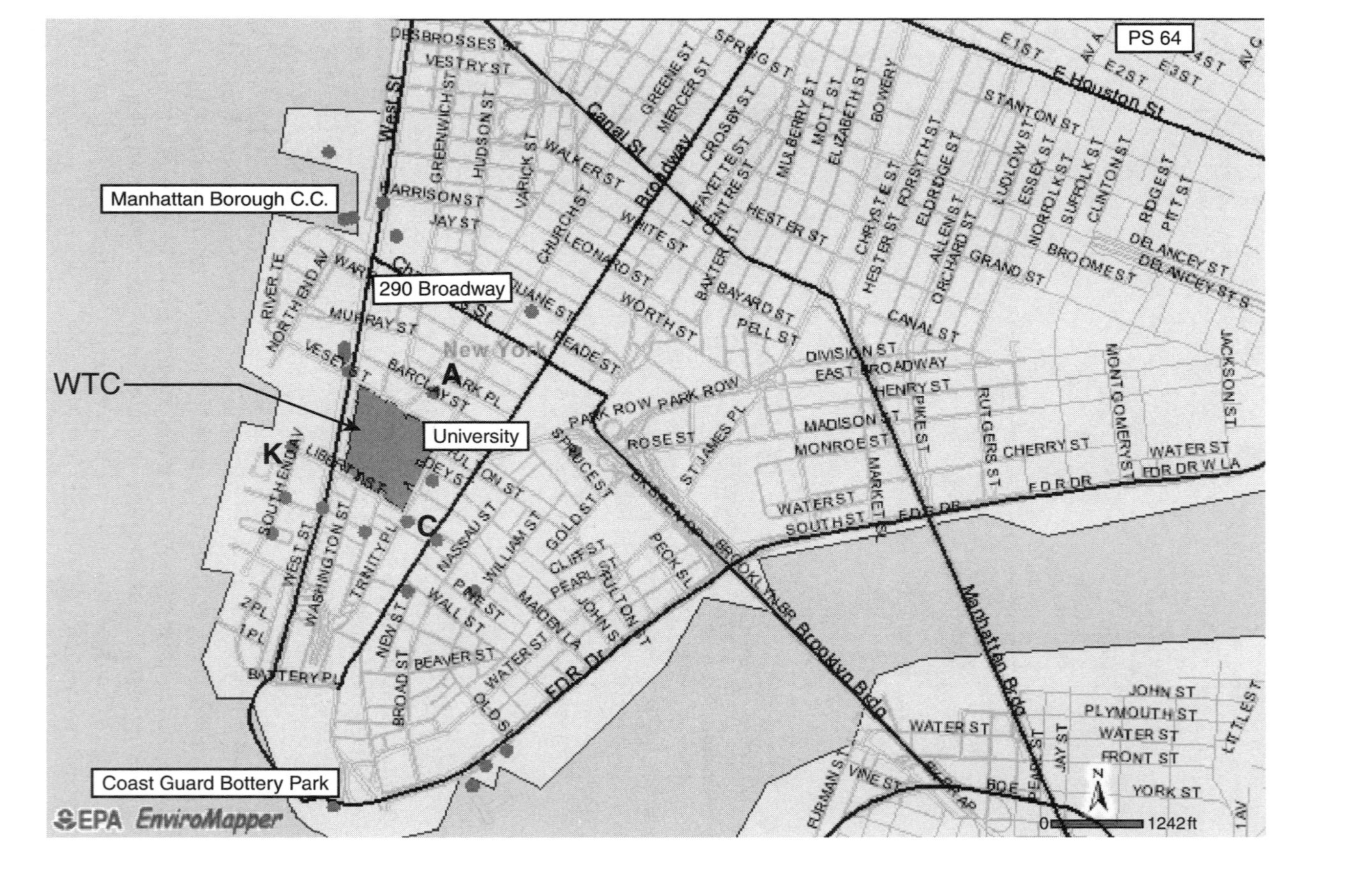

FIGURE 7-3. Location of Monitoring Sites Referenced in Figure 7-2. (*Source:* EnviroMapper, U.S. Environmental Protection Agency.)

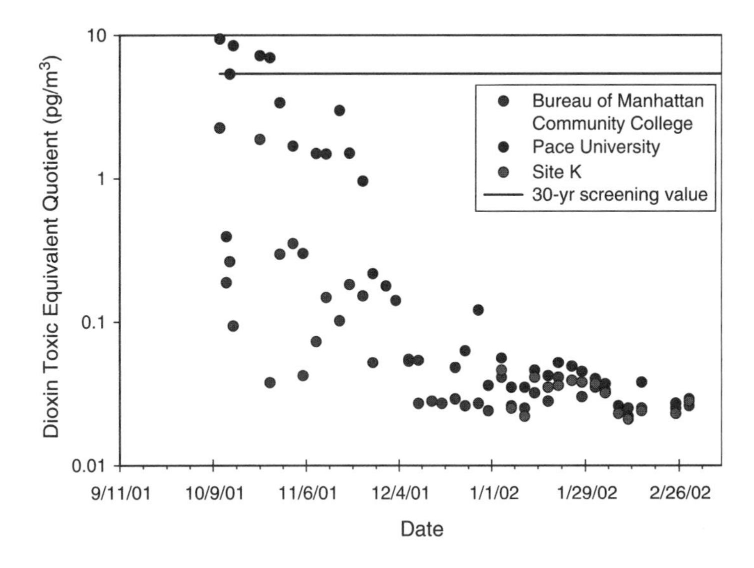

FIGURE 7-4. Measurements of airborne dioxins and furans at outermost monitoring sites shown in Figure 7-3. The measurements were 24-hour integrated samples. The concentrations of dioxin/furans (17 congeners in TEQ) were below all action levels (1 ppb). All concentrations after October 2001 were below the screening level.

that were used to arrive at this risk are dated, and there appears to be a greater number of immuno-compromised individuals in the United States than when the studies were conducted, so the risks may be greater. But, assuming that risk is close to correct and everyone in the United States is vaccinated, that would mean about 300 fatalities.

Pity the policy makers. Since the U.S. government has decided to have only limited numbers of vaccinations (e.g., emergency response personnel and health care providers), a massive contingency plan will likely have to be developed in the event of outbreak. The risks associated with the implementation of this plan, such as rapid mobilization and dangerous logistics to get the vaccinations where they are needed, may be greater than those of the vaccinations.

Often, the public must make so-called "risk tradeoffs."[12] Nosocomial infections, those that the patient contracts *after* being admitted to the hospital, is a serious health care problem, which may become more serious with any increases in microbial resistance to antibiotics. Law enforcement risk tradeoffs include the crime and corruption associated with attempts to reduce the risks associated with illegal drugs, and the benefits of deterring break-ins by keeping a gun at home associated with a three-fold increase in gunshot mortalities for gun owners (in large part the result of the increased availability of the weapon during domestic disputes). Environmental examples include the use of asbestos in building materials to reduce the risk of fires, but increasing the risk of lung diseases. Even, in hazardous waste cleanups, the risk of chemical exposure to the nearby population is hopefully decreased, but other risks, such as accidents in construction activities, are increased.

This is the stuff of ethical decision makers. These are classic cases of trying to weigh perceived risks, not only against benefits, but against the risks from the chosen action or inaction.

CHAPTER 8

Closing Thoughts on the Future of Hazardous Waste Engineering

In the 1970s, the phrase *hazardous waste* was rarely used. The next few decades saw infamous chemical catastrophes in Love Canal, Times Beach, the Valley of the Drums, Seveso, Italy, and Bhopal, India, to name just a few. These incidents were simply the tip of the iceberg. For each of these notorious cases, hundreds of smaller sites and facilities were being threatened by the release of toxic contaminants. In the relatively short span of time since Love Canal, engineers have stepped up to the challenge of protecting the health and environment from risks posed by hazardous wastes. The public has demanded answers and has been willing to use public funds to deal with these risks. Engineers have opened whole new frontiers in designing treatment, storage, and disposal facilities; have helped to ensure that hazardous substances are transported safely; and have found new methods for measuring and modeling the fate of these substances after their entry into the environment. Civil engineering has been at the forefront of this revolution. Environmental engineers, with their particularly strong emphasis on risk assessment and management, have led the way.

The environmental engineer's risk management roles will continue to grow in importance. We have offered several ways for the engineer to incorporate an appreciation for risks into every facet of hazardous waste management. The engineering profession has progressed in its understanding of the physical, chemical, and biologic principals needed to confront the wastes that have been released and to find new ways to prevent releases and exposures in the future. Beyond this, we have incorporated the social sciences and humanities into our approaches for managing hazardous wastes, as evidenced by the success of waste exchanges and risk communications.

The learning curve has been uneven. Like the rest of the civil engineering profession, hazardous waste engineers have learned many lessons from September 11, 2001. Many apply to engineering the risks of hazardous wastes. We can engender trust in those who rely on us as professionals. We are up to the challenge of making our environment a safer and healthier place. How thoroughly would one approach a project if it affected one's own family and neighborhood? The answer to that question finds its way into all plans and actions addressing hazardous wastes.

Engineering is a high calling. The public has rightfully placed a great amount of trust in the engineering profession. We as engineers have the ability to place ourselves in the shoes of the people affected by our work. Our capacity to appreciate and even empathize with a company or community threatened by releases of hazardous wastes is among our greatest strengths. We have taken our public trust to heart. *Credat emptor!* "The client can trust us."

Perhaps it is appropriate to end this book with a reminder of what it means to be an engineer. We are a helping profession. In a sense, all hazardous waste engineering is "value engineering."[1] That is, we must obtain the maximum per-unit value from every step in identifying and providing remedies to hazardous wastes problems; however, value is more than dollars and cents. It is even more than credible applications of sound science. Value is a human construct.

As we strive to solve the hazardous wastes problems, let us keep in mind canons of the civil engineer.[2] First, we must protect the safety, health, and welfare of the public. This is the driving force behind hazardous waste engineering. Second, we must be competent. This means we must understand the sciences underpinning our environmental recommendations. We must be objective and truthful, and we must fairly represent those who have entrusted the profession to us. We must have integrity, honor, and dignity in the performance of our duties. Finally, we must recognize our need for continual growth as scientists and engineers. Environmental challenges are a certainty of modern life. They are also a certainty for the engineering profession. New and bold approaches are the province of hazardous wastes engineering.

Ironically, the call to reduce and manage the risks posed by hazardous wastes is risky to the engineer. Unlike in the laboratory, managing hazardous wastes in the real world does not give the engineer the luxury of controlling all the variables; however, an engineer who is well informed with a sound science underpinning is more likely to be successful. The rewards are a cleaner environment and improved public health. In the words of St. Thomas Aquinas, "If the primary aim of a captain were to preserve his ship, he would keep it in port forever." The hazardous waste engineering ship has already left port. Let us ensure that its journey is successful.

APPENDIX 1

Glossary of Hazardous Waste Engineering Terminology

Abandoned waste site A hazardous waste site that has been closed and is no longer in operation and the original owner/operator is no longer in business.

Abandoned well A permanently discontinued well or one that is in a state of such disrepair that it cannot be used for its intended purpose.

Abatement Amelioration or reduction of pollution.

Abatement debris Waste from remediation activities.

Abiotic Nonliving components of the environment (opposite of biotic).

Abiotic factors Nonliving influences on an organism's functions.

Absolute error In statistics and quality assurance, the difference between the measured value and the true value.

Absorbed dose In exposure assessment, the amount of a substance that penetrates an exposed organism's absorption barriers (e.g., skin, lung tissue, gastrointestinal tract) through physical or biologic processes. The term is synonymous with internal dose.

Absorption 1. Penetration and collection of a chemical within the surface of a body (contrast with adsorption). A form of sorption. 2. The uptake of chemicals by an organism, making the chemical available to metabolic processes. 3. In soil science, the movement of ions and water into plants. Active soil absorption uses the plant's metabolic processes to remove chemicals, while passive absorption depends on chemical diffusion.

Absorption barrier Any of the exchange sites of the body that permit uptake of various substances at different rates (e.g., skin, lung tissue, and gastrointestinal tract wall).

Absorption factor The fraction of a chemical that reaches the cells and tissues of an organism. This is one of the factors of the exposure calculation.

Acceptable daily intake (ADI) The daily dose of a chemical that has been determined by research and scientific investigation to be free from adverse effects in the general human population after a lifetime of exposure.

Accident site Location of an unexpected occurrence resulting in a release of hazardous materials.

Acclimatization An organism's physiologic and behavioral adjustments to changes in its environment.

Accuracy The degree to which a measurement or statistic reflects the true value. More accurate measurements have lower absolute error.

Acid Corrosive compound with the following characteristics: (1) reacts with metals, yielding hydrogen; (2) reacts with a base, forming a salt; (3) dissociates in water, yielding hydrogen or hydronium ions; (4) pH less than 7.0; and (5) neutralizes bases or alkaline media.

Acid-catalyzed hydrolysis Enhanced hydrolysis resulting from protonation.

Action level (AL) The concentration of a contaminant where some intervention is required by a regulatory body. For example, the Occupational Safety and Health Administration's AL for an air pollutant is the concentration above which exposed workers must undergo medical monitoring. The Food and Drug Administration sets ALs to protect public health, for example, fish advisories would be issued when thresholds are exceeded in water bodies for either naturally occurring substances, such as paralytic shellfish toxins and trace elements like cadmium or mercury are geologically leached from the environment, or anthropogenically enhanced concentrations of substances, such as pesticides and combustion by-products, like polycyclic aromatic hydrocarbons or dioxins.

Activated carbon Highly adsorbent form of carbon with a great amount of particle surface area; used to collect toxic substances from gaseous emissions and to remove dissolved organic matter from wastewater.

Activated sludge The suspended solids, predominantly composed of the living biomass of microbes, found in the wastewater treatment plant's aeration tanks.

Activation The toxicological term for rendering a substance more toxic after being transformed biochemically after entering an organism. The transformations are mediated by biological catalysts, i.e. enzymes. An example of a compound being activated is the metabolism of the polycyclic aromatic hydrocarbon, benzo(a)pyrene (B(a)P). The compound becomes more toxic and carcinogenic when it is metabolized to an epoxide form, when an oxygen atom joins with two of the B(a)P carbon atoms.

Active chemical A chemical that readily combines with other chemicals.

Active ingredients The chemical(s) in a pesticide or pharmaceutical formulation that is responsible for the target effect. The other ingredients are inert.

Acute effect A disease or other adverse health outcome wherein symptoms occur shortly after exposure to a chemical.

Acute exposure Usually, the amount of an exposure received in one day or less.

Acute toxicity The ability of a chemical to cause adverse effects from acute exposure.

Addition reaction Chemical reaction, in which a molecule is added to another molecule that contains a double bond, and converts the double bond into a single bond.

Adiabatic process A change where no loss or gain in heat is involved.

Administered dose The amount of a chemical given to an organism.

Adsorption Collection of a chemical on the surface of a body. A form of sorption.

Aerobic Requiring the presence of molecular oxygen (O_2). Term is applied to microbial processes that decompose organic wastes and, if complete, yield carbon dioxide (CO_2) and water.

A horizon The uppermost layer of a mineral soil containing organic matter. This layer has the soil's largest amount of biologic activity and removal of soil material by chemical suspension and solution.

Air sparging Hazardous waste treatment technology that introduces air or other gases beneath the water table. Sparging combines volatilization (see *Stripping*) and bioremediation processes. Examples include in-well aeration and aquifer air injection.

Alkalinity Capacity of water to neutralize acids, attributable to the water's ionic strength from carbonates, bicarbonates, hydroxides, borates, silicates, and phosphates. Units are equivalents (eq) or microequivalents (μeq).

Anaerobic Microbial processes that do not use molecular oxygen (O_2). Anaerobic microbes usually cannot live in the presence of O_2.

Anion A negatively charged ion.

Applied dose Amount of a chemical given to an organism to determine dose-response relationships. Applied dose does not differentiate the amount of the chemical administered versus the amount absorbed by the organism.

Aquifer Underground layer of the earth that can transmit a sufficient amount of water as a source for water supply.

Base Substance usually capable of freeing OH^- anions when dissolved in water. Can weaken a strong acid, reacts with an acid, forms a salt, and has pH higher than 7.

Base-catalyzed hydrolysis Enhanced hydrolysis resulting from the attacks of hydroxyl ions.

Benchmark Chemical concentration used to calculate the hazard quotient—the likelihood that a chemical will be associated with an adverse health or environmental outcome. Examples include quality criteria for water, sediment, and air, and thresholds, such as the no observable adverse effect level (NOAEL).

Bioaccumulation Rate of increase of a chemical in an organism resulting from an excess of chemical intake versus the organism's ability to detoxify and eliminate the chemical.

Bioconcentration Buildup of a chemical in organisms, reaching concentrations above that found in the surrounding environment.

Biologically effective dose Amount of chemical taken up by a person to cause an adverse effect. This is usually organ-specific (e.g., dose leading a specific type of liver damage).

Biomarker Measurement of an interaction between a biologic system and a chemical. Exposure biomarkers of a chemical are measurements of the chemical itself (e.g. lead levels in hair) or its metabolite indicating that the organism has been exposed to a chemical (an example would be the concentration of blood level cotinine, a metabolite of nicotine, indicating that a person has been exposed to nicotine). Effect biomarkers are measurements of biochemical, behavioral, or physiologic changes in an organism resulting from exposure to a chemical (an example is the reduction in sperm count and lower testosterone blood levels following an exposure to an antiandrogen, such as DDE).

Bioremediation Deploying organisms to decontaminate waste sites. For example, phytoremediation has taken advantage of certain plant life at Superfund sites to extract, sequester, and biotransform toxic compounds into less toxic forms.

Biosolids Sludge.

Biotransformation Conversion of a chemical within an organism after absorption. This usually leads to a less toxic compound compared to the compound to which the organism was exposed. However, the biotransformation compounds may be more toxic than the parent (See Activation).

Body burden Total amount of a chemical to which a person has been exposed from all sources over time.

Bulk density Density/volume ratio for a solid, especially a soil, not corrected for the voids contained in the bulk of material. Units are kg m^{-3}.

Cadmium (Cd) A heavy metal (atomic number 48, mean atomic weight 112.4) considered to be carcinogenic, nepatotoxic, and neurotoxic. Formerly used as a pigment in paints, coatings, and other materials. Still used in galvanizing metal processing and batteries. Sequestered in kidneys via metallothionein binding.

Cancer New, malignant growth. Carcinoma is a malignant epithelial tumor that affects surrounding tissue and can lead to metastases (movement of the cancer cells to other parts of the body). Sarcoma is a malignant, connective tissue tumor of anaplastic cells that resembles the supporting tissues.

Cancer Risk Evaluation Guides (CREGs) Estimated contaminant concentrations in air, soil, or water expected to cause no greater than one excess cancer in a million (risk $\leq 10^{-6}$) persons exposed over a lifetime. CREGs are calculated from the EPA's cancer slope factors.

Cap The final, permanent layer of impermeable material, such as compacted clay or synthetics, on top of a landfill. Part of a closure.

Carcinogen Substance that causes, or is suspected of causing, cancer.

Carotenoids Labile, oxidizable red, purple, orange or yellow pigments distributed by plants and animals. Carotenoids are usually lipophilic.

Cation A positively charged ion.

Characteristic waste Substances defined as hazardous wastes because of their ignitability, corrosivity, reactivity, or toxicity (from 40 CFR, Part 261, Subpart C).

Chelating agents Organic compounds that are able to withdraw ions from their water solutions into soluble complexes.

Chelation Formation of a heterocyclic ring that contains at least one metal ion (cation) or hydrogen ion.

Chemical bond Force holding atoms together. Ionic bonds transfer electrons from one atom to another. Covalent bonds share electrons among atoms. The covalent bond is stronger than the ionic bond, so covalently bonded compounds are more difficult to degrade.

Chemical equilibrium The equality of chemical reactions in forward and reverse directions.

Chemical treatment Technology using abiotic, chemical processes to treat waste.

Chemicals of potential concern Chemicals at a site about which available data indicate the need for conducting a quantitative risk assessment.

Chemisorption Sorption process of integrating a chemical into porous materials surface via chemical reaction.

Chlorinated hydrocarbon Class of persistent pesticides, notably DDT, that linger in the environment and have a strong ability to bioaccumulate. Examples include DDT, aldrin, dieldrin, heptachlor, chlordane, lindane, endrin, mirex, benzene, hexachloride, and toxaphene.

Chronic exposure Exposures lasting more than six months.

Chronic reference dose (RfD) Estimated lifetime daily exposure level for the human population, likely to be without an appreciable risk of deleterious effects, protecting the population from long-term exposure to a chemical (more than 7 years).

Clay Soil particles with grain sizes less than 0.002 mm in diameter.

Cleanup Set of steps taken to address a release or threat of release of a hazardous substance. Synonyms include *remedial action, removal action, response action,* or *corrective action.*

Closed system Processes designed and used so that chemicals are not released. Closed systems are measures to control exposures to hazardous materials in industrial operations.

Comparative biology Use of animal testing to determine toxicity of a chemical and to develop models of how the chemical may behave in humans.

Comparison values Screening values in the preliminary identification of contaminants of concern at a hazardous waste site.

Composite sample Environmental sample made up of a combination of several samples taken over a specified time period (e.g., a 24-hour composite air sample).

Composting Engineered degradation of organic wastes using aerobic microbes to generate a high-nutrient substance (i.e., compost).

Compound of Potential Concern (COPC) A chemical that the U.S. Environmental Protection Agency associates with data of sufficient quality to be used to screen for exposure and risk using quantitative analysis, for example, in assessing hazardous air emissions or in calculating "ecological screening quotients" (ESQs).

Concentration gradient Change in concentration of a chemical ([C]) over a unit length. Diffusion occurs from higher to lower [C] (law of potentialities). Rates of diffusion increase with increasing [C] in a medium.

Congener Compounds possessing similar structures to another compound.

Consent decree Binding agreement by both parties that settles a lawsuit (this can be between commercial parties, between commercial and governmental parties, or between private citizens and governmental and/or commercial parties).

Contaminants of concern Chemical found at the site that the health professionals select to be analyzed for potential human health effects.

Contingency plan Document laying out the organized, planned, and coordinated courses of action to be taken in the event of an accident where toxic chemicals, hazardous wastes, or radioactive materials are released.

Convection Transfer by a moving fluid, such as air or water.

Corrective action Required cleanup of hazardous substance releases. Often refers to releases before passage of the Resource Conservation and Recovery Act.

Corrosivity Hazardous waste characteristic for any waste with pH of 2.0 or less or 12.5 or greater.

Cost-benefit analysis Approach for determining whether a project is worth pursuing, taking into consideration expected costs and benefits. The costs and benefits can be either monetized (based on dollar value), or nonmonetized (some other system of value). The analysis should yield a cost-benefit ratio. If the ratio is greater than 1, the project may be worth pursuing, but engineers usually seek projects where the ratio is much larger than 1.

Cradle-to-grave Requirement of the Resource Conservation and Recovery Act that a substance be accounted for from generation, through transport, to its ultimate treatment and disposal.

Darcy's law Expression of the laminar flow of water through porous media. Darcy's law states that the velocity of the water through porous media is proportional to the hydraulic gradient (i.e., $V = Ki$).

De minimus risk A risk not deemed to be important or of no public health concern. For carcinogens, this has recently been deemed to be a one-in-a-million risk (1×10^{-6}).

Dehalogenation Removal of chlorine, bromine, or other halogen atoms from a molecule. This process usually reduces the toxicity and persistence of chemical contaminants, so it can be an important part of cleanup efforts at sites contaminated with halogenated wastes.

Delisting Formally removing a substance from the U.S. EPA's listing of regulated materials, as new data are made available.

Denitrifying bacteria Soil bacteria that can convert nitrogenous compounds to gaseous nitrogen (N_2).

Dense nonaqueous phase liquid (DNAPL) Organic compound that sinks in groundwater.

Dermal Relating to the skin.

Designated facility Entity that treats, stores, or disposes of hazardous wastes.

Detection limit Minimum amount of a chemical that can be measured (specific to instruments and laboratories).

Developmental toxicity Impairment of the developing organism. Often linked in risk assessment to "reproductive toxicity" and, recently, to "endocrine disruption."

Diffusion Transport of a chemical through an environmental medium based on a concentration gradient of the chemical. See *Fick's Law*.

Dispersion model A mathematical approach to predict the transport of chemicals from one site to another. Models may be stochastic (statistically based) or deterministic (including the scientific attributes and parameters expected to drive the movement of chemicals).

Dose-response Relationship of adverse effects in an organism to the amount of chemical to which the organism is exposed.

Dosimetry Measurement of amount of toxic substance.

Downgradient well Well used to sample groundwater that has passed beneath a facility that may release contaminants, such as a landfill.

Dredging Removal of sediments and accumulated material from surface waters.

Dyes Chemicals that are used to color a substance. Dyes are generally used to impart permanent color change. They may be organic (such as the aromatic amine, analine, or the hydrocarbon, carotene), metallic (such as cadmium compounds), or inorganic (such as cyanine or quinoline blue dye used in film).

Ecologic risk assessment Estimating the contributing factors and the effects of human activities on ecosystems.

Ecological Screening Quotient (ESQ) A quotient used by the U.S. Environmental Protection Agency to assess risk during the risk assessment when default assumptions are used. The ESQ's numerator is the reasonable worst-case "compound of potential concern" (COPC) concentration at the point of exposure. The denominator is the chemical's "no observable adverse effects level" (NOAEL).

Electrical resistivity Noninvasive, geophysical measurement method for determining types of underlying strata and their characteristics. Resistivity measurements are made by placing electrodes on the ground surface, sending through an electric current, and determining the decrease in potential between the electrode locations. Inverse of conductivity. In general, dry, coarse-grained matrices have higher resistance than finer-grained, moisture-laden matrices. Such relationships can be quantified and used as input for inverse models to estimate permeability, porosity, and chemical concentrations in groundwater. Units of resistance are mohs (reciprocal of conductivity units, ohms).

Emergency response team (ERT) Group of experts and responders who have been trained to assist following a spill or release of contaminants. Tasks include measuring the extent of contamination and implementing decontamination efforts.

Endocrine disruption Dysfunction of normal hormonal processes in humans and wildlife. Three types of endocrine disruption can occur: agonism, antagonism, and indirect. Endocrine *agonists* are chemicals that mimic estrogen, testosterone, and other hormones because the chemicals are in some ways similar in chemical structure (e.g., functional groups) to that of the natural hormone and are able to bind with the receptors of a cell. *Antagonists* are chemicals that interfere or block normal hormone receptors, so that the cell cannot produce sufficient amounts of a hormone (e.g., antiandrogens shut off testosterone receptors, causing a net increase in estrogens and feminizing the organism). *Indirect disruptors* may be agonists or antagonists for nonendocrine systems, such as the neurologic or immune systems, but the changes to these other systems leads to endogenous changes that ultimately affect the normal hormonal functions (i.e., reproductive, developmental, and physiologic homeostasis).

Endpoint Disease or harmful outcome associated with exposures to a chemical that is the focus of a study.

Environmental audit Investigation of a company's compliance with a range of environmental regulations or an assessment of potential environmental liabilities, such as a condition of a real estate transaction.

Environmental engineering Application of the principles of physics, chemistry, and biology to address and to prevent environmental and public health problems. The profession has evolved from the more confined field of sanitary engineering, which was concerned with the design and operation of environmental facilities (e.g., drinking water plants, wastewater treatment facilities, and air pollution abatement equipment). The field presently addresses hazardous waste management, risk assessment, and ecosystem protection.

Environmental justice Inclusion of race and social issues in environmental decisions. Combination of environmental and social justice.

Environmental Media Evaluation Guides (EMEGs) Concentrations of a chemical in the air, soil, or water below which noncancer effects are not expected to be associated with exposures over a specified duration of exposure.

Epidemiology Study of the occurrence and distribution of diseases in humans or adverse effects in ecosystems. Epidemiology considers the factors that influence the distribution of these effects. Descriptive epidemiology is concerned with the delineation of diseases and adverse effects in populations and subpopulations (known as *polymorphs*). Analytic epidemiology delves further into the potential reasons for such occurrences.

Equilibrium Steady-state condition where no net gain or net less occurs (i.e., inflow equals outflow).

Eulerian model Pollutant dispersion model that applies fluid mechanics to estimate and characterize flow across fixed locations in space. These fixed locations are referred to as a grid.

Exogenous Taking place outside of an organism.

Exposure Contact of a person (or organism) with a chemical agent, quantified as the mass of the chemical available at the exchange boundaries (e.g., lungs, skin, digestive tract) and available to be absorbed. If the organism is exposed in the medium of release (e.g., a person breathes air that contains a chemical released from a stack), such exposure is considered direct. If the exposure occurs only after the chemical has moved through various media (e.g., the chemical is deposited onto a water body and is taken up by a fish eaten by a person), then the exposure is indirect. Units are in mass of chemical per mass of body weight per time ($mg/kg^{-1}/d^{-1}$).

Exposure pathway Physical course taken by a chemical to reach the exposed organism.

Extraction procedure (EP) test Leach test required by the U.S. Environmental Protection Agency to estimate the likelihood that a waste would transport 15 toxic metals and organic compounds into groundwater. Modified and expanded to address 40 chemicals in 1990 (renamed the toxicity characteristic leaching procedure).

Fate The ultimate site of a pollutant following its release. The pollutant will undergo numerous stages before reaching its fate. These include physical transport, chemical (including photochemical and biochemical) and biologic transformations, sequestration, and storage. Fate is often described according to environmental media or compartments. For example, a chemical may have an affinity for sediments, so its physicochemical may make for low residence time in water and air, driving its fate toward the sorption onto sediment particles.

Fick's law Fick's first law of diffusion states that the rate of diffusion of one material through a different material is proportional to the cross-sectional area of diffusion, the concentration gradient of the first material, and a coefficient of diffusion. The law is expressed as $M/A = -D(d[C]/dX)$, where M is the mass transfer rate, A is the cross-sectional area, D is the coefficient of diffusion, and $d[C]/dX$ is the concentration gradient.

Fixation Immobilization of wastes by combining them with relatively inert and stable materials, such as fly ash, concrete, or refractory clays.

Flux Flow rate through a cross-sectional area (mass or volume per unit area per time).

Forcing functions In environmental models, variables from outside the system being studied that are used in predicting potential changes with time. For example, rainfall and sunlight will change the rates of degradation of chemical compounds in soil.

Free radical Molecule with an unshared electron.

Fugacity The propensity of a chemical to move out of a compartment, such as from the water to the air. Expressed by Henry's law constants.

Gaussian model Pollutant dispersion model that assumes the time-averaged concentration of a chemical released from a point has a Gaussian (normal) distribution about the mean centerline. This distribution may be either two-dimensional (along x and y axes) or three-dimensional (along x, y and z axes). These models are based upon probabilities, and the standard deviations are calculated for the location of the plume along each axis. The Gaussian model is a type of Lagrangian model.

Generator Producer of hazardous wastes.

Groundwater Water that has infiltrated through the soil and is stored underground, usually for long time periods.

Half-life ($T^{1/2}$) Time needed to decrease the concentration or mass of a chemical by one-half. Half-lives may be biologic (by metabolism and elimination processes), chemical (by transformation and reactions), and radiologic (instability of the atom's nucleus).

Hazard Ranking System (HRS) Process that screens the threats of each site to determine if the site should be on the National Priority Listing (NPL) of most serious sites identified for possible long-term cleanup, and what the rank of a listed site should be.

Hazardous substance Chemical that will threaten human health and the environment if released in sufficient quantities. Specifically defined by various agencies, including the Department of Transportation and the Occupational Safety and Health Administration.

Hazardous waste Substance that has been produced as a by-product of human activities with the potential of harming human health or environmental resources. Hazardous wastes must possess at least one of four characteristics (i.e., ignitable, corrosive, reactive, or toxic) or they must appear on a U.S. Environmental Protection Agency special list.

Hazardous waste management Comprehensive approach for dealing with hazardous wastes, including pollution prevention, exchanges, and engineering approaches.

Headspace Zone above the contents of a closed container. Three-dimensional space in a container above a liquid that receives gases that have partitioned from the liquid.

Henry's law constant (K_H) Ratio of a chemical compound's mass in the gas phase to its mass in the aqueous phase. An expression of fugacity.

Homeostasis Body's ability to maintain a relatively consistent internal environment.

Horizon Horizontal soil layer from top: A horizon (highest organic material content); B horizon (nutrients from leaching); and C horizon (partially weathered parent rock).

Hydraulic conductivity (K) Coefficient expressing the permeability of an aquifer.

Hydraulic gradient (i) Rate of change in hydraulic head over a unit distance. Also, the head loss over a horizontal distance (dimensionless).

Hydraulic head (i or h) Height of the water column. Elevation of the water surface above a plane of reference (e.g., mean sea level). Expressed in units of length.

Hydraulic head loss (Δh) Decrease in the height of a water column.

Hydrolysis Chemical reaction in which a molecule of water or a hydroxide ion replaces an atom or group of atoms of another molecule, making the transformation products more polar.

Hydrophilic Propensity to dissolve in water.

Hydrophobic Resistance to dissolving in water. Usually synonymous with lipophilic (fat soluble).

Ignitability Hazardous characteristic of a chemical pertaining to its likelihood to catch fire.

Impermeability Resistant to passage of a liquid.

In situ "In place." For example, in situ remediation or treatment occurs where the contamination exists, rather than being removed and treated elsewhere.

In utero exposure Contact with a chemical through the placenta, during an organism's gestation period.

In vitro "In glass." Experiments that are performed in test tubes and other laboratory apparatus.

In vivo "In a living organism." Experiments that are performed on living organisms.

Incineration Combustion of organic materials.

Interstices Void spaces between particles of unconsolidated materials, such as soil.

Ion exchange Surface exchange process by which positively charged ions (cations) are attracted to negatively charged particle surfaces or negatively charged ions (anions) are attracted to positively charged particle surfaces, causing ions on the particle surfaces to be displaced. Particles undergoing ion exchange can include soils, sediment, airborne particulate matter, or even biota, such as pollen particles. The cation exchange capacity (CEC), and to a lesser extent the anion exchange capacity (AEC) in tropical soils are the means by which nutrients are made available to plant roots.

Isomer Chemical compound with the same molecular formula of another, but where the atoms are arranged differently, making for different chemical and physical characteristics, such as solubility and vapor pressure, and different biological characteristics, such as the potential for biological accumulation and cellular receptor binding.

Labile Easily reactive. A labile compound enters into reactions, such as oxidation and thermal reactions, readily with other compounds. Opposite of inert.

Lagrangian model Pollutant dispersion model that applies fluid mechanics to estimate and characterize flow by simulating the movement of a point (or "particle") moving with a plume. Lagragian models are also called "large-particle" or "macro-particle" models.

Leachate collection system Arrangement of catchments and piping underlying a landfill or other waste site that is engineered to catch and remove water that migrates through the site.

Lifetime average daily dose (LADD) Total dose that a person receives over a lifetime. A measure of chronic exposure.

Lowest observable adverse effects level (LOAEL) Lowest level of exposure to a chemical where an adverse effect in the exposed population increases significantly (statistically and biologically) compared to the unexposed population.

Maximum daily dose (MDD) Highest dose received in a 24-hour period during an exposure period.

Metallothionein (MT) Derivative of thionein, a sulfur-rich protein, that is a key part of metal metabolism in animals. MT binds metals into ligands at the cellular level, and is part of the detoxification and excretion processes.

Microenvironmental exposure An estimate of a person's potential contact with a chemical agent measured from the immediate local environment (e.g., indoor air in a home or a vehicle). Units are in mass of chemical per mass of body weight per time ($mg\,kg^{-1}\,d^{-1}$).

Microstructures The structure that is part of a larger structure, that renders certain physical characteristics to the larger structure. Microstructures are usually too small to view with the naked eye. An example is the microstructures of a feather that are able to diffuse light to give the feather coloring.

Minimal risk levels (MRLs) Estimates of daily human exposure to a chemical agent ($mg\,kg^{-1}\,d^{-1}$), which are not expected to be associated with any appreciable risk of noncancer effects over a specified duration of exposure.

Modifying factor Factor that reflects the results of qualitative assessments of the studies used to determine the threshold values (dimensionless, usually factors of 10).

National Priority Listing (NPL) Annual list compiled by the U.S. Environmental Protection Agency of the hazardous wastes sites in most need of cleanup.

No observable adverse effect level (NOAEL) Highest dose where no adverse effects are seen.

No-action alternative Status quo. No additional intervention.

Nonaqueous phase liquid (NAPL) Organic compound that floats within the groundwater.

Nondietary ingestion Exposures through the mouth to the digestive tract from sources other than eating food, including pica (e.g., children eating paint chips or dirt) and contaminants transferred by hand to mouth from surfaces.

Octanol-water partition constant (K_{ow}) Expression of a compound's affinity for an organic medium (less polar) versus its affinity for the aqueous medium (more polar).

Optics The study of the generation, transmission, and change of light.

Organic carbon-normalized sorption coefficients (K_{oc}) Measure of extent of adsorption of an organic compound to soil or sediment particles. Expressed as the ratio of mass of adsorbed carbon per unit mass of total organic carbon.

Parameters In environmental models, coefficients that mathematically express various processes. Parameters may be constants or constants ranging in value, including Henry's Law constants, solubilities, vapor pressures, and cationic exchange capacities. Single value constants include the universal gas constant and atomic weights of chemical compounds.

Personal exposure Actual contact of a person with a chemical agent, quantified as the mass of the chemical available at the exchange boundaries (e.g., lungs, skin, digestive tract) and available to be absorbed. Personal exposure can be measured directly (see PEMs) or modeled from chemical measurements in the ambient environment or in microenvironments (e.g., indoor air). Units are in mass of chemical per mass of body weight per time ($mg\,kg^{-1}d^{-1}$).

Personal exposure monitors (PEMs) Devices placed on or carried by people to determine actual, personal exposures (as contrasted with ambient measurements and microenvironmental exposures). These can be active monitors (those that include a pump to gather samples) or passive monitors (those that are based on diffusion and Fick's law).

pH Expression of the molar hydrogen ion concentration $[H^+]$ of a solution. Calculated as the negative logarithm of $[H^+]$. Thus a pH 5 solution has two orders of magnitude or 100 times the $[H^+]$ of a pH 7 (neutral) solution. The $[H^+]$ and the hydroxide ion concentration $[OH^-]$, have equal molar concentrations in a pH 7 solution. Measures of pH are often erroneously used synonymously with alkalinity/acidity; however, the latter include numerous other ions, besides H^+ and OH^-. The pOH is the negative log of the molar hydroxide ion concentration.

Pharmacokinetic (PK) modeling Means of estimating the temporal and spatial distribution of a chemical in a system. A general expression of how a chemical moves and changes in an organism. Types of PK models include the fixed effect model, maximum exposure models, and physiological-based PK models. These models may be single compartmental, when a compound equilibrates within an organism in a linear manner compared to elimination, or multicompartmental when the compound's concentration in an organism is curvilinear with respect to its elimination from the organism. Parameters include physiology (e.g. blood flow), thermodynamics (free concentrations of a "xenobiotic"), transport (transport across membranes), and amount (mass or concentration of the xenobiotic in a compartment).

Physiologically based pharmacokinetic (PBPK) modeling Simulation of absorption, distribution, metabolism, and excretion of chemical agents in a biologic system over time based on physiological factors and variables, such as respiration rates, blood flow, and endocrine processes.

Pica Ingestion of nonfood substances, such as paint chips and dirt.

Pigment Substance that changes the tint, tone or color of another material. Pigments are found in living tissues, such as melanin in skin as a protection for ultraviolet radiation and "carotenoids" in plant tissue.

Postclosure plan Steps to be taken by a hazardous waste facility to protect groundwater and to prevent exposures following cleanup. Requires environmental monitoring, reporting, waste containment, security, and other actions to prevent exposures for 30 years after site closure.

Preliminary assessment/site inspection (PA/SI) First stage of collecting data and evaluating a site that contains a hazardous waste. Required under the Comprehensive Environmental Response, Compensation, and Liability Act.

Products of incomplete combustion (PICs) Compounds formed from thermal processes of incineration and manufacturing. Includes dioxins, furans, hydrocarbons, and polycyclic aromatic hydrocarbons.

Protonation Attack of chemical compound by hydrogen ions.

Quality assurance/quality control (QA/QC) Approaches and procedures employed to ensure accurate and reliable results from environmental studies. Includes field and laboratory protocols for sample collection and handling, and blanks, duplicates, and split samples in the laboratory.

Rate constant Proportionality constant for the rate of a chemical reaction.

Reactivity Hazardous property of a chemical caused by the chemical's high likelihood to react chemically with other substances in the environment.

Receptor 1. Person, organism, or material that is exposed to a contaminant. 2. Location on a cell wall or within a cell that binds to a chemical.

Receptor binding The ability of a chemical to act as a ligand and bind to a cell in an organism. For example, some endocrine disruptors are estrogenic, meaning they bind to the estrogen receptors of a cell.

Record of decision (ROD) Document that contains the selected remedial action to be taken at a site, based on the results of the remedial investigation/feasibility study.

Reference concentration (RfC) Estimate of the daily inhalation exposure to a human population that is likely to be without appreciable risk of adverse effects during a lifetime.

Reference dose (RfD) Estimate of the daily exposure to a human population that is likely to be without adverse effects during a lifetime.

Remedial design/remedial action (RD/RA) Specifies remedies that will be undertaken at the site and lays out all plans for meeting cleanup standards for all environmental media.

Remedial investigation/feasibility study (RI/FS) Formal study following the initial investigation of a hazard to assess the nature and the extent of contamination.

Remediation Process of cleaning up a contaminated site.

Risk Probability of an adverse outcome resulting from exposure to a chemical.

Risk assessment paradigm Scientific framework for assessing risks. The paradigm usually includes hazard identification, dose-response determinations, exposure assessments, and risk characterization.

Risk management Set of engineering and policy approaches that are employed to prevent, remove, treat, exchange, and recycle wastes identified and characterized by a risk assessment.

Risk tradeoff The process of selecting an approach, intervention, or action by comparing and weighing various countervailing risks associated with each option. This is a common, everyday approach such as when decisions are made as to whether side-effects of medical treatment are worth the possible cure of a disease, whether possible "unintended consequences" from public policies are worthwhile, and perceived or actual safety of larger vehicles is worth the decrease in fuel efficiencies. Hazardous waste engineering deals with risk tradeoffs when comparing natural attenuation to more invasive remediation practices, when deciding how best to reduce exposures during design and construction, and when selecting possible remedial measures.

Route of exposure How a chemical enters an organism after contact (e.g., by dermal, ingestion, or inhalation).

Sand A soil or detritus particle with a grain size larger than silt and smaller than gravel. Generally, sand grains range between 0.07 mm to less than 5 mm diameter.

Screening level The concentration of a chemical contaminant that may be associated with potential risks to human populations and the environment. Models and other analytical tools use these levels to assess "worst case" exposures, erring in the interests of safety (i.e., they estimate high or perhaps higher than actual values of exposure). These artificially high estimates will likely mean that some substances will have exposure concerns where there actually are none; but the use of screening levels provides some confidence that substances with exposure estimates indicating no concern are in fact not a concern. Put another way, screening levels are more likely to give "false positives," but are much less likely to give "false negatives." An example of a screening level tool is the U.S. EPA's Ecological Screening Quotient (ESQ).

Sediment Matter that was once suspended in a liquid and has subsequently settled to the bottom as a multiphase system of organic (including living microbes) and inorganic matter.

Semivolatile organic compound (SVOC) Compound that may exist in various physical states, depending on environmental conditions. Generally includes compounds with vapor pressures less than 10^{-2} kilopascals, but greater than 10^{-5} kilopascals.

Silt Grains of soil finer than sand and coarser than clay (commonly between 0.07 mm and 0.002 mm grain diameter).

Sorption Solid-liquid distribution of a chemical in a given volume of an aquatic compartment (see *Absorption* and *Adsorption*).

Specific gravity Ratio of the weight of a given volume of a substance to that of a given volume of water.

Specific yield Ratio of water volume that a given mass of saturated soil will yield by gravity.

Stratigraphy Arrangement of geologic strata.

Stripping Removal of organic compounds (usually volatile organic compounds) from a soil or other contaminated matrix. The compounds are volatilized and transferred into a gas flow. The gas is then collected and treated.

Surface tension Force holding a liquid in pore spaces of a matrix preventing flow due to gravity.

Texture Grain size of soil particles.

Toxic Ability to cause adverse effects.

Toxic equivalency factor (TEF) Aggregate means of estimating the risks associated with exposure to chemical classes of highly toxic groups, such as the chlorinated dioxins and furans and polycyclic aromatic hydrocarbons. Usually compared to the most toxic isomer (e.g., tetrachorodibenzo-*p*-dioxin for dioxins and furans).

Toxicity Characteristic of chemical wherein it can cause acute or chronic adverse effects in humans or wildlife.

Toxicity characteristic leaching procedure (TCLP) Test designed to provide replicable results for organic compounds and to yield the same type of results for inorganic substances as those from the original EP test. Twenty-five organic compounds were added to the original EP list.

Toxicity slope factor Dimensionless slope factors used to calculate the estimated probability of increased occurrence of adverse outcome over a person's lifetime (for cancer, this is the so-called excess lifetime cancer risk, or ELCR). Like the reference doses, slope factors follow exposure pathways.

Toxicologic profile Document prepared for a specific substance in which scientists interpret available information on the chemical to specify hazardous exposure levels. The profile also identifies knowledge gaps and uncertainties about the chemical.

Transformation Change in the chemical form of a substance after its release. This transformation can take place in the ambient environment (e.g., via hydrolytic processes) or within an organism via metabolism (known as *biotransformation*).

Transgenerational effects Adverse effects in the progeny or later generations of individuals who were actually exposed to a contaminant. DES is a classic example: Women were prescribed DES with little or no effect on them, but their daughters developed cervical cancers, which have been linked to the DES exposures during their mothers' pregnancies.

Transmissivity Rate at which water moves through a unit width of an aquifer under a unit hydraulic gradient.

Uncertainty factor Adjustment to toxicity data to set acceptable human dose levels to protect against noncancer adverse outcomes. Designed to account for the large amounts of uncertainty from animal testing and other health data.

Unconsolidated aquifer Underground water bearing stratum composed of loose geologic materials, such as gravel or sand.

Uncontrolled site Abandoned hazardous waste site where wastes are being released or may be released.

Vadose zone Underground strata above the water table (i.e., unsaturated zone). Also called the *zone of aeration.*

Vapor pressure (P^0) Pressure exerted by a gaseous substance in equilibrium with its liquid or solid phase. Units are pascals, atmospheres, and other units of pressure.

Volatile organic compound (VOC) Compound with an affinity for the gas phase, usually with vapor pressures greater than 10^{-2} kilopascals.

Waste exchange Practice of matching the chemicals considered wastes of companies, laboratories, government agencies, and other entities with entities where those same chemicals are needed. Active waste exchange makes use of an organization (e.g., a clearinghouse) to arrange the transfer of waste chemicals from a waste generator to an entity needing the chemicals, whereas a passive waste exchange is one where information is made available more generally and the interested parties are responsible for working together (e.g., an adopt-a-chemical program that advertises available chemicals on the Internet).

Water-filled pore space (WFPS) An expression of the amount of water that can be stored by an unconsolidated material in its interstices.

Xenobiotic A substance that is foreign to an organism. Usually applied to toxic compounds that enter an organism from other than natural pathways, such as a pesticide or product of incomplete combustion.

Zone of saturation Underground stratum or strata with all pore spaces (i.e., interstices) filled with water that is under higher pressure than that of the atmosphere. The saturation zone is below the vadose zone.

In addition to the author's working definitions, the sources for the terms in this glossary include, as well as others cited elsewhere in the text:

1. U.S. Environmental Protection Agency, 2002, *Terms of Environment*, http://www.epa.gov/OCEPAterms/intro.htm.
2. B. Wyman and L.H. Stevenson, 2001, *The Facts on File Dictionary of Environmental Science*, Checkmark Books, New York.
3. Vincent Covello, 1992, "Risk Comparisons and Risk Communications," in *Communicating Risk to the Public*, edited by Roger E. Kasperson and P. Stallen, Kluwer, New York.
4. U.S. Environmental Protection Agency, 1997, *Exposure Factors Handbook, Volume 1*; EPA/600/P-95/002FA; Washington, DC.
5. C. D. Klaasssen, 1996, *Casarett & Doull's Toxicology: The Basic Science of Poisons*, 5th edition, McGraw-Hill, New York.
6. Michael J. Derelanko, 1999, *CRC Handbook of Toxicology*, "Risk Assessment," M. J. Derelanko and M.A. Hollinger, editors, CRC Press, Boca Raton, FL.
7. G.M. Rand, editor, 1995, *Fundamentals of Aquatic Toxicology*, Taylor & Francis, Washington, DC.

APPENDIX 2

Minimum Risk Levels for Chemicals

TABLE A2-1

Chemical	*Route*	*Duration*	*MRL**	*Factors*	*Endpoint*
ACENAPHTHENE	Oral	Intermittent	$0.6\ mg\,kg^{-1}\,d^{-1}$	300	Liver
ACETONE	Inhalation	Acute	26 ppm	9	Nervous System
		Intermittent	13 ppm	100	Nervous System
		Chronic	13 ppm	100	Nervous System
	Oral	Intermittent	$2\ mg\,kg^{-1}\,d^{-1}$	100	Blood
ACROLEIN	Inhalation	Acute	0.00005 ppm	100	Eye
		Intermittent	0.000009 ppm	1,000	Lung
	Oral	Chronic	$0.0005\ mg\,kg^{-1}\,d^{-1}$	100	Blood
ACRYLONITRILE	Inhalation	Acute	0.1 ppm	10	Nervous System
	Oral	Acute	$0.1\ mg\,kg^{-1}\,d^{-1}$	100	Developmental
		Intermittent	$0.01\ mg\,kg^{-1}\,d^{-1}$	1,000	Reproductive
		Chronic	$0.04\ mg\,kg^{-1}\,d^{-1}$	100	Blood
ALDRIN	Oral†	Acute	$0.002\ mg\,kg^{-1}\,d^{-1}$	1,000	Developmental
		Chronic	$0.00003\ mg\,kg^{-1}\,d^{-1}$	1,000	Liver
ALUMINUM	Oral	Intermittent	$2.0\ mg\,kg^{-1}\,d^{-1}$	30	Nervous System
AMMONIA	Inhalation	Acute	0.5 ppm	100	Lung
		Chronic	0.3 ppm	10	Lung
	Oral	Intermittent	$0.3\ mg\,kg^{-1}\,d^{-1}$	100	Other
ANTHRACENE	Oral	Intermittent	$10\ mg\,kg^{-1}\,d^{-1}$	100	Liver
ARSENIC	Oral	Acute	$0.005\ mg\,kg^{-1}\,d^{-1}$	10	Gastrointestinal
		Chronic	$0.0003\ mg\,kg^{-1}\,d^{-1}$	3	Dermal
ATRAZINE	Oral‡	Acute	$0.01\ mg\,kg^{-1}\,d^{-1}$	100	Body Weight
BENZENE	Inhalation	Acute	0.05 ppm	300	Immune System
		Intermittent	0.004 ppm	90	Nervous System

BERYLLIUM	Oral†	Chronic	0.001 $mg\,kg^{-1}\,d^{-1}$	100	Gastrointestinal
BIOALLETHRIN	Oral†	Acute	0.0007 $mg\,kg^{-1}\,d^{-1}$	300	Developmental
BIS(CHLOROMETHYL) ETHER	Inhalation	Intermittent	0.0003 ppm	100	Lung
BIS(2-CHLOROETHYL) ETHER	Inhalation	Intermittent	0.02 ppm	1,000	Body Weight
BORON	Oral	Intermittent	0.01 $mg\,kg^{-1}\,d^{-1}$	1,000	Developmental
BROMODICHLOROMETHANE	Oral	Acute	0.04 $mg\,g^{-1}\,d^{-1}$	1,000	Liver
		Chronic	0.02 $mg\,kg^{-1}\,d^{-1}$	1,000	Kidney
BROMOFORM	Oral	Acute	0.6 $mg\,kg^{-1}\,d^{-1}$	100	Nervous System
		Chronic	0.2 $mg\,kg^{-1}\,d^{-1}$	100	Liver
BROMOMETHANE	Inhalation	Acute	0.05 ppm	100	Nervous System
		Intermittent	0.05 ppm	100	Nervous System
		Chronic	0.005 ppm	100	Nervous System
	Oral	Intermittent	0.003 $mg\,kg^{-1}\,d^{-1}$	100	Gastrointestinal
CADMIUM	Oral	Chronic	0.0002 $mg\,kg^{-1}\,d^{-1}$	10	Kidney
CARBON DISULFIDE	Inhalation	Chronic	0.3 ppm	30	Nervous System
	Oral	Acute	0.01 $mg\,kg^{-1}\,d^{-1}$	300	Liver
CARBON TETRACHLORIDE	Inhalation	Acute	0.2 ppm	300	Liver
		Intermittent	0.05 ppm	100	Liver
	Oral	Acute	0.02 $mg\,kg^{-1}\,d^{-1}$	300	Liver
		Intermittent	0.007 $mg\,kg^{-1}\,d^{-1}$	100	Liver
CESIUM	Radiation	Acute	4 mSv	3	Developmental
		Chronic	1 $mSv\,yr^{-1}$	3	Other
CHLORDANE	Inhalation	Intermittent	0.0002 $mg\,m^{-3}$	100	Liver
		Chronic	0.00002 $mg\,m^{-3}$	1,000	Liver
	Oral	Acute	0.001 $mg\,kg^{-1}\,d^{-1}$	1,000	Liver
		Intermittent	0.0006 $mg\,kg^{-1}\,d^{-1}$	100	Liver
		Chronic	0.0006 $mg\,kg^{-1}\,d^{-1}$	100	Liver

(continued)

TABLE A2-1 *(continued)*

Chemical	*Route*	*Duration*	*MRL**	*Factors*	*Endpoint*
CHLORDECONE	Oral	Acute	0.01 $mg\,kg^{-1}\,d^{-1}$	100	Nervous System
		Intermittent	0.0005 $mg\,kg^{-1}\,d^{-1}$	100	Kidney
		Chronic	0.0005 $mg\,kg^{-1}\,d^{-1}$	100	Kidney
CHLORFENVINPHOS	Oral	Acute	0.002 $mg\,kg^{-1}\,d^{-1}$	1,000	Nervous System
		Intermittent	0.002 $mg\,kg^{-1}\,d^{-1}$	1,000	Immune System
		Chronic	0.0007 $mg\,kg^{-1}\,d^{-1}$	1,000	Nervous System
CHLOROBENZENE	Oral	Intermittent	0.4 $mg\,kg^{-1}\,d^{-1}$	100	Liver
CHLORODIBROMOMETHANE	Oral	Acute	0.04 $mg\,kg^{-1}\,d^{-1}$	1,000	Kidney
		Chronic	0.03 $mg\,kg^{-1}\,d^{-1}$	1,000	Liver
CHLOROETHANE	Inhalation	Acute	15 ppm	100	Developmental
CHLOROFORM	Inhalation	Acute	0.1 ppm	30	Liver
		Intermittent	0.05 ppm	100	Liver
		Chronic	0.02 ppm	100	Liver
	Oral	Acute	0.3 $mg\,kg^{-1}\,d^{-1}$	100	Liver
		Intermittent	0.1 $mg\,kg^{-1}\,d^{-1}$	100	Liver
		Chronic	0.01 $mg\,kg^{-1}\,d^{-1}$	1,000	Liver
CHLOROMETHANE	Inhalation	Acute	0.5 ppm	100	Nervous System
		Intermittent	0.2 ppm	300	Liver
		Chronic	0.05 ppm	1,000	Nervous System
CHLORPYRIFOS	Oral	Acute	0.003 $mg\,kg^{-1}\,d^{-1}$	10	Nervous System
		Intermittent	0.003 $mg\,kg^{-1}\,d^{-1}$	10	Nervous System
		Chronic	0.001 $mg\,kg^{-1}\,d^{-1}$	100	Nervous System
CHROMIUM(VI), AEROSOL MISTS	Inhalation	Intermittent	0.000005 $mg\,m^{-3}$	100	Lung
CHROMIUM(VI), PARTICULATES	Inhalation	Intermittent	0.001 mg,m^{-3}	30	Lung

COBALT	Inhalation[†]	Chronic	0.0001 $mg\,m^{-3}$	10	Lung
	Oral	Intermittent	0.01 $mg\,kg^{-1}\,d^{-1}$	100	Blood
	Radiation	Acute	4 mSv	3	Developmental
		Chronic	1 $mSv\,yr^{-1}$	3	Other
CRESOL, META-	Oral	Acute	0.05 $mg\,kg^{-1}\,d^{-1}$	100	Lung
CRESOL, ORTHO-	Oral	Acute	0.05 $mg\,kg^{-1}\,d^{-1}$	100	Nervous System
CRESOL, PARA-	Oral	Acute	0.05 $mg\,kg^{-1}\,d^{-1}$	100	Nervous System
CYANIDE, SODIUM	Oral	Intermittent	0.05 $mg\,kg^{-1}\,d^{-1}$	100	Reproductive
CYCLOTETRAMETHYLENE	Oral	Acute	0.1 $mg\,kg^{-1}\,d^{-1}$	1,000	Nervous System
TETRANITRAMINE (HMX)		Intermittent	0.05 $mg\,kg^{-1}\,d^{-1}$	1,000	Liver
CYCLOTRIMETHYLENE	Oral	Acute	0.06 $mg\,kg^{-1}\,d^{-1}$	100	Nervous System
TRINITRAMINE (RDX)		Intermittent	0.03 $mg\,kg^{-1}\,d^{-1}$	300	Reproductive
DDT, P,P′-	Oral[†]	Acute	0.0005 $mg\,kg^{-1}\,d^{-1}$	1,000	Developmental
		Intermittent	0.0005 $mg\,kg^{-1}\,d^{-1}$	100	Liver
DELTAMETHRIN	Oral[†]	Acute	0.002 $mg\,kg^{-1}\,d^{-1}$	300	Developmental
DI(2-ETHYLHEXYL)PHTHALATE	Oral[†]	Intermittent	0.01 $mg\,kg^{-1}\,d^{-1}$	300	Developmental
DI-N-BUTYL PHTHALATE	Oral[‡]	Acute	0.5 $mg\,kg^{-1}\,d^{-1}$	100	Developmental
DI-N-OCTYL PHTHALATE	Oral	Acute	3 $mg\,kg^{-1}\,d^{-1}$	300	Liver
		Intermittent	0.4 $mg\,kg^{-1}\,d^{-1}$	100	Liver
DIAZINON	Inhalation	Intermittent	0.009 $mg\,m^{-3}$	30	Nervous System
	Oral	Intermittent	0.0002 $mg\,kg^{-1}\,d^{-1}$	100	Nervous System
DICHLORVOS	Inhalation	Acute	0.002 ppm	100	Nervous System
		Intermittent	0.0003 ppm	100	Nervous System
		Chronic	0.00006 ppm	100	Nervous System
	Oral	Acute	0.004 $mg\,kg^{-1}\,d^{-1}$	1,000	Nervous System

(continued)

TABLE A2-1 *(continued)*

Chemical	*Route*	*Duration*	*MRL**	*Factors*	*Endpoint*
		Intermittent	0.003 $mg\,kg^{-1}\,d^{-1}$	10	Nervous System
		Chronic	0.0005 $mg\,kg^{-1}\,d^{-1}$	100	Nervous System
DIELDRIN	Oral†	Intermittent	0.0001 $mg\,kg^{-1}\,d^{-1}$	100	Nervous System
		Chronic	0.00005 $mg\,kg^{-1}\,d^{-1}$	100	Liver
DIETHYL PHTHALATE	Oral	Acute	7 $mg\,kg^{-1}\,d^{-1}$	300	Reproductive
		Intermittent	6 $mg\,kg^{-1}\,d^{-1}$	300	Liver
DIISOPROPYL	Oral	Intermittent	0.8 $mg\,kg^{-1}\,d^{-1}$	100	Blood
METHYLPHOSPHONATE (DIMP)		Chronic	0.6 $mg\,kg^{-1}\,d^{-1}$	100	Blood
DISULFOTON	Inhalation	Acute	0.006 $mg\,m^{-3}$	30	Nervous System
		Intermittent	0.0002 $mg\,m^{-3}$	30	Nervous System
	Oral	Acute	0.001 $mg\,kg^{-1}\,d^{-1}$	100	Nervous System
		Intermittent	0.00009 $mg\,kg^{-1}\,d^{-1}$	100	Developmental
		Chronic	0.00006 $mg\,kg^{-1}\,d^{-1}$	1,000	Nervous System
ENDOSULFAN	Oral	Intermittent	0.005 $mg\,kg^{-1}\,d^{-1}$	100	Immune System
		Chronic	0.002 $mg\,kg^{-1}\,d^{-1}$	100	Liver
ENDRIN	Oral	Intermittent	0.002 $mg\,kg^{-1}\,d^{-1}$	100	Nervous System
		Chronic	0.0003 $mg\,kg^{-1}\,d^{-1}$	100	Nervous System
ETHION	Oral	Acute	0.002 $mg\,kg^{-1}\,d^{-1}$	30	Nervous System
		Intermittent	0.002 $mg\,kg^{-1}\,d^{-1}$	30	Nervous System
		Chronic	0.0004 $mg\,kg^{-1}\,d^{-1}$	150	Nervous System
ETHYLBENZENE	Inhalation	Intermittent	1.0 ppm	100	Developmental
ETHYLENE GLYCOL	Inhalation	Acute	0.5 ppm	100	Kidney
	Oral	Acute	2.0 $mg\,kg^{-1}\,d^{-1}$	100	Developmental
		Chronic	2.0 $mg\,kg^{-1}\,d^{-1}$	100	Kidney

ETHYLENE OXIDE	Inhalation	Intermittent	0.09 ppm	100	Kidney
FLUORANTHENE	Oral	Intermittent	$0.4\ mg\,kg^{-1}\,d^{-1}$	300	Liver
FLUORENE	Oral	Intermittent	$0.4\ mg\,kg^{-1}\,d^{-1}$	300	Liver
FLUORIDE, SODIUM	Oral†	Chronic	$0.06\ mg\,kg^{-1}\,d^{-1}$	10	Musculoskeletal
FLUORINE	Inhalation†	Acute	0.01 ppm	10	Lung
FORMALDEHYDE	Inhalation	Acute	0.04 ppm	9	Lung
		Intermittent	0.03 ppm	30	Lung
		Chronic	0.008 ppm	30	Lung
	Oral	Intermittent	$0.3\ mg\,kg^{-1}\,d^{-1}$	100	Gastrointestinal
		Chronic	$0.2\ mg\,kg^{-1}\,d^{-1}$	100	Gastrointestinal
FUEL OIL NO. 2	Inhalation	Acute	$0.02\ mg\,m^{-3}$	1,000	Nervous System
HEXACHLOROBENZENE	Oral†	Acute	$0.008\ mg\,kg^{-1}\,d^{-1}$	300	Developmental
		Intermittent	$0.0001\ mg\,kg^{-1}\,d^{-1}$	90	Reproductive
		Chronic	$0.00002\ mg\,kg^{-1}\,d^{-1}$	1,000	Developmental
HEXACHLOROBUTADIENE	Oral	Intermittent	$0.0002\ mg\,kg^{-1}\,d^{-1}$	1,000	Kidney
HEXACHLOROCYCLOHEXANE, ∀-	Oral	Chronic	$0.008\ mg\,kg^{-1}\,d^{-1}$	100	Kidney
HEXACHLOROCYCLOHEXANE, ∃-	Oral	Acute	$0.2\ mg\,kg^{-1}\,d^{-1}$	100	Nervous System
		Intermittent	$0.0006\ mg\,kg^{-1}\,d^{-1}$	300	Liver
HEXACHLOROCYCLOHEXANE, Δ-†	Oral	Acute	$0.01\ mg\,kg^{-1}\,d^{-1}$	100	Nervous System
		Intermittent	$0.00001\ mg\,kg^{-1}\,d^{-1}$	1,000	Immune System
HEXACHLOROCYCLOPENTADIENE	Inhalation	Intermittent	0.01 ppm	30	Lung
		Chronic	0.0002 ppm	90	Lung
	Oral	Intermittent	$0.1\ mg\,kg^{-1}\,d^{-1}$	100	Kidney
HEXACHLOROETHANE	Inhalation	Acute	6 ppm	30	Nervous System
		Intermittent	6 ppm	30	Nervous System

(continued)

TABLE A2-1 *(continued)*

Chemical	*Route*	*Duration*	*MRL**	*Factors*	*Endpoint*
	Oral	Acute	$1\ mg\ kg^{-1}\ d^{-1}$	100	Liver
		Intermittent	$0.01\ mg\ kg^{-1}\ d^{-1}$	100	Liver
HEXAMETHYLENE DIISOCYANATE	Inhalation	Intermittent	0.00003 ppm	30	Lung
		Chronic	0.00001 ppm	90	Lung
HEXANE, N-	Inhalation	Chronic	0.6 ppm	100	Nervous System
HYDRAZINE	Inhalation	Intermittent	0.004 ppm	300	Liver
HYDROGEN FLUORIDE	Inhalation§	Acute	0.03 ppm	30	Lung
		Intermittent	0.02 ppm	30	Lung
HYDROGEN SULFIDE	Inhalation	Acute	0.07 ppm	30	Lung
		Intermittent	0.03 ppm	30	Lung
IODIDE	Oral†	Acute	$0.01\ mg\ kg^{-1}\ d^{-1}$	1	Endocrine System
		Chronic	$0.01\ mg\ kg^{-1}\ d^{-1}$	1	Endocrine System
ISOPHORONE	Oral	Intermittent	$3\ mg\ kg^{-1}\ d^{-1}$	100	Other
		Chronic	$0.2\ mg\ kg^{-1}\ d^{-1}$	1,000	Liver
JP-4 (Jet Fuel)	Inhalation	Intermittent	$9\ mg\ m^{-3}$	300	Liver
JP-5/JP-8 (Jet Fuel)	Inhalation	Intermittent	$3\ mg\ m^{-3}$	300	Liver
JP-7 (Jet Fuel)	Inhalation	Chronic	$3\ mg\ m^{-3}$	300	Liver
KEROSENE	Inhalation	Intermittent	$0.01\ mg\ m^{-3}$	1,000	Liver
MALATHION	Inhalation†	Acute	$0.2\ mg\ m^{-3}$	100	Nervous System
		Intermittent	$0.02\ mg\ m^{-3}$	1,000	Lung
	Oral	Intermittent	$0.02\ mg\ kg^{-1}\ d^{-1}$	10	Nervous System
		Chronic	$0.02\ mg\ kg^{-1}\ d^{-1}$	100	Nervous System
MANGANESE	Inhalation	Chronic	$0.00004\ mg\ m^{-3}$	500	Nervous System

MERCURIC CHLORIDE	Oral	Acute	$0.007\ mg\,kg^{-1}\,d^{-1}$	100	Kidney
		Intermittent	$0.002\ mg\,kg^{-1}\,d^{-1}$	100	Kidney
MERCURY	Inhalation	Chronic	$0.0002\ mg\,m^{-3}$	30	Nervous System
METHOXYCHLOR	Oral†	Intermittent	$0.005\ mg\,kg^{-1}\,d^{-1}$	1,000	Reproductive
METHYL PARATHION	Oral	Intermittent	$0.0007\ mg\,kg^{-1}\,d^{-1}$	300	Nervous System
		Chronic	$0.0003\ mg\,kg^{-1}\,d^{-1}$	100	Blood
METHYL-Tertiary-BUTYL ETHER	Inhalation	Acute	2 ppm	100	Nervous System
		Intermittent	0.7 ppm	100	Nervous System
		Chronic	0.7 ppm	100	Kidney
	Oral	Acute	$0.4\ mg\,kg^{-1}\,d^{-1}$	100	Nervous System
		Intermittent	$0.3\ mg\,kg^{-1}\,d^{-1}$	300	Liver
METHYLENE CHLORIDE	Inhalation	Acute	0.6 ppm	100	Nervous System
		Intermittent	0.3 ppm	90	Liver
		Chronic	0.3 ppm	30	Liver
	Oral	Acute	$0.2\ mg\,kg^{-1}\,d^{-1}$	100	Nervous System
		Chronic	$0.06\ mg\,kg^{-1}\,d^{-1}$	100	Liver
METHYLMERCURY	Oral	Chronic	$0.0003\ mg\,kg^{-1}\,d^{-1}$	4	Developmental
MIREX	Oral	Chronic	$0.0008\ mg\,kg^{-1}\,d^{-1}$	100	Liver
MUSTARD GAS	Inhalation†	Acute	$0.0002\ mg\,m^{-3}$	900	Lung
	Oral	Acute	$0.5\ \mu g\,kg^{-1}\,d^{-1}$	1,000	Developmental
		Intermittent	$0.02\ \mu g\,kg^{-1}\,d^{-1}$	1,000	Gastrointestinal
N-NITROSODI-N-PROPYLAMINE	Oral	Acute	$0.095\ mg\,kg^{-1}\,d^{-1}$	100	Liver
NAPHTHALENE	Inhalation	Chronic	0.002 ppm	1,000	Lung
	Oral	Acute	$0.05\ mg\,kg^{-1}\,d^{-1}$	1,000	Nervous System
		Intermittent	$0.02\ mg\,kg^{-1}\,d^{-1}$	300	Liver
NICKEL	Inhalation	Chronic	$0.0002\ mg\,m^{-3}$	30	Lung

(continued)

TABLE A2-1 *(continued)*

Chemical	*Route*	*Duration*	*MRL**	*Factors*	*Endpoint*
PENTACHLOROPHENOL	Oral	Acute	0.005 $mg\,kg^{-1}\,d^{-1}$	1,000	Developmental
		Intermittent	0.001 $mg\,kg^{-1}\,d^{-1}$	1,000	Reproductive
		Chronic	0.001 $mg\,kg^{-1}\,d^{-1}$	1,000	Endocrine System
PHOSPHORUS, WHITE	Inhalation	Acute	0.02 $mg\,m^{-3}$	30	Lung
	Oral	Intermittent	0.0002 $mg\,kg^{-1}\,d^{-1}$	100	Reproductive
POLYBROMINATED BIPHENYLS (PBBs)	Oral	Acute	0.01 $mg\,kg^{-1}\,d^{-1}$	100	Endocrine System
POLYCHLORINATED BIPHENYLS	Oral	Intermittent	0.03 $\mu g\,kg^{-1}\,d^{-1}$	300	Nervous System
(PCBs) Aroclor 1254		Chronic	0.02 $\mu g\,kg^{-1}\,d^{-1}$	300	Immune System
PROPYLENE GLYCOL DINITRATE	Inhalation	Acute	0.003 ppm	10	Nervous System
		Intermittent	0.00004 ppm	1,000	Blood
		Chronic	0.00004 ppm	1,000	Blood
PROPYLENE GLYCOL	Inhalation	Intermittent	0.009 ppm	1,000	Lung
SELENIUM	Oral†	Chronic	0.005 $mg\,kg^{-1}\,d^{-1}$	3	Skin
STRONTIUM	Oral†	Intermittent	2 $mg\,kg^{-1}\,d^{-1}$	30	Musculoskeletal
STYRENE	Inhalation	Chronic	0.06 ppm	100	Nervous
	Oral	Intermittent	0.2 $mg\,kg^{-1}\,d^{-1}$	1,000	Liver
SULFUR DIOXIDE	Inhalation	Acute	0.01 ppm	9	Lung
TETRACHLOROETHYLENE	Inhalation	Acute	0.2 ppm	10	Nervous System
		Chronic	0.04 ppm	100	Nervous System
	Oral	Acute	0.05 $mg\,kg^{-1}\,d^{-1}$	100	Developmental
TITANIUM TETRACHLORIDE	Inhalation	Intermittent	0.01 $mg\,m^{-3}$	90	Lung
		Chronic	0.0001 $mg\,m^{-3}$	90	Lung
TOLUENE	Inhalation	Acute	1 ppm	10	Nervous System
		Chronic	0.08 ppm	100	Nervous System
	Oral	Acute	0.8 $mg\,kg^{-1}\,d^{-1}$	300	Nervous System

		Intermittent	0.02 $mg\,kg^{-1}\,d^{-1}$	300	Nervous System
TOXAPHENE	Oral	Acute	0.005 $mg\,kg^{-1}\,d^{-1}$	1,000	Liver
		Intermittent	0.001 $mg\,kg^{-1}\,d^{-1}$	300	Liver
TRICHLOROETHYLENE	Inhalation	Acute	2 ppm	30	Nervous System
		Intermittent	0.1 ppm	300	Nervous System
	Oral	Acute	0.2 $mg\,kg^{-1}\,d^{-1}$	300	Developmental
URANIUM, HIGHLY SOLUBLE SALTS	Inhalation	Intermittent	0.0004 $mg\,m^{-3}$	90	Kidney
		Chronic	0.0003 $mg\,m^{-3}$	30	Kidney
	Oral	Intermittent	0.002 $mg\,kg^{-1}\,d^{-1}$	30	Kidney
URANIUM, INSOLUBLE COMPOUNDS	Inhalation	Intermittent	0.008 $mg\,m^{-3}$	30	Kidney
VANADIUM	Inhalation	Acute	0.0002 $mg\,m^{-3}$	100	Lung
	Oral	Intermittent	0.003 $mg\,kg^{-1}\,d^{-1}$	100	Kidney
VINYL ACETATE	Inhalation	Intermittent	0.01 ppm	100	Lung
VINYL CHLORIDE	Inhalation	Acute	0.5 ppm	100	Developmental
		Intermittent	0.03 ppm	300	Liver
	Oral	Chronic	0.00002 $mg\,kg^{-1}\,d^{-1}$	1,000	Liver
XYLENE, meta-	Oral	Intermittent	0.6 $mg\,kg^{-1}\,d^{-1}$	1,000	Liver
XYLENE, para-	Oral	Acute	1 $mg\,kg^{-1}\,d^{-1}$	100	Nervous System
XYLENES, total	Inhalation	Acute	1 ppm	100	Nervous System
		Intermittent	0.7 ppm	300	Developmental
		Chronic	0.1 ppm	100	Nervous System
	Oral	Intermittent	0.2 $mg\,kg^{-1}\,d^{-1}$	1,000	Kidney
ZINC	Oral	Intermittent	0.3 $mg\,kg^{-1}\,d^{-1}$	3	Blood
		Chronic	0.3 $mg\,kg^{-1}\,d^{-1}$	3	Blood
1-METHYLNAPHTHALENE	Oral	Chronic	0.07 $mg\,kg^{-1}\,d^{-1}$	1,000	Lung

(continued)

TABLE A2-1 *(continued)*

Chemical	*Route*	*Duration*	*MRL**	*Factors*	*Endpoint*
1,1-DICHLOROETHENE	Inhalation	Intermittent	0.02 ppm	100	Liver
	Oral	Chronic	$0.009\ mg\,kg^{-1}\,d^{-1}$	1,000	Liver
1,1-DIMETHYLHYDRAZINE	Inhalation	Intermittent	0.0002 ppm	300	Liver
1,1,1-TRICHLOROETHANE	Inhalation	Acute	2 ppm	100	Nervous System
		Intermittent	0.7 ppm	100	Nervous System
1,1,2-TRICHLOROETHANE	Oral	Acute	$0.3\ mg\,kg^{-1}\,d^{-1}$	100	Nervous System
		Intermittent	$0.04\ mg\,kg^{-1}\,d^{-1}$	100	Liver
1,1,2,2-TETRACHLOROETHANE	Inhalation	Intermittent	0.4 ppm	300	Liver
	Oral	Intermittent	$0.6\ mg\,kg^{-1}\,d^{-1}$	100	Body Weight
		Chronic	$0.04\ mg\,kg^{-1}\,d^{-1}$	1,000	Lung
1,2-DIBROMO-3-CHLOROPROPANE	Inhalation	Intermittent	0.0002 ppm	100	Reproductive
	Oral	Intermittent	$0.002\ mg\,kg^{-1}\,d^{-1}$	1,000	Reproductive
1,2-DICHLOROETHENE, cis-	Oral	Acute	$1\ mg\,kg^{-1}\,d^{-1}$	100	Blood
		Intermittent	$0.3\ mg\,kg^{-1}\,d^{-1}$	100	Blood
1,2-DICHLOROETHANE	Inhalation	Chronic	0.6 ppm	90	Liver
	Oral	Intermittent	$0.2\ mg\,kg^{-1}\,d^{-1}$	300	Kidney
1,2-DICHLOROPROPANE	Inhalation	Acute	0.05 ppm	1,000	Lung
		Intermittent	0.007 ppm	1,000	Lung
	Oral	Acute	$0.1\ mg\,kg^{-1}\,d^{-1}$	1,000	Nervous System
		Intermittent	$0.07\ mg\,kg^{-1}\,d^{-1}$	1,000	Blood
		Chronic	$0.09\ mkg^{-1}\,d^{-1}$	1,000	Blood
1,2-DICHLOROETHENE, trans-	Inhalation	Acute	0.2 ppm	1,000	Liver
		Intermittent	0.2 ppm	1,000	Liver
	Oral	Intermittent	$0.2\ mg\,kg^{-1}\,d^{-1}$	100	Liver

1,2-DIMETHYLHYDRAZINE	Oral	Intermittent	$0.0008\ mg\,kg^{-1}\,d^{-1}$	1,000	Liver
1,2,3-TRICHLOROPROPANE	Inhalation	Acute	0.0003 ppm	100	Lung
	Oral	Intermittent	$0.06\ mg\,kg^{-1}\,d^{-1}$	100	Liver
1,3-DICHLOROPROPENE	Inhalation	Intermittent	0.003 ppm	100	Lung
		Chronic	0.002 ppm	100	Lung
1,3-DINITROBENZENE	Oral	Acute	$0.008\ mg\,kg^{-1}\,d^{-1}$	100	Reproductive
		Intermittent	$0.0005\ mg\,kg^{-1}\,d^{-1}$	1,000	Blood
1,4-DICHLOROBENZENE	Inhalation	Acute	0.8 ppm	100	Developmental
		Intermittent	0.2 ppm	100	Blood
		Chronic	0.1 ppm	100	Blood
	Oral	Intermittent	$0.4\ mg\,kg^{-1}\,d^{-1}$	300	Liver
2-BUTOXYETHANOL	Inhalation	Acute	6 ppm	9	Blood
		Intermittent	3 ppm	9	Blood
		Chronic	0.2 ppm	3	Blood
	Oral	Acute	$0.4\ mg\,kg^{-1}\,d^{-1}$	90	Blood
		Intermittent	$0.07\ mg\,kg^{-1}\,d^{-1}$	1,000	Blood
2,3,4,7,8-PENTACHLORO-DIBENZOFURAN	Oral	Acute	$0.001\ \mu g\,kg^{-1}\,d^{-1}$	3,000	Immune System
		Intermittent	$0.00003\ \mu g\,kg^{-1}\,d^{-1}$	3,000	Liver
2,3,7,8-TETRACHLORODIBENZO-para-DIOXIN	Oral	Acute	$0.0002\ \mu g\,kg^{-1}\,d^{-1}$	21	Immune System
		Intermittent	$0.00002\ \mu g\,kg^{-1}\,d^{-1}$	30	Lymph Nodes
		Chronic	$0.000001\ \mu g\,kg^{-1}\,d^{-1}$	3,000	Developmental
2,4-DICHLOROPHENOL	Oral	Intermittent	$0.003\ mg\,kg^{-1}\,d^{-1}$	100	Immune System
2,4-DINITROPHENOL	Oral	Acute	$0.01\ mg\,kg^{-1}\,d^{-1}$	100	Body Weight
2,4-DINITROTOLUENE	Oral	Acute	$0.05\ mg\,kg^{-1}\,d^{-1}$	100	Nervous System
		Chronic	$0.002\ mg\,kg^{-1}\,d^{-1}$	100	Blood

(continued)

TABLE A2-1 *(continued)*

Chemical	*Route*	*Duration*	*MRL**	*Factors*	*Endpoint*
2,4,6-TRINITROTOLUENE	Oral	Intermittent	0.0005 $mg\,kg^{-1}\,d^{-1}$	1,000	Liver
2,6-DINITROTOLUENE	Oral	Intermittent	0.004 $mg\,kg^{-1}\,d^{-1}$	1,000	Blood
4-CHLOROPHENOL	Oral	Acute	0.01 $mg\,kg^{-1}\,d^{-1}$	100	Liver
4,4′-METHYLENEBIS (2-CHLOROANILINE)	Oral	Chronic	0.003 $mg\,kg^{-1}\,d^{-1}$	3,000	Liver
4,4′-METHYLENEDIANILINE	Oral	Acute	0.2 $mg\,kg^{-1}\,d^{-1}$	300	Liver
		Intermittent	0.08 $mg\,kg^{-1}\,d^{-1}$	100	Liver
4,6-DINITRO-O-CRESOL	Oral	Acute	0.004 $mg\,kg^{-1}\,d^{-1}$	100	Nervous System
		Intermittent	0.004 $mg\,kg^{-1}\,d^{-1}$	100	Nervous System

*All MRLs are established as final by the Agency for Toxic Substance and Disease Registry, except where otherwise noted.
†Draft MRL
‡Provisional oral MRL
§Known commercially as "lindane"

APPENDIX 3

What to Do If a Company Produces Only a Small Amount of Hazardous Waste

First, one must define small. The rules under the Resource Conservation and Recovery Act (RCRA) define a "small-quantity generator" (SQG) as a facility that generates less than 1,000 kg but more than 100 kg of hazardous waste per month (or 1 kg of extremely hazardous waste per month). This allows local governments, such as cities and counties, to offer businesses the opportunity to dispose of their hazardous waste in a less restrictive manner than is required of large generators.

The State of Washington, for example (www.co.pacific.wa.us/dcd/SmallQuantityGenPrgm.htm), has established a program in which small-quantity generators of dangerous wastes (DW) or extremely hazardous wastes (EHW) can simplify the way they must deliver their regulated wastes to a state/federally authorized treatment, storage, and disposal facility (TSD). Even if a local government's moderate-risk waste facility is not a TSD, it is permitted to receive this waste if the local government has a contract with a state/federally licensed TSD hazardous waste disposal contractor.

Businesses can qualify for this program and can be exempt from the full hazardous waste regulations that apply to generators of larger quantities of hazardous wastes if a company's total quantity of dangerous waste generated in one month (including both DW and EHW) does not equal or exceed 9 kg (20 lbs); however, companies do not qualify as a SQG if they accumulate more than 1,000 kg of dangerous waste on site at any time.

What Steps Must the Company Take?

The first step in the SQG process is to determine if the company is in fact producing hazardous waste. If so, the actual amount of hazardous waste generated per month must be calculated. The business may either document this information itself or use a solid waste coordinator to do so.

Upon concluding that the company is a SQG, the local government's solid waste coordinator will conduct an inventory of the waste via telephone or the company may link to a website to fill out an SQG Waste Inventory form (see Figure A3-1).

The company must then schedule an appointment to deliver the waste to the local government's disposal facility, keeping waste in its original container whenever possible. If the waste is not in its original container, the delivered container must clearly show the chemical contents.

All waste must be clearly and completely described. It is best to keep the waste in original containers, since the labeling provides important information on handling.

Copies of each documented chemical's Material Safety Data Sheet (MSDS) should also be provided. A MSDS describes possible hazards associated with a product. Information about the MSDS for a chemical can be found at www.siri.org.

To be considered for SQG designation, the company must provide each waste's chemical or trade name, the waste's physical and chemical characteristics, the percentages of each chemical constituent comprising the waste, and a description of the process that leads to the generation of the waste, including how the material was used. Certain wastes must be segregated because mixtures are more difficult to manage and may chemically react with one another, creating explosion and toxic gas release hazards at the local facility.

There are many SQGs. Some of the more common businesses are:

- Dry cleaners and laundry plants (wastes from solvent distillation, spent filter cartridges, and cooked powder)
- Furniture and wood manufacturing and refinishing operations (wastes from solvents, paints, used oil, acids, and bases)
- Printing businesses (acids/bases, heavy metals, solvents, and inks)
- Motor vehicle maintenance (acids/bases, wastes from solvents, paints, and batteries)

Pacific County Small Quantity Generator Program

Waste Collection Worksheet

Company Name: ____

Company Phone Number: ____________________________________

Material – *chemical and trade name*:
Characteristics – *solid, liquid, sludge, pH, etc.*
Chemical Constituents – *by percentage*:
What process generates this waste?
How much is generated each month?
How much do you store?
Disposal Quantity – *both gallons and pounds*:
Container – *type, size and quantity*:

Material – *chemical and trade name*:
Characteristics – *solid, liquid, sludge, pH, etc.*
Chemical Constituents – *by percentage*:
What process generates this waste?
How much is generated each month?
How much do you store?
Disposal Quantity – *both gallons and pounds*:
Container – *type, size and quantity*:

FIGURE A3-1. Pacific County, Washington's Waste Collection Worksheet. Local governments must gather information from companies to determine whether the business is a small-quantity generator.

(continued)

Material – *chemical and trade name*:
Characteristics – *solid, liquid, sludge, pH, etc.*
Chemical Constituents – *by percentage*:
What process generates this waste?
How much is generated each month?
How much do you store?
Disposal Quantity – *both gallons and pounds*:
Container – *type, size and quantity*:

Material – *chemical and trade name*:
Characteristics – *solid, liquid, sludge, pH, etc.*
Chemical Constituents – *by percentage*:
What process generates this waste?
How much is generated each month?
How much do you store?
Disposal Quantity – *both gallons and pounds*:
Container – *type, size and quantity*:

I certify that the above information is correct and true to the best of my knowledge. I further certify that I am a Small Quantity Generator as defined by Washington State Regulations WAC 173.303 and this quantity of waste does not exceed the specified limits for the type of waste being disposed. If this waste is later found to exceed SQG limits or contain materials not accepted under this program, I agree to complete a hazardous waste manifest and comply with other state regulations as appropriate. I accept full liability for my waste and will not hold Pacific County liable in any way through this program.

Signature: ____

Title: ____

Date Submitted: ____

FIGURE A3-1. *(continued)*

APPENDIX 4

Safety, Health, and Environmental Management Protocol for Field Activities[1]

Title of Project or Study:

____

____

Name of Site(s): ______________________________

Location of Site(s): ____________________________

Duration of Field Activity: ______________________

Principal Investigator (PI): ______________________

Laboratory, Division, Branch: ____________________

Location: Office ________________ Lab ________________

Phone: Office ________________ Lab ________________

Principal Investigator Signature ____________ Date ____________

APPROVALS

Supervisory ________________ Date ________________

Managerial ________________ Date ________________

Safety Office ________________ Date ________________

[1]Source: Adapted from the U.S. EPA Safety Protocol, 2002.

PART I. PERSONNEL POTENTIALLY EXPOSED TO HAZARDS DURING FIELD ACTIVITIES

A. Personnel

Personnel Authorized to Perform Field Work: Each authorized person must complete and sign a Personnel Qualification form.

1. ______________________ 6. ______________________
2. ______________________ 7. ______________________
3. ______________________ 8. ______________________
4. ______________________ 9. ______________________
5. ______________________ 10. ______________________

Are all personnel working with this study participants in a medical monitoring program (including baseline health status and periodic health evaluations)?

Yes ____ No ____. If yes, describe type of program. If no, explain.

Have all personnel working on this study successfully completed the required initial Field Activity Safety Training and/or Annual Refresher Training? **Note**: This information is available on the SHEM website at: http://intranet.epa.gov/nerlintr/shem/subpages/gd.htm

Yes ____ No ____. If no, explain.

B. Location(s) where work will be conducted (include site name and address).

Site Name: ____

Address: ____

Is this site a remote location ________ or an urban setting ________? If site is in a remote location, include a map and global positioning system (GPS) or longitude/latitude coordinates.

C. Contact personnel representing the site (include name, title, and phone number).

Contact Name: ____

Title: ______________________ Phone #: ______________________

D. Description of study (Research or Monitoring Protocol should be attached if applicable).

E. Describe in detail all potentially hazardous operations and duration

1. Identify the type(s) of environments the study will be conducted in:

☐ Mobile Laboratory
☐ Non-EPA Laboratory
☐ Terrestrial Ecology
☐ Aquatic Ecology
☐ Industry
☐ Other ____________________
☐ Other ____________________

2. Identify **physical** hazards (noise, heat, electrical, climbing/falling).

Hazard(s): ____

Anticipated Exposure: Hours: ________/day Total Days: ________

3. Identify **chemical** hazards (those that exist at the site and those EPA will transport to the site).

Hazard(s): ____

Anticipated Exposure: Hours: ________/day Total Days: ________

4. Identify **biological** hazards (pathogen, poison plants, wastewater, polluted stream, poisonous insects/snakes, human blood/fluids).

Hazard(s): ____

Anticipated Exposure: Hours: ________/day Total Days: ________

5. Identify any locations on the site that EPA personnel are restricted from entering. (Note: Employees are not authorized to enter confined spaces.)

6. Identify any pre-field visit vaccines that are necessary.

   ☐ Tetanus
   ☐ Hepatitis A (wastewater)
   ☐ Hepatitis B (blood, body fluids)
   ☐ Other ____________________
   ☐ None required

7. Describe the level of physical exertion required:

   ☐ Low (Office work)
   ☐ Moderate (Frequent walking)
   ☐ High (Frequent climbing, lifting)

F. Provide the following information for any hazardous agent that will be taken into the field by EPA personnel.

Hazardous Agent:

1. Common name:
2. Chemical name (and/or scientific name):
3. Quantity (to be taken into the field):
4. Condition/method of storage (in transport and when in use in the field):
5. Physical/chemical properties (form, solubility, volatility, vapor pressure, stability, flash point, reactivity):
6. Department of Transportation labeling requirements (Contact ORD SHEM Office for assistance x-2613):

G. Toxicity of materials to be used

1. Will any chemical materials be used that are considered hazardous agents by the EPA/RTP/ORD SHEM Office?

   A hazardous agent, as defined by the RTP Health and Safety office, is one that has:

   - An LD_{50} (oral, rat) < 50 mg/kg body weight
   - An inhalation LC_{50} toxicity (rat) of < 2 mg/liter
   - A dermal LD_{50} toxicity (rabbit) of < 200 mg/kg

- One that causes carcinogenic effects (*confirmed or suspect in humans and/or confirmed in animals*)
- One that causes teratogenic or mutagenic effects (*in humans or animals*)
- Any infectious biological agent (*as defined by CDC and/or NIH*)
- Any explosive or violently reactive agent (*shock sensitive, peroxide forming, and/or violently reactive with moisture/air*)
- Is a sensitizing agent.

Yes ______ No ______. If yes, describe:

H. Is Personnel Protective Equipment required during this visit?
☐ Yes ☐ No. Please select type below?

Face/Eye Protection	*Hand Protection (gloves)*	*Protective Clothing/ Footwear*	*Respiratory Protection*	*Hearing Protection*
☐ Safety Glasses ☐ Splash Goggles ☐ Face Shield ☐ Other	☐ Chemical (Butyl, Viton, Nitrile, ______) ☐ Latex ☐ Cotton ☐ Leather ☐ Thermal ☐ Double gloves	☐ Lab Coat (tyvek, cotton) ☐ Lab Apron ☐ Jumpsuit ☐ Snake Chaps ☐ Flotation devices ☐ Safety Shoes/Boots ☐ Hard Hat	☐ Air Purifying-full face ☐ Air Purifying-half mask ☐ Surgical mask ☐ Dust mask-not true respirator	☐ Single (plugs or muffs) ☐ Double (plugs and muffs)

I. Precautionary procedures to be used (e.g. controlled access, covered work surfaces, etc.).

J. Hazardous Waste Disposal

(Fill out the following information only if you are taking materials into the field and anticipate generating waste materials that must be returned to an EPA facility.)

Type of waste anticipated (exact chemical name and concentrations):

Volume of waste (*Provide time period, for example: 1 Liter/week solvent waste*)

Unused stock (to be disposed of at site or kept):

K. Attach copy of Hazardous Material Safety Data Sheet (MSDS), or a copy of information found in NIOSH *Registry of Toxic Effects of Chemical Substances.*

PART II EMERGENCY PROCEDURES

This information must be coordinated with representatives from the field site. This is referring to the emergency procedures dictated by the site personnel.

A. In the event of an accident or chemical/biological spill:

1. Describe procedures in event of overt personnel exposure (inhalation, ingestion, inoculation, asphyxiates, flammables, corrosives, etc.):

2. Describe plans for containment to prevent spread of the agent from the immediate area, decontamination procedures and monitoring methods to assure decontamination.

3. Describe the procedures for emergency evacuation of the facility (include diagram).

B. In the event of a medical emergency:

1. Emergency phone number (Is 911 available or does facility have its own medical emergency number?)
2. Is response by EMS available?
3. First response hospital (attach map of hospital location relative to site).

   Hospital: ____________________

   Address: ____________________

   Phone #: ____________________

4. Is hospital equipped to handle:

   —Burns?

—Chemical splashes (skin, eye, respiratory)?
—Chemical burns?
—Severe trauma?

If the answer to any of the above is no, designate an alternate facility that can handle these injuries. Include the hospital name, address, phone number, and location relative to the site.

C. Safety Checklist

Refer to the attached Safety Checklist and identify who will be responsible for completing this list.

Once the project ends this checklist should be sent to the Safety Office (MD-50) to be incorporated into the Protocol File for this project.

Travel to remote locations:

☐ Airplane ____________________ ☐ Boat ____________________

☐ Train ____________________ ☐ Bus ____________________

☐ Other ____________________

Hotel Information:

Hotel Name: ____

Phone #: ____

Nights of stay: ____

Satellite Phones

Will you have access to a satellite telephone? Yes ______ No ______.

If yes, telephone number: ______________________________

PERSONNEL QUALIFICATIONS FOR WORKING WITH HAZARDOUS AGENTS

(Complete this page for all personnel working on the project.)

Name: ____

Number of years related experience: __________

Research Specific Formal Training (include all health and safety courses applicable to this type of work):

Previous on-the-job training (work with specific hazardous agents related to this job, quantities worked with, and training received on these hazardous materials).

Restrictions (to be completed by appropriate safety office):

Protocol Title: ____

I have read the Health and Safety Research Protocol and agree to comply with all procedures and protective measures outlined in the Protocol.

____

Signature Date

Field Activity Protocol Checklist

Employees Participating in Field Activity:

____

____

____

Dates of Field Activity:

____

Location of Field Activity:

____

____

____

Designate whether or not the following items have been completed or obtained for each participating employee

ITEMS	COMPLETE
Field Activity Training (*includes having current certificate or card*)	
Medical Monitoring Program Participation	
Personnel Protective Equipment	
Safety Glasses or Goggles	
Safety Shoes/ Over Boots	
Hard Hat	
Respirator and Cartridges (appropriate to potential hazards)	
Appropriate Gloves (appropriate to potential hazards)	
Chemical Protective Clothing (appropriate to potential hazards)	
Cell Phone(s)	
Accident Report Forms (Injury, Vehicle)	
Health & Safety Protocol if Required	
Government Vehicle to Be Taken? (*Does it contain the following items*) First Aid Kit Fire Extinguisher Other Supplies	
Site Contact (*list name and phone number*)	
Itinerary Left with (Name of Person at Permanent Office):	
Verification of Certifications *Signature of Safety Office Representative*	

APPENDIX 5

Fundamentals of Chemical Equilibria

Equilibrium is both a physical and chemical concept. It is the state of a system where the energy and mass of that system are distributed in a statistically most probable manner, obeying the laws of conservation of mass, conservation of energy (first law of thermodynamics), and efficiency (second law of thermodynamics).

In environmental situations, we are mainly concerned with thermodynamic and chemical equilibria. That is, we must ascertain whether a system's influences and reactions are in balance. We know from the conservation laws that everything is balanced eventually, but since we only observe systems within finite time frames and confined spatial frameworks, we may only be seeing some of the steps in reaching equilibrium. Thus, for example, it is not uncommon in the environmental literature to see nonequilibrium constants.

To understand the concepts of environmental equilibria, let us begin with some fundamental chemical concepts. The first is that chemical reactions depend on "colligative" (collective) relationships between reactants and products. Colligative properties are expressions of the number of solute particles available for a chemical reaction. So, in a liquid solvent, like water, the number of solute particles determines the property of the solution. This means that the concentration of solute determines the colligative properties of a chemical solution. These solute particle concentrations for pollutants are expressed as either mass-per-mass (e.g., mg kg^{-1}) and, most commonly, as mass-per-volume (e.g., mg l^{-1}) concentrations. In gas solutions, the concentrations are expressed as mass-per-volume (mg m^{-3}). Colligative properties may also be expressed as mole fractions, where the sum of all mole fractions in any solution equals 1.

A simple example is 1 g sucrose dissolved in 9 g water. The total mass of this solution would be 10 g. If a given sugar solution contains 240 g sucrose per 1,000 g water, and the molecular weight of sucrose ($C_6H_{12}O_6$) is 180 g, we would have $240/180 = 1.3$ moles sucrose in 1,000 g water. Since the molecular weight of H_2O is 18, the mole-fraction of our sugar solute =

$$\frac{moles(solute)}{moles(solute) + moles(solvent)} = \frac{240/180}{240/180 + 1000/18} = 1.3/56.9 = 0.02.$$

And, the mole fraction of water is $\frac{1000/18}{1000/18 + 240/180} = 0.98.$

Thus, the mole percent of our solute is approximately 2% and the mole percent of our solvent is about 98%. The sum of all mole percentages is 100% because the sum of all mole fractions is 1.

Colligative properties depend directly upon concentration. One important property is vapor pressure, which is decreased with increased temperature. This is why water will require higher temperatures to boil when a solute is present. For example, pure water will boil at 100°C because one atmosphere (760 mm Hg) of pressure, and escapes as water vapor. In this case, all of the molecules are water (100% mole fraction). By adding solute to the pure water, we change the mole fraction. For example, if we heat the example solution above (2% sucrose), our vapor pressure is lowered by 2%, so that rather than 760 mm Hg, our vapor pressure = (0.98 water mole fraction)(760 mm Hg) = 745 mm Hg. Thus, the vapor pressure of the solvent (P) in any solution is found by:

$$P = X_A P^0 \quad \text{(A5-1)}$$

Where,

X_A = Mole fraction of solvent
P^0 = Vapor pressure of 100% solvent

Solution Equilibria

A body is considered to be in thermal equilibrium if there is no heat exchange within the body and between that body and its environment. Analogously, a system is said to be in chemical equilibrium when the forward and reverse reactions proceed at equal rates. Again, since we are looking at finite space and time, such as a spill or an emission, or movement through the environment, reactions within that time and space may be either nonequilibrium ($xA + yB \rightarrow zC + wD$) or equilibrium ($xA + yB \Leftrightarrow zC + wD$) chemical reactions. The x, y, z, and w terms are the stoichiometric coefficients, which represent the relative number of molecules of each reactant (A and B) and each product (C and D) involved in the reaction. So, to have chemical equilibrium, the reaction must be reversible, so that the concentrations of the reactants and the concentrations of the products are constant with time.

The Law of Concentration Effects states that the concentration of each reactant in a chemical reaction dictates the rate of the reaction. So, using our equilibrium reaction ($xA + yB \Leftrightarrow zC + wD$), we see that the rate of the forward reaction, i.e., the rate that the reaction moves to the right, is most often dictated by the concentrations of A and B. So, we can express the forward reaction as:

$$r_1 = k_1[A]^x[B]^y \tag{A5-2}$$

The brackets indicate molar concentrations of each chemical species (i.e., all products and reactants). Further, the rate of the reverse reaction can be expressed as:

$$r_2 = k_2[C]^z[D]^w \tag{A5-3}$$

Since at equilibrium, $r_1 = r_2$ and $k_1[A]^x[B]^y = k_2[C]^z[D]^w$ we can rearrange the terms to find the equilibrium constant K_{eq} for the reversible reaction:

$$\frac{k_1}{k_2} = \frac{[C]^z[D]^w}{[A]^x[B]^y} = K_{eq} \tag{A5-4}$$

The equilibrium constant for a chemical reaction depends upon the environmental conditions, especially temperature and ionic strength of the solution.

An example of a thermodynamic equilibrium reaction is chemical precipitation water treatment process. This is a heterogeneous reaction in that it involves more than one physical state. For an equilibrium reaction to occur between solid and liquid phases the solution must be saturated and undissolved solids must be present. So, at a high hydroxyl ion concentration (pH = 10), the solid phase calcium carbonate ($CaCO_3$) in the water reaches equilibrium with divalent calcium (Ca^{2+}) cations and divalent carbonate (CO_3^{2-}) anions in solution. So, when a saturated solution of $CaCO_3$ contacts solid $CaCO_3$, the equilibrium is:

$$CaCO_3(s) \Leftrightarrow Ca^{2+}(aq) + CO_3^{2-}(aq) \tag{A5-5}$$

The (s) and (aq) designate that chemical species are in solid and aqueous phases respectively.

Thus, applying the equilibrium constant relationship in Equation A5-3, the dissolution (precipitation) of calcium carbonate is:

$$K_{eq} = \frac{[Ca^{2+}] + [CO_3^{2-}]}{[CaCO_3]} \tag{A5-6}$$

The solid phase concentration is considered to be a constant K_s. In this instance, the solid $CaCO_3$ is represented by K_s, so:

$$K_{eq}K_s = [Ca^{2+}] + [CO_3^{2-}] = K_{sp} \tag{A5-7}$$

K_{sp} is known as the solubility product constant. These K_{sp} constants for inorganic compounds are published in engineering handbooks (e.g., in Part 1, Appendix C of the *Handbook of Environmental Engineering Calculations*).

Other equilibrium constants, such as the Freundlich Constant (K_d) discussed in Chapter 3, are also published for organic compounds (e.g., in Part 1, Appendix D of the *Handbook of Environmental Engineering Calculations*).

Gas Equilibria

For gases, the thermodynamic "equation of state" expresses the relationships of pressure (p), volume (V), and thermodynamic temperature (T) in a defined quantity (n) of a substance. For gases, this relationship is defined most simply in the ideal gas law:

$$pV = nRT \tag{A5-8}$$

where,

R = the universal gas constant or molar gas constant = $8.31434\,\mathrm{J\,mol^{-1}\,{}^{\circ}K^{-1}}$

It should be noted that the ideal gas law only applies to ideal gases, those that are made up of molecules taking up negligible space, with negligible spaces between the gas molecules. So, for real gases, the equilibrium relationship is:

$$(p + k)(V - nb) = nRT \tag{A5-9}$$

where,

k = factor for the decreased pressure on the walls of the container due to gas particle attractions
nb = volume occupied by gas particles at infinitely high pressure.

Further, the van der Waals equation of state is:

$$k = \frac{n^2 a}{V^2} \tag{A5-10}$$

where, a is a constant.

The van der Waals equation generally reflects the equilibria of real gases. It was developed in the early 20th century and has been updated, but these newer equations can be quite complicated.

Gas reactions, therefore, depend upon partial pressures. The gas equilibrium K_p is quotient of the partial pressures of the products and reactants, expressed as:

$$K_p = \frac{p_C^z p_D^w}{p_A^x p_B^y} \tag{A5-11}$$

And from Equations A5-1, 5, 6, and 7, K_p can also be expressed as:

$$K_{\mathrm{p}} = K_{\mathrm{eq}}(RT)^{\Delta v} \tag{A5-12}$$

where Δv is defined as the difference in stoichiometric coefficients.

Equilibrium constants can be ascertained thermodynamically by employing the Gibbs free energy (G) change for the complete reaction. Free energy is the measure of a system's ability to do work, in this case to drive the chemical reactions. This is expressed as:

$$G = H - TS \tag{A5-13}$$

where G is the energy liberated or absorbed in the equilibrium by the reaction at constant T. H is the system's enthalpy and S is its entropy.

Enthalpy is the thermodynamic property expressed as:

$$H = U + pV \tag{A5-14}$$

where U is the system's internal energy.

Entropy is a measure of a system's energy that is unavailable to do work. Numerous handbooks explain the relationship between Gibbs free energy and chemical equilibria. However, the relationship between a change in free energy and equilibria can be expressed by:

$$\Delta G^* = \Delta G_f^{*0} + RT \ln K_{\mathrm{eq}} \tag{A5-15}$$

where,

ΔG_f^{*0} = Free energy of formation at steady state (kJ gmol^{-1}).

Importance of Free Energy in Microbial Metabolism

Microbes play a large role in hazardous waste degradation, whether in natural attenuation, where the available microbial populations adapt to the hazardous wastes as an energy source, or in engineered systems that do the same in a more highly concentrated substrate. In either case the waste is degraded as an energy source or through co-metabolic processes. Free energy is an important factor in microbial metabolism.

The reactant and product concentrations and pH of the substrate affect the observed free energy values. If a reaction's ΔG^* is a negative value the free energy is released and the reaction is exergonic. If a reaction's ΔG^* is positive, the reaction required energy input, so the reaction is endergonic. In this case, the reverse reaction is favored.

Time and energy are limiting factors in whether a microbe can efficiently mediate a chemical reaction. An enzyme is a biological catalyst. These proteins speed up the chemical reactions of degradation without themselves being used up, by helping to break chemical bonds in the reactant

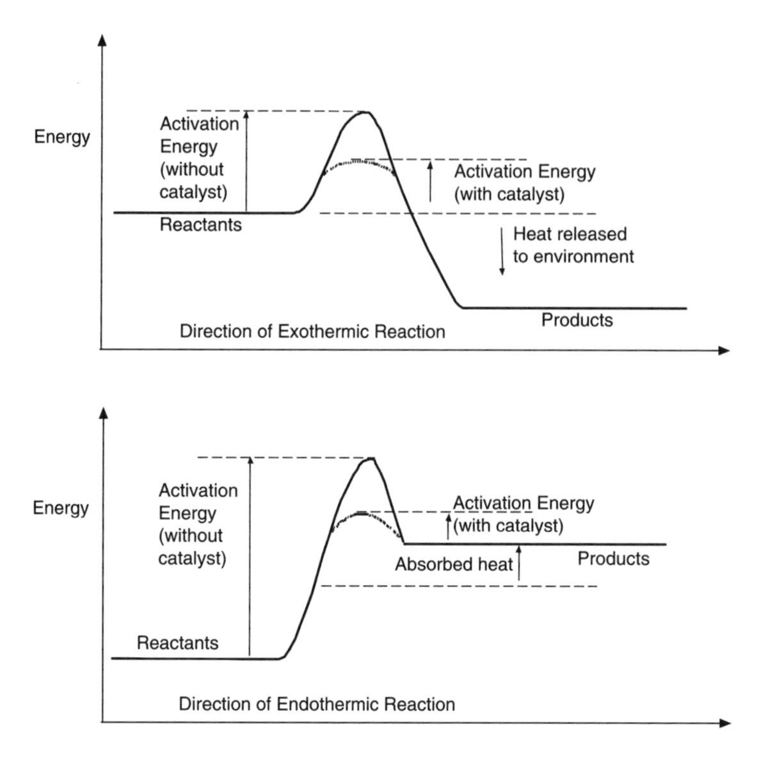

FIGURE A5-1. Effect of catalyst on exothermic and endothermic reactions.

molecules (see Figure A5-1). Enzymes are essential to microbial metabolism. They reduce the reaction's activation energy, which is the minimum free energy required for a molecule to undergo a specific reaction. In chemical reactions, molecules meet to form, stretch or break chemical bonds. During this process, the energy in the system is maximized, then decreases to the energy level of the products. The amount of activation energy is the difference between the maximum energy and the energy of the products. This difference is the energy barrier that must be overcome for a chemical reaction to take place. Catalysts speed up and increase the likelihood of a reaction by reducing the amount of energy, i.e., the activation energy, needed for the reaction.

The most common microbial coupling of exergonic and endergonic reactions by means of high-energy molecules to yield a net negative free energy is that of the nucleotide, adenosine triphosphate (ATP) with $\Delta G^* = -12$ to -15 kcal mole^{-1} (standard free energies). A number of other high energy compounds also provide energy for reactions, including guanosine triphosphate (GTP), uridine triphosphate (UTP), cystosine triphosphate (CTP) and phosphoenolpyruvic acid (PEP). These molecules store their energy using high-energy phosphate bonds. An example of free energy in microbial degradation is the possible first step in acetate metabolism by bacteria shown in Equation A5-16:

$$\text{Acetate} + \text{ATP} \rightarrow \text{acetyl} - \text{coenzyme A} + \text{ADP} + \pi \qquad \text{(A5-16)}$$

where π is a phosphate molecule.

Endnotes and Commentary

Preface

1. This quote comes from P. Aarne Vesilind, J. Jeffrey Peirce, and Ruth F. Weiner, *Environmental Engineering*, 3rd edition (Boston, MA: Butterworth-Heinemann, 1993). The text is an excellent introduction to the field of environmental engineering and one of the sources of inspiration for this book.
2. For example, see Michael LaGrega, Phillip Buckingham, and Jeffrey Evans, *Hazardous Waste Management*, 2nd edition (Boston, MA: McGraw-Hill, 2001).
3. For example, a recent issue of *Civil Engineering* (Vol. 71, No. 11, Nov. 2001, pp. 36–49) discussed the important roles of civil engineers in the aftermath of attacks on the World Trade Center and the Pentagon, but failed even to mention the critical roles of environmental engineers in the emergency response, public health protection, and environmental monitoring.
4. William W. Lowrance, *Of Acceptable Risk: Science and the Determination of Safety* (Los Altos, CA: William Kaufmann, 1976).

Chapter 1

1. The National Academy of Engineering dedicated its entire Spring 2002 issue of "The Bridge" (Vol. 32, No. 1) to "Engineering and Homeland Defense," including the article "Reflections of the World Trade Center" by its design engineer, Leslie E. Robertson. However, most of the issue ignored the roles of environmental engineering.
2. Some may argue that engineering is really what philosophers and ethicists refer to as *teleology*. The term is derived from the Greek, *telos*, meaning "far." So, what engineers do is based on something far "out there" (i.e., an outcome that the engineer desires). Further, some argue that engineering is utilitarian (i.e., we are called on to produce the most good for the most people). (See Mike Martin and Ronald Schintzing's *Ethics in Engineering* [New York: McGraw-Hill, 1996] for an excellent discussion of the roles of moral reasoning and ethical theories

in engineering decision making). John Stuart Mill is most famous for espousing this decision theory; however, engineers are duty-bound to our codes of ethics, design criteria, regulations, and standards of practice. In this way, engineering is a type of *deontology* (Greek, *deon* for "duty"). This theory was formally articulated by Immanuel Kant. In practice, the engineer is both outcome-based and duty-bound. Risk assessment is a factor for both frameworks. For a more extensive, recent discussion of basic philosophical principles in engineering, see Edmund Seebauer and Robert Barry, *Fundamentals for Ethics for Scientists and Engineering* (New York: Oxford University Press, 2001).

3. National Academy of Sciences, *Risk and Decision Making: Perspectives and Research* (Washington, DC: NAS, 1981).
4. Abraham Maslow, *Motivation and Personality*, 2nd edition (New York: Harper & Row, 1970). Maslow presents a hierarchy of needs consisting of two classes of needs: basic and growth. The basic needs must first be satisfied before a person can progress toward higher-level growth needs. Within the basic needs classification, Maslow separated the most basic physiologic needs, such as water, food, and oxygen, from the need for safety. Therefore, one must first avoid starvation and thirst, satisfying minimum caloric and water intake, before being concerned about the quality of the air, food, and water. The latter is the province of environmental protection.
5. However, the issue of antibiotic resistance is a looming problem today, and it has an environmental engineering component. For example, some of the cross–resistance of bacteria is being accelerated by the large and widespread use of antibiotics in confined animal feeding operations (so called CAFOs), such as hog and poultry farms.
6. D.G. Wilson, "History of Solid Waste Management," in *Handbook of Solid Waste Management* (New York: Van Nostrand Reinhold, 1997), 1–9.
7. J.G. Landels, *Engineering in the Ancient World* (New York: Barnes & Noble Books, 1978). This book contains an excellent discussion of water supplies as recorded by the Roman architect and engineer, Vitruviusm in his eighth book of *De Architectura* in the 1st century B.C.
8. Landels, *Engineering in the Ancient World.* The two major challenges to delivering water in sprawling, ancient Rome were pressure and sediment. The large head needed to transport water over distances would split the lead and earthenware pipes, so the harder bronze was recommended. The siltation in the aqueducts was addressed using cisterns similar to sedimentation basins in modern wastewater treatment plants.
9. H. Tammenagi, *The Waste Crisis: Landfills, Incinerators, and the Search for a Sustainable Future* (Oxford, England: Oxford University Press, 1999), 22–24.
10. The source of this information is the EPA Public Information Office, Carborundum Center-Room 530, Niagara Falls, New York 14303.
11. "Superfund" is actually a woefully inadequate nickname for the federal Comprehensive Environmental Response, Compensation and Liability Act (CERCLA) passed in 1980 (42 U.S.C.A. Sections 9601 to 9675) and reauthorized as the Superfund Amendments and Authorization Act (SARA) in 1986. The law

does not simply provide cleanup monies. As the full name implies, there are three parts of the law: (1) rapid response to emergencies; (2) funding to remediate abandoned sites; and (3) the legal means to seek damages to pay for cleanups.

12. Henry Petroski, *To Engineer Is Human: The Role of Failure in Successful Design* (New York: Vintage Books, 1992). Although the examples of failures are mainly those of design and construction, they hold equally to the other aspects of waste engineering, including operation and maintenance, environmental monitoring, and exposure reduction. Professor Petroski has recently discussed the role of the engineer in failure analysis and remediation in the context of the World Trade Center collapse. He surmises that many factors beyond those within the realm of engineering will be prominent in design decisions. The way that society reacts to the terrorist threat will determine our role. He expects that advances in "micro-miniaturization, telecommunications, information technology, business practice, management science, economics, psychology and politics" will be more important than engineering in what society demands of its urban landscape [H. Petroski, "The Fall of Skyscrapers," *American Scientist*, 90(1): 16–21, January-February 2002]. The lessons remind us that engineering depends on a myriad of societal variables. The environmental engineer is constantly reminded of these same variables when proposing hazardous waste plans.
13. Threshold levels include no observable adverse effect levels (NOAELs) and lowest observable effect levels (LOAELs), discussed in Chapter 2.
14. J. Duffus and H. Worth provide an excellent introduction to the concepts of dose, hazards, and risk in their 2001 training program, "The Science of Chemical Safety: Essential Toxicology—4; Hazard and Risk," *IUPAC Educators' Resource Material*, International Union of Pure and Applied Chemistry.
15. The importance of the so-called "pharmacokinetic (PK) modeling" was recently brought home to the author. As mentioned in Chapter 6, the compound diphenyl propane was detected in the air in Manhattan following the collapse of the WTC towers. In an attempt to assess the hazard, the author attempted to use the results of a tumor cell study to determine if endocrine effects in humans were potentially caused by the compound. However, the absence of any PK models rendered this effort fruitless.
16. Mandated by 40CFR261 (U.S. Code of Federal Regulations).
17. One should consult with legal experts in such matters as confidential business information, as regulated under federal and state laws, such as the Toxic Substances Act and The Federal Insecticide, Fungicide, and Rodenticide Act.
18. U.S. Environmental Protection Agency, 1995, *Test Methods for Evaluating Solid Waste*, Volumes I and II (SW-846), 3rd edition, November 1986.
19. U.S. Environmental Protection Agency, 1976, National Interim Primary Drinking Water Regulations, EPA-570/9-76-003.
20. The federal laboratories in Research Triangle Park, North Carolina, have instituted an adopt-a-chemical program that not only includes chemicals, but also has subsequently been extended to share glassware and other laboratory apparatus. Although much of the apparatus does not qualify as hazardous waste per se,

it may contain hazardous substances that would have to be disposed of properly if they were not adopted. One example would be the chromatographs and detectors, such as the electron capture detectors (ECDs) that contain radioactive nickel.

Chapter 2

1. National Research Council, *Risk Assessment in the Federal Government* (Washington, DC: National Academy of Sciences, 1983).
2. National Academy of Sciences, *Risk and Decision Making: Perspectives and Research* (Washington, DC: National Academy of Sciences, 1981).
3. In fact, the touchstone of William Ruckelshaus's return to be Administrator of the U.S. EPA was to disentangle risk assessment from risk management. He saw that the simultaneous attention to assessing and managing risks put science and engineering at a disadvantage to the political, economic, and other social factors embodied in risk management. So, upon his return, he declared that the science (risk assessment) would be done as independently as possible. This science, in turn, would be one of the crucial components of managing risks, but not the only one.
4. In 1982, for example, the U.S. government purchased all the homes in the town of Times Beach, Missouri, after finding tetrachorodibenzo-*para*-dioxin (TCDD) in soil throughout the town. Times Beach became a test case for the recently enacted hazardous waste laws: emergency response was needed to prevent the exposure of humans and wildlife to TCDD; the potentially responsible parties who dumped the waste had to be found and prosecuted; and the mess had to be cleaned up. These issues reflect the three mandates of Superfund.
5. Polychlorinated biphenyls were highly efficient for insulating energy, especially electrical and heat, so by the 1960s and 1970s they had become ubiquitous across the environmental landscape. The Toxic Substances Control Act (TSCA, 15 U.S.C.A. Sections 2601 to 2692) called for the EPA to promulgate rules for PCB disposal and labeling, and virtually banned the manufacturing, processing, and distribution of these substances [Section 2605(e)]. This put tremendous pressure on businesses to find safe disposal approaches and opened the door to shoddy profiteers who accepted the PCB wastes, but who lacked the expertise and wherewithal to dispose of and to decontaminate them.
6. Also in 1982, a major protest took place against a landfill in North Carolina that was burying PCBs. This event has been recognized as one of the major events of the environmental justice movement. Environmental justice has fairness as its objective in environmental decisions. It has also been referred to as environmental equity, and some advocacy groups conversely address environmental racism. The common thread of the justice movements is that the segments of society with the least economic and political power have been most likely to be exposed to pollutants. For example, the advocates argue that the likelihood of siting power plants, heavy manufacturing facilities, and waste

disposal operations increases with decreasing average incomes, so that low-income neighborhoods have the highest likelihood of being exposed to toxic substances. Another possible explanation for some of the sites is that the existence of the polluting facilities caused the income levels to drop because of the decrease in desirability of the neighborhoods and accompanying precipitous fall in real estate value. In the past 50 years in the United States, factory workers have generally lived closer to the factory than have the company executives. Before this time, and especially before the general availability of the automobile, executives and workers were both likely to live near the factory. In fact, some of the most valuable residential properties were those nearest and most accessible to the factory. Culture also plays a role in siting. For example, some Latin American cities' low-income neighborhoods and barrios are located away from central business districts in what would be suburbia in Canada and the United States.

7. National Research Council, *Risk Assessment in the Federal Government* (Washington, DC: National Academy of Sciences, 1983).
8. The methods for deriving these values are provided by the U.S. Environmental Protection Agency, *Risk Assessment Guidance for Superfund: Volume 1, Human Health Evaluation Manual*, Part A (EPA/540/1-89/002), 1989, and the U.S. Environmental Protection Agency, *Dermal Exposure Assessment: Principles and Applications* (EPA/600/8-91/011B), 1992.
9. A. Bradford Hill, "The Environment and Disease: Association or Causation?" Proceedings of the Royal Society of Medicine, *Occupational Medicine*, 58:295, 1965.
10. See the ATSDR's website at www.atsdr.cdc.gov.
11. Sources of the cancer slope factor, RfD, and RfC values used in risk assessments are (1) the Integrated Risk Information System (IRIS), an online database run by the U.S. EPA that provides internally peer-reviewed toxicity data for many commonly detected chemicals, and (2) the U.S. EPA's (1994) Health Effects Assessment Summary Tables (HEAST), which include information from the literature compiled for use in the Superfund program.
12. Michael J. Derelanko, "Risk Assessment," in M.J. Derelanko and M.A. Hollinger, eds., *CRC Handbook of Toxicology* (Boca Raton, FL: CRC Press, 1999).
13. D. Kincaid, "Toxicity Code" (Bronx, NY: Lehman College, City University of New York, 2001).
14. The limit of detection (LOD) is both an analytical and a sampling threshold. If an instrument can detect only down to 1 part per billion (ppb), then this is an analytical limitation; however, in reality, if the sample has been held for some time, or the sample must be extracted from the soil or trapping device in the field, then this is a limit, even if the laboratory can detect down to 1 ppb. In our discussion, such a laboratory would report the nondetects as 500 parts per trillion (ppt, i.e., one-half of 1 ppb). This is certainly not the only means of dealing with nondetects. Other statistical methods for dealing with nondetects are used, but a nondetect should never be reported as 0 because one can only

say with confidence that it was not seen. It may not be present, but we can only report what we know, and that is dictated by the LOD.

15. For example, the World Health Organization (WHO) has recently assigned the value of 1 to the pentachlorodibenzo-para-dioxin, which means that the organization considers the toxicity equal to that of TCDD.
16. For detailed information on how exposures can be calculated and the derivation of exposure factors, see the U.S. Environmental Protection Agency, *Exposure Factors Handbook*, 1999, EPA/600/C-99/001.
17. American Society of Heating, Refrigerating, and Air-Conditioning Engineers, *ASHRAE Handbook: Equipment* (Atlanta, GA: ASHRAE, 1988). The differences among indoor and outdoor and room-to-room make for density differences that are used to determine basic patterns of air motion. For example, in winter, warmer air indoors tends to rise and to exit the building structure at upper stories by stack action. The exiting air is displaced at lower levels by an influx of colder air from the outside. During the summer, the movement is reversed, so that stack forces during the summer are generally not as strong as they are in the winter because the indoor–outdoor temperature differentials are less pronounced. Also, the position where the indoor–outdoor pressures are equal (i.e., neutral pressure level) depends on the leakage configuration of the building envelope. When there is unrestricted connection between floors, the building behaves as a unit volume that is influenced by a current that, in general, rises in winter and falls in the summer.
18. For sampling and analyzing dioxins and furans in soil and water, a good place to start is the U.S. EPA "Method 1613," Tetra-through octa-chlorinated dioxins and furans by isotope dilution HRGC/HRMS(Rev. B) (Washington, DC: Office of Water, Engineering and Analysis Division, 1994), as well as "RCRA SW846 Method 8290," Polychlorinated dibenzodioxins (PCDDs) and polychlorinated dibenzofurans (PCDFs) by high-resolution gas chromatograph/high-resolution mass spectrometry (HRGC/HRMS), Office of Solid Waste, U.S. EPA (September 1994). For air, the best method is the PS-1 high-volume sampler system described in the U.S. EPA "Method TO-9A" in Compendium of Methods for the Determination of Toxic Organic Compounds in Ambient Air, 2nd edition, EPA/625/R-96/010b, 1999.
19. See M. Ekhtera, G. Mansoori, M. Mensinger, A. Rehmat, and B. Deville, "Supercritical Fluid Extraction for Remediation of Contaminated Soil," in M. Abraham and A. Sunol, eds., *Supercritical Fluids: Extraction and Pollution Prevention*, ACSSS Vol. 670, 280–298 (Washington, DC: American Chemical Society, 1997).
20. See N. Duan, "Models for Human Exposure to Air Pollution," *Environmental International*, 8: 305–309, 1982; and N. Duan, "Stochastic Microenvironment Models for Air Pollution Exposure," *Journal of Experimental Analytic Epidemiology*, 1: 235–257, 1991.
21. See M. Derelanko, "Risk Assessment," in M.J. Derelanko and M.A. Hollinger, eds., *CRC Handbook of Toxicology* (Boca Raton, FL: CRC Press, 1999) for equations related to all routes and pathways of exposure.

22. The default value for absorption equals 1. That is, unless otherwise specified, one can assume that all of the contaminant is absorbed.
23. The Hazard Ranking System (HRS) is the principal mechanism used by the U.S. EPA to place uncontrolled waste sites on the National Priority List (NPL). It is a numerically based screening system that incorporates information from initial, limited investigations in the preliminary assessment, as well as from the site inspection. The combined information is then used to assess and rank the relative potential of a site to pose a threat to human health or the environment.
24. See G. Suter, *Ecological Risk Assessment* (Boca Raton, FL: Lewis Publishers, 1993); the U.S. Environmental Protection Agency, Framework for Ecological Risk Assessment. Washington, DC, EPA/630/R-92/001, 1992; and the Federal Register 63(93):26846-26924.
25. Source: Westfall, R. 1993, *The Life of Isaac Newton*, Cambridge University Press, Boston, MA.
26. I first was made aware of this whole new paradigm driving to Greensboro from Durham, NC, heading to the National Environmental Science and Technology Conference. A radio program was dealing with new ways to think about materials and how we have relied upon old, inefficient means of using technology. Unfortunately, I do not know the name of the expert who was being interviewed, nor even that of the radio program (I tuned in mid-interview), but the discussion was intriguing. I am indebted to this anonymous expert, including his insights about birds and nature.

    However, one of the postulations of the expert may not hold. He contended that a reason that nature would not select coloring by heavy metals like cadmium is that flight depends upon lift exceeding drag, so why would nature use something heavy for breeding that hurts the essence of flight? I contacted Professor Geoffrey Hill, a respected ornithologist at Auburn University, who informed me that it is not unusual for large molecules to be used as pigments, because they are readily available in the birds' habitats. He cited the example of carotenoid pigments manufactured by carrots and other orange and red plants. Birds ingest the plants and translocate the carotenoids to their feathers. So, it may cost the birds something in flight, but the availability of the pigments can override this need.

    By coincidence, the keynote speaker at the conference was Joseph DeSimone, Professor of Chemistry and Chemical Engineering at the University of North Carolina. Professor DeSimone has been recognized a leader in sustainable industry. He has used supercritical carbon dioxide, for example, as a substitute for hazardous solvents in the dry cleaning industry. Such cleaners are now found throughout the U.S. and increasingly around the globe. Over 100,000 plants use pressure and polymers to make the CO_2 supercritical. In this form, CO_2 is very efficient at dissolving most organic compounds. Thus, a new application of a well understood physical concept is preventing pollution.
27. See R.O. Prum, R.H. Torres, S. Williamson, and J. Dyck, 1998, "Coherend Light Scattering by Blue Feather Barbs." *Nature*, Volume 396, 28–29.

28. For more information, see Jack Shipley, "Waste Reduction through Better Management Program Helps Installations Cut Costs, Hazardous Materials Use," U.S. Army Environmental Center, Winter 1998.

Chapter 3

1. United Nations Environmental Programme, *Chemicals in the European Environment: Low Doses, High Stakes?* Annual Message on the State of Europe's Environment, UNEP/ROE/97/16, 1997.
2. For a more extensive discussion of the organic chemistry of environmental systems, see the excellent reference text by R.P. Schwarzenbach, P.M. Gschwend, and D.M. Imboden, *Environmental Organic Chemistry* (New York: John Wiley & Sons, 1993). For a discussion of the process chemistry associated with waste systems, see Chapter 5 of M.D. LaGrega, P.L. Buckingham, and J.C. Evans, *Hazardous Waste Management*, 2nd edition (Boston: McGraw-Hill, 2001)
3. For background on the biochemistry and degradation of vinclozolin, see W. Kelce, E. Monosson, and M. Gamcsik, "Environmental Hormone Disruptors: Evidence That Vinclozolin Developmental Toxicity Is Mediated by Anti-Androgenic Metabolites," *Toxicology and Applied Pharmacology,* 126: 275–285, 1994.
4. See, for example, J. Weber and C. Miller, "Organic Chemical Movement over and through Soil," and D. Glotfelty and C.Schomburg, "Volatilization of Pesticides from Soil," both in *Reactions and Movement of Organic Chemicals in Soils,* Monograph 22 (Madison, WI: American Society of Agronomy, Soil Science Society of America, 1989), 305–334.
5. The presentation, "Groundwater Modelling: Theory of Solute Transport" by Professor W. Schneider, Technische Unversität Hamburg-Harburg, was a rich resource for this discussion. I am very grateful for Dr. Schneider's sharing this presentation via the Internet.
6. Cation exchange has been characterized as being the second most important chemical process on earth, after photosynthesis. This is because the cation exchange capacity (CEC), and to a lesser degree anion exchange capacity (AEC) in tropical soils, is the means by which nutrients are made available to plant roots. Without this process, the atmospheric nutrients and the minerals in the soil would not come together to provide for the abundant plant life on planet earth. Professor Daniel Richter of Duke University's Nicholas School of the Environment has waxed eloquently on this subject.
7. This value is taken from W.A. Tucker and L.H. Nelkson, 1982, "Diffusion Coefficients in Air and Water." *Handbook of Chemical Property Estimation Techniques.* McGraw-Hill, New York. Flows this low are not uncommon in some ground water systems or at or in clay liners in landfills.
8. Science is not always consistent with its terminology. The term "particle" is used in many ways. In dispersion modeling, the term particle usually means a

theoretical point that is followed in a fluid. The point represents the path that the pollutant is expected to take. Particle is also used to mean aerosol in atmospheric sciences. Particle is also commonly used to describe unconsolidated materials, such as soils and sediment. The present discussion, for example, accounts for the effects of these particles (e.g., frictional) as the fluid moves through unconsolidated material. The pollutant PM, particle matter, is commonly referred to as "particles." Even the physicist's particle-wave dichotomy comes into play in environmental analysis, as the behavior of light is important in environmental chromatography.

9. Excellent sources on this topic include J. Pignatello, "Sorption Dynamics of Organic Compounds in Soils and Sediments," in *Reactions and Movement of Organic Chemicals in Soils,* Monograph 22 (Madison, WI: American Society of Agronomy, Soil Science Society of America, 1989), 305–334; and V. Evangelou, *Environmental Soil and Water Chemistry: Principles and Applications* (New York: John Wiley & Sons, 1998).
10. Theophrastus Philippus Aureolus von Hohenheim Paracelsus, *Operum Medico-Chimicaorum,* 3 vols. (Geneva, Switzerland: 1605).
11. F. Wöhler, "On the Artificial Production of Urea," Leipzig, Germany: *Annalen der Physik und Chemie,* 88: 1828.
12. Not all compounds containing carbon are organic. For example, the chemistry of calcium carbonate ($CaCO_3$), carbonic acid (H_2CO_3), and other carbonic compounds is an important part of environmental inorganic chemistry because it must be understood to explain environmental conditions such as hardness, ionic strength, buffering capacity, and metabolism of aquatic organisms. Likewise, inorganic atmospheric chemistry is concerned about cyanide compounds, which contain the radical, CN (e.g., the toxic fumigant cyanogen bromide, CNBr), as well as oxides of carbon, especially carbon monoxide (CO) and carbon dioxide (CO_2), none of which are organic compounds. All of these inorganic carbon compounds lack a carbon bonded to another carbon.
13. Substitution is often used to render organic compounds more useful, or even more "safe," such as when methane and ethane are chlorinated. The chlorinated methanes and ethanes are much less flammable than when unsubstituted. This change allowed for these compounds to be very useful in dry cleaning and solvent cleaning process. Unfortunately, they were susbsequently found to be more toxic than their unsubstituted counterparts.
14. The sources for the discussions regarding dioxin and dioxin-like compounds are Ontario, Canada's Ministry of Environment and Energy, "Green Facts," July 1997 (including the tables and figures); and personal conversations with Linda Birnbaum of the U.S. EPA's National Health and Environmental Effects Laboratory in Research Triangle Park, NC. The source of the PCB discussion is reported by the North American Free Trade Agreement's Commission for Environmental Cooperation, *Status of PCB Management in North American,* Montreal, 1996.
15. Resource Planning Corporation, *Appendix A: Estimated 1988 PCB Equipment Inventory* (Final Report), 1988.

16. Federal Register, 59 FR 62791.
17. For discussion of the transport of dioxins, see C. Koester and R. Hites, "Wet and Dry Deposition of Chlorinated Dioxins and Furans," *Environmental Science and Technology*, 26: 1375–1382, 1992; and R. Hites, "Atmospheric Transport and Deposition of Polychlorinated Dibenzo-p-dioxins and Dibenzofurans" (Research Triangle Park, NC: 1991), EPA/600/3-91/002.
18. Nonmetallic ions can also indicate reduction of environments. For example, oxided forms of nitrogen (e.g., nitrites NO_2^{-1}) and reduced nitrogen (e.g., ammonium NH^{+1}).
19. D. Vallero, J. Farnsworth, and J. Peirce, "Degradation and Migration of Vinclozolin in Sand and Soil," *Journal of Environmental Engineering*, 127(10): 952–957, 2001.
20. E. Nyer, *Groundwater Treatment Technology* (New York: Van Nostrand Reinhold, 1985), 188.
21. For a discussion of the properties of volatility of organic compounds, see R. Lewis and S. Gordon, "Sampling for Organic Chemicals in Air," in *Principles of Environmental Sampling* (Washington, DC: American Chemical Society, 1996), 401–470. For information regarding the measurement of semivolatile organic compounds, see R. Williams, R. Watt, R. Stevens, C. Stone, and J. Lewtas, "Evaluation of Personal Air Sampler for Twenty-Four Hour Collection of Fine Particles and Semivolatile Organics," *Journal of Exposure Analysis and Environmental Epidemiology*, 2: 158–166, 1999.
22. This equation is derived from H. Hemond and E. Fechner, *Chemical Fate and Transport in the Environment* (San Diego, CA: Academic Press, 1996).
23. D. Vallero and J. Peirce, "Transformation and Transport of Vinclozolin from Soil to Air," *Journal of Environmental Engineering*, 128(3): 261–268, 2002.
24. This case study is taken from K. Walker, D. Vallero, and R. Lewis, "Factors Influencing the Distribution of Lindane and Other Hexachlorocyclohexanes in the Environment," *Environmental Science and Technology*, 33(24): 4373–4378, 1999.
25. F. Kutz, P. Wood, and D. Bottimore, *Review of Environmental Contamination Toxicology*. 120: 1–82, 1991. "Organochlorine Pesticides and Polychlorinated Biphenyls in Human Adipose Tissue."
26. Agency for Toxic Substances and Disease Registry (ATSDR), *Toxicological Profile for Alpha-, Beta-, Gamma- and Delta-Hexachlorocyclohexane*, 205-93-0606 (Research Triangle Park, NC: Research Triangle Institute, 1997), 1–239.
27. The term *aquifer* is usually reserved for those underground strata that can provide sufficient yield to serve as a source of water supply; however, in hazardous waste engineering, it is better to consider almost any strata that holds or could hold water to be an aquifer because the contamination of even a low-yielding aquifer can be a source of exposure and risk.
28. This case study is taken from work conducted by a team led by our Duke colleague, Professor Miguel Medina, at a waste site in Duke Forest, North Carolina, who has graciously permitted its reproduction here. The full paper is M.A. Medina, Jr., W. Thomann, J.P. Holland, and Y-C. Lin, "Integrating

Parameter Estimation, Optimization and Subsurface Solute Transport," *Hydrological Science and Technology*, 17: 259–282, 2001.

29. Generally, the only groundwater movement that moves by turbulent flow is that of Karst topography, wherein massive limestone or dolomite is fractured and eroded to form caverns and caves. Flow within these systems is often found to occur in underground streams.
30. P. Domenico and F. Swartz, *Physical and Chemical Hydrogeology* (New York: John Wiley & Sons, 1990).
31. Paradioxane's carcinogenesis was established by the National Toxicology Program, *Bioassay of 1,4-Dioxane for Possible Carcinogenicity* (CAS No. 123-91-1), Technical Report No. 80, 1978.
32. The source for this case study is the description of the National Priority Listing sites in Oklahoma (www.health.state.ok.us/PROGRAM/envhlth/sites/).
33. See S. Van der Zee's discussion regarding qualitative descriptions of organic liquids in groundwater in Chapter 7 of *Pollution Risk Assessment and Management*, P. Douben, ed. (Chicester, England: John Wiley & Sons, 1998).
34. This structure explains why household dishwashing liquids are so effective in removing the lipophilic substances, such as fats and grease, compared to water alone. The water is able to remove the hydrophilic substances (such as foods predominantly composed of sugars and starches) but is not effective for the fatty wastes. Other physical factors such as water pressure (e.g., physical removal of egg yolks by a high-velocity spray) and temperature (increased temperature increases the water solubility of the organic compounds, such as the bacon grease stuck on the breakfast plate) also come into play in dishwashing. These culinary principles also factor in hazardous waste engineering. My son, Daniel Joseph Vallero, a recently minted engineer, and I have had many such conversations at the dinner table (to the chagrin of his mother and his sister Amelia).
35. See J. Leete, "Groundwater Modeling in Health Risk Assessment," in S. Benjamin and D. Belluck, eds., *A Practical Guide to Understanding, Managing and Reviewing Environmental Risk Assessment Reports* (Boca Raton, FL: Lewis Publishers, 2001).
36. The compartmental or box models such as the one in Figure 3-30 are being enhanced by environmental scientists and chemical engineers. Much of the information in this figure can be attributed to discussions with Yoram Cohen, a chemical engineering professor at UCLA, and Ellen Cooter, a National Oceanic and Atmospheric Administration modeler on assignment to the U.S. EPA's National Exposure Research Laboratory in Research Triangle Park, North Carolina.

Chapter 4

1. An interesting recent publication that will introduce the reader to life cycle analysis is J.K. Smith and J.J. Peirce, "Life Cycle Assessment Standards: Industrial Sectors and Environmental Performance," *International Journal of Life Cycle Assessment*, 1(2): 115–118, 1996.

2. Numerous textbooks address the topic of incineration in general and hazardous waste incineration in particular. For example, see C.N. Haas and R.J. Ramos, *Hazardous and Industrial Waste Treatment* (Englewood Cliffs, NJ: Prentice-Hall, 1995); C.A. Wentz, *Hazardous Waste Management* (New York: McGraw-Hill, 1989); and J.J. Peirce, R.F. Weiner, and P.A. Vesilind, *Environmental Pollution and Control* (Boston, MA: Butterworth–Heinemann, 1998).
3. For decades books have been published that focus on the current understandings of the science, engineering, and technology of biologic waste treatment. See, for example, Metcalf and Eddy as revised by G. Tchobanoglous and F.L. Burton, *Wastewater Engineering* (New York: McGraw-Hill, 1991); A.F. Gaudy and E.T. Gaudy, *Elements of Bioenvironmental Engineering* (San Jose, CA: Engineering Press, 1988); and J.J. Peirce, R.F. Weiner, and P.A. Vesilind, *Environmental Pollution and Control* (Boston, MA: Butterworth–Heinemann, 1998). For a particular focus on the biotreatment of hazardous wastes, see for example, C.N. Haas and R.J. Ramos, *Hazardous and Industrial Waste Treatment* (Englewood Cliffs, NJ: Prentice-Hall, 1995); and C.A. Wentz, *Hazardous Waste Management* (New York: McGraw-Hill, 1989).
4. Reported by the *Environmental News Service*, "Altered Algae Soaks up Toxic Metals," May 14, 2002.
5. "Remeditation: Research Could Enhance PCB Cleanup Efforts," *Civil Engineering*, 72(3): 24, 2002.
6. A more complete discussion of hazardous waste storage facilities appears in a wide range of textbooks, including C.N. Haas and R.J. Ramos, *Hazardous and Industrial Waste Treatment* (Englewood Cliffs, NJ: Prentice-Hall, 1995); and C.A. Wentz, *Hazardous Waste Management* (New York: McGraw-Hill, 1989).
7. Developing research in the area of nitric oxide emissions from soil includes F.E. Chase, C.T. Corke, and J.B. Robinson, "Nitrifying Bacteria in the Soil," in T.R.G. Gray and D. Parkinson, eds., *Ecology of Soil Bacteria* (Liverpool, England: University of Liverpool Press, 1968); H. Christensen, M. Hansen, and J. Sorensen, "Counting and Size Classification of Active Soil Bacteria by Fluorescence In Situ Hybridization with an rRNA Oligonucleotide Probe," *Applied and Environmental Microbiology*, 65(4): 1753–1761, 1999; I. E. Galbally, "Factors Controlling NO Emissions from Soils," in M.O. Andreae and D.S. Schimel, eds., *Exchange of Trace Gases between Terrestrial Ecosystems and the Atmosphere: The Dahlem Conference* (New York: Wiley, 1989); S. Jousset, R.M. Tabachow, and J.J. Peirce, "Nitrification and Denitrification Contributions to Soil Nitric Oxide Emissions," *Journal of Environmental Engineering*, 127(4): 222–238, 2001; J.J. Peirce and V.P. Aneja, "Laboratory Study of Nitric Oxide Emissions from Sludge Amended Soil," *Journal of Environmental Engineering*, 126(3): 225–232, 2000; and D. Rammon and J.J. Peirce, "Biogenic Nitric Oxide from Wastewater Land Application," *Atmospheric Environment*, 33: 2115–2121, 1999.
8. Developing research in the area of FISH applications to the microbial populations in water and soil includes G.A. Kowalchuk, J.R. Stephen, W. De Boer,

J.I. Prosser, T.M. Embley, and J.W. Woldendorp, "Analysis of *B*-Proteobacteria Ammonia-Oxidising Bacteria in Coastal Sand Dunes Using Denaturing Gradient Gel Electrophoresis and Sequencing of PCR Amplified 16S rDNA Fragments," *Applied and Environmental Microbiology,* 63: 1489–1497, 1997; W. Manz, R. Amann, M. Wagner, and K.-H. Schleifer, "Phylogenetic Oligonucleotide Probes for the Major Subclasses of Proteobacteria: Problems and Solutions," *Systems of Applied Microbiology,* 15: 593–600, 1992; B. Nogales, E.R.B. Moore, E. Llobet-Brossa, R. Rossello-Mora, R. Amann, and K.N. Timmis, "Combined Use of 16S Ribosomal DNA and 16S RNA to Study the Bacterial Community of Polychlorinated Biphenyl-polluted Soil," *Applied and Environmental Microbiology,* 67(4): 1874–1884, 2001; and M. Wagner, G. Rath, H.-P. Koops, J. Flood, and R. Amann, "In Situ Analysis of Nitrifying Bacteria in Sewage Treatment Plants," *Water and Science Technology,* 34(1-2): 237–244, 1996.

Chapter 5

1. The site manager must have an acceptable health and safety plan before beginning remediation. The United States E.P.A. safety protocol (Appendix 4) is a good place to start, but individual sites may have been unique threats to worker safety that must be accounted for in the site safety plan.
2. G.W. Ware, *The Pesticide Handbook*, 3rd edition (Fresno, CA: Thomson, 1999).
3. Agency for Toxic Substances and Disease Registry, *Toxicological Profile for Alpha-, Beta-, Gamma- and Delta-Hexachlorocyclohexane,* 205-93-0606 (Research Triangle Park, NC: 1997).
4. The term *inert* when applied to pesticide chemical composition can be misleading. It does not mean that the so-called inert materials are necessarily chemically nonreactive or even that they have low toxicity. Ingredients in pesticides that are classified as inert are simply contrasted with those ingredients classified as active. So the engineer must take great care to identify all substances used in the manufacturing processes, whether active or inert. This will help to characterize the site, decide on the type of field measurements needed, structure the analytical chemistry program, and develop the remediation plan.
5. In addition to walking the site, it may be helpful to ask neighbors and local businesses about previous activities. These oral histories may uncover practices that could lead to discoveries of additional sites, such as the observation of neighbors of past movement of vehicles, memories of former workers of tasks that may have required the burial and other disposal practices of hazardous materials, and nearby farmers and ranchers who accepted barrels and drums for "rip-rapping" and erosion control in ditches and gulches. Although such information can be subjective and sometimes unreliable compared to chemical and physical measurements, it can provide many insights.

6. Michael J. Derelanko, "Risk Assessment," in M.J. Derelanko and M.A. Hollinger, eds., *CRC Handbook of Toxicology* (Boca Raton, FL: CRC Press, 1999).
7. The default value for absorption is 1. That is, unless otherwise specified, one can assume that all of the contaminant is absorbed.
8. U.S. Environmental Protection Agency, *Water Quality Criteria Documents; Availability*. Federal Register, 45(231): 79318–79379, November 28, 1980; and U.S. Environmental Protection Agency, *National Primary Drinking Water Regulation; Final Rule*. Federal Register, 56(20): 3526–3597, January 30, 1991.
9. National Academy of Sciences, *Recommended Dietary Allowances*, 8th edition (Washington, DC: National Academy of Sciences, National Research Council, 1974).
10. The default value for absorption is 1.
11. U.S. EPA, *Dermal Exposure Assessment: Principles and Applications*, EPA/600/8-9-91, (Washington, DC: U.S. EPA, 1992).
12. For an extensive discussion, see U.S. EPA, *Guidelines for Exposure Assessment*. Federal Register, 57(104): 22888–22938, May 29, 1992. Also, the basis for these guidelines can be found in U.S. EPA, *Development of Statistical Distributions or Ranges of Standard Factors Used in Exposure Assessments*, EPA 600/8-85-010 (Washington, DC: U.S. EPA, 1985).
13. The default value for absorption is 1.
14. For example, *The Precautionary Principle* by Indur Goklany (Washington, DC: Cato Institute, 2001) refers to Carolyn Raffensperger and Joel Tickner's definition of *precautionary principle*: "When an activity raises threats of harm to human health or the environment, precautionary measures should be taken even if some cause and effect relationships are not established scientifically. In this context the proponent of the activity, rather than the public, should bear the burden of proof." Within the framework of our example here, we may agree that it is physiologically impossible for one to maintain a heavy ventilation rate for an entire workday, but it does provide a margin of safety, which is a vital role of the on-site engineer. Some have argued that if the principle is carried to an extreme, however, it could severely reduce technological advancement. For example, see Julian Morris, *Rethinking Risk and the Precautionary Principle* (Boston, MA: Butterworth–Heinemann, 2000).
15. For example, refer to C.D. Klaassen, *Casarett & Doull's Toxicology: The Basic Science of Poisons*, 5th edition (New York: McGraw-Hill, 1996), especially the discussion on toxicokinetics and mechanisms of toxicity.
16. This does not, however, necessarily mean that the cost for risk reduction will be less to clean up the air than the soil. As was found in the 1980s and 1990s with tetrachlorodibenzo-*para*-dioxin (TCDD) and other dioxins and furans, certain compounds have a strong affinity for soil, making treatment by extraction very difficult.

Chapter 6

1. The source of this WTC dust discussion is P. Lioy et al., "Characterization of the Dust/Smoke Aerosol That Settled East of the World Trade Center (WTC) in Lower Manhattan After the Collapse of September 11, 2001," *Environmental Health Perspectives,* 110(7): 703–714, 2002.
2. For an early assessment of the environmental impacts from the attacks on the WTC towers, see the article "Environmental Aftermath," *Environmental Health Perspectives,* 109: A528–A537, 2001.
3. "Environmental Aftermath," *Environmental Health Perspectives,* 109: A528–A537, 2001
4. E. Swartz, L. Stockburger, and D. Vallero, "Preliminary Data of Polyaromatic Hydrocarbons (PAHs) and Other Semi-Volative Organic Compounds Collected in New York City in Response to the Events of September 11, 2001," Report of NERL, RTP, NC.
5. The levels of polychlorinated dioxins and furans found at the WTC are compared to those found in sludges by R. Hale, M. LaGuardia, E. Harvey, M. Gaylor, T. Mainor, and W. Duff, "Flame Retardants: Persistent Pollutants in Land-Applied Sludges," *Nature,* 412: 141–142, 2001.
6. Federal Register, 40CFR, Part 745, 66: 1206–1240 (Washington, DC: U.S. Environmental Protection Agency, 2001).
7. The study by K.Ohyama, F. Nagai, and Y. Tsuchiya, in the *Journal of Health Science,* volume 109, pp. 699–703, found that diphenyl propane binds to estrogen receptors in tumor. However, since the study was conducted using the solvent DMSO and factors such as absorption factors are unknown, there is no way to extrapolate doses for humans.
8. IRIS is found at www.epa.gov/iriswebp/iris/index.html. It is an electronic database that is updated and maintained by the U.S. EPA, containing information on human health effects that may result from chemical exposures. It is a key source of data for risk assessments that is used by government agencies in decision making and regulatory activities. The IRIS files contain descriptive and quantitative information regarding oral reference doses and inhalation reference concentrations (RfDs and RfCs, respectively) for chronic, noncarcinogenic health effects. IRIS also contains information about a chemicals hazard identification, oral slope factors, and oral and inhalation unit risks for carcinogenic effects.
9. U.S. Environmental Protection Agency, *Exposure Factors Handbook, Volume 1.* EPA/600/P-95/002FA (Washington, DC: U.S. EPA, 1997).
10. When a substance in a plume is uniquely associated with a compound, such as retene wood smoke, it can be used as a marker in so-called "receptor models." These models, such as the Chemical Mass Balance (CMB) models, mathematically link sources to downwind sites (i.e., receptors) and are used by regulatory agencies as evidence that a source is contributing pollution to the plume.
11. The opposite error (i.e., false positive) can also result when the test shows the presence of the chemical, but in fact there is none. This can result from artifacts in the laboratory, resulting from poor laboratory practice, such as residual

dioxins left on glassware from a prior analysis. It may also result from misreading peaks on the chromatogram. For example, a particular dioxin congener comes off the column at the same time (i.e., retention time or RT) as another compound. A false positive would occur if the chromatographer identifies this other compound as the dioxin compound. A false negative would occur if the chromatographer identifies an actual dioxin peak as a nondioxin compound.

12. A dramatic example of the importance of method selection was demonstrated in measurements of asbestos at WTC. The scientists on the emergency response team decided to use a conservative method to measure the airborne asbestos fibers, which uses an electron microscope to count the number of particles. Unfortunately, the unique situation of the collapsing buildings created a situation where very short fibers were produced, likely the result of breaking from pulverization during the collapses. Therefore, one 10 μm fiber that had broken into 100 fibers of 100 nm length would be counted as 100 fibers. The respiratory toxicology community has no unanimity of thought regarding whether small fibers are more dangerous than large fibers. (This is the case for particle matter.) That is, the smaller particles (< 2.5 μm) are generally considered to cause the most health problems. In fact, some say that the very short fibers are *less dangerous* than the longer fibers. This conclusion is primarily based on the etiology of asbestosis and lung cancer, which may be caused by ineffective phagocytosis. In other words, the longer the fiber, the less likely the cell will be able to surround it and be able to eliminate the fiber from the lung tissue. The contrary argument is that small fibers, like small particles, are able to penetrate lung tissue more deeply, leading to respiratory illness. It is beyond our purposes here to decide which argument is correct, but this discussion does point to the importance of knowing the contaminants of concern as soon as possible and choosing the correct means for measuring these contaminants.
13. U.S. EPA, Method 1613A, Report No. 821/R-93-017, September 1, 1993.
14. U.S. EPA, Compendium Method TO-9A, Determination of Polychlorinated, Polybrominated and Brominated/Chlorinated Dibenzo-p-Dioxins and Dibenzofurans in Ambient Air, Report No. EPA/625/R-96/010b, January 1999.

Chapter 7

1. William D. Ruckelshaus, "Risk, Science, and Democracy," in Theodore S. Glickman and Michael Gough, eds., *Readings in Risk* (Baltimore, MD: Resources for the Future, 1990).
2. Vincent Covello, "Risk Comparisons and Risk Communications," in Roger E. Kasperson and P. Stallen, eds., *Communicating Risk to the Public* (New York: Kluwer, 1992).
3. This is also an argument for better science education for nonscientists. For example, North Carolina Central University has instituted a program called Critical Foundations in Art and Sciences (CFAS). Part of CFAS is a required course for nonscience majors, Science Odyssey, where they learn about the

physics, chemistry, and biology of everyday life. Similarly, Duke University has recently revised its curriculum to ensure that all Duke undergraduate students complete courses in the sciences. Rather than "general studies," each student must complete courses to fill in a curriculum matrix of both "Areas of knowledge" (i.e., Arts and Literature; Civilization; Social Sciences; Natural Sciences; and Mathematics) and "Inquiries and Competencies" (i.e., Quantitative Reasoning; Interpretive and Aesthetic Approaches; Cross-Cultural Inquiry; Science, Technology, and Society; Ethical Inquiry; and Communication Competencies).

4. For example, detection limits continue to fall so more recent data appear to show that things are getting worse. In reality the old "nondetects" may have been just as high or higher than more recent analyses.
5. The FQPA was enacted on August 3, 1996, to amend the Federal Insecticide, Fungicide, and Rodenticide Act (FIFRA) and the Federal Food, Drug, and Cosmetics Act (FFDCA). Especially important to risk assessment, the FQPA established a health-based standard to provide for a reasonable certainty of no harm from pesticide residues in foods. This new provision was enacted to ensure protection from unacceptable pesticide exposure and to strengthen the health protection measure for infants and children from pesticide risks.
6. A very interesting development over the past decade has been the increasing awareness that health research has often ignored several polymorphs or subpopulations, such as women and children, and is plagued by the so-called healthy worker effect. Much occupational epidemiology has been based on a tightly defined population of relatively young and healthy adult white males who had already been screened and selected by management and economic systems put in place during the 20th century. Also, health studies have tended to be biased toward adult white males even when the contaminant or disease of concern was distributed throughout the general U.S. population. For example, much of the cardiac and cancer risk factors for women and children have been extrapolated from studies of adult white males. Pharmaceutical efficacy studies had also been targeted more frequently toward adult white males. This approach has been changing recently, but the residual uncertainties are still problematic.
7. For an excellent and thorough introduction to emerging paradigms for dealing with environmental problems in disadvantaged communities, see Chapter 6, "Communities of Color Respond to Environmental Threats to Health: The Environmental Justice Framework," in R. Braithwaite, S. Taylor, and J. Austin, *Building Health Coalitions in the Black Community* (London: Sage, 1995).
8. Commission for Racial Justice, *Toxic Wastes and Race in the United States* (United Church of Christ, 1987).
9. It has been two or three decades since I heard the expression "a word will mean what it *can* mean." I've since forgotten the name of the instructor, but he said this in a Technical Writing course in Kansas City, Missouri. I often remind students of this sage advice: Take care to ensure *no ambiguity* in what you say and write as a professional, and make certain that the only possible interpretation of your words is what you intend them to be.

10. R.M. Hograth, *Educating Intuition*, Chapter 1, "The Sixth Sense." (Chicago: University of Chicago Press, 2001)
11. Department of Materials Science and Engineering State University of New York at Stony Brook.
12. For an excellent volume on the subject of risk tradeoffs, see J.D. Graham and J.B. Wiener. 1995, *Risk versus Risk*, Harvard University Press, Cambridge, MA. The book includes the examples mentioned here, as well as other examples of the complex nature of comparing risks.

Chapter 8

1. Value engineering (VE) was developed by L.D. Miles of General Electric in the 1950s and has been employed by engineers for the last 50 years to reduce costs by eliminating inefficiencies in design and operations. It is also the basis for multidisciplinary teams of engineers and nonengineers to solve problems.
2. Taken from the Code of Ethics of the American Society of Civil Engineers.

Appendix 5

1. For the calculations and discussions of solubility equilibrium, including this example, see C.C. Lee and S.D. Lin (eds.), 2000, *Handbook of Environmental Engineering Calculations*, pp. 1.368–1.373 (New York: McGraw-Hill Professional, 2000).
2. See Michael LaGrega, Phillip Buckingham, and Jeffrey Evans, *Hazardous Waste Management*, 2nd Edition (New York: McGraw-Hill, 2001).

Index